AIDS-ASSOCIATED VIRAL ONCOGENESIS

Cancer Treatment and Research

Steven T. Rosen, M.D., *Series Editor*

Sugarbaker, P. (ed.): *Peritoneal Carcinomatosis: Principles of Management.* 1995. ISBN 0-7923-3727-1.
Dickson, R.B., Lippman, M.E. (eds): *Mammary Tumor Cell Cycle, Differentiation and Metastasis.* 1995. ISBN 0-7923-3905-3.
Freireich, E.J., Kantarjian, H. (eds): *Molecular Genetics and Therapy of Leukemia.* 1995. ISBN 0-7923-3912-6.
Cabanillas, F., Rodriguez, M.A. (eds): *Advances in Lymphoma Research.* 1996. ISBN 0-7923-3929-0.
Miller, A.B. (ed.): *Advances in Cancer Screening.* 1996. ISBN 0-7923-4019-1.
Hait, W.N. (ed.): *Drug Resistance.* 1996. ISBN 0-7923-4022-1.
Pienta, K.J. (ed.): *Diagnosis and Treatment of Genitourinary Malignancies.* 1996. ISBN 0-7923-4164-3.
Arnold, A.J. (ed.): *Endocrine Neoplasms.* 1997. ISBN 0-7923-4354-9.
Pollock, R.E. (ed.) *Surgical Oncology.* 1997. ISBN 0-7923-9900-5.
Verweij, J., Pinedo, H.M., Suit, H.D. (eds): *Soft Tissue Sarcomas: Present Achievements and Future Prospects.* 1997. ISBN 0-7923-9913-7.
Walterhouse, D.O., Cohn, S.L. (eds): *Diagnostic and Therapeutic Advances in Pediatric Oncology.* 1997. ISBN 0-7923-9978-1.
Mittal, B.B., Purdy, J.A., Ang, K.K. (eds): *Radiation Therapy.* 1998. ISBN 0-7923-9981-1.
Foon, K.A., Muss, H.B. (eds): *Biological and Hormonal Therapies of Cancer.* 1998. ISBN 0-7923-9997-8.
Ozols, R.F. (ed.): *Gynecologic Oncology.* 1998. ISBN 0-7923-8070-3.
Noskin, G.A. (ed.): *Management of Infectious Complications in Cancer Patients.* 1998. ISBN 0-7923-8150-5.
Bennett, C.L. (ed.): *Cancer Policy.* 1998. ISBN 0-7923-8203-X.
Benson, A.B. (ed.): *Gastrointestinal Oncology.* 1998. ISBN 0-7923-8205-6.
Tallman, M.S., Gordon, L.I. (eds): *Diagnostic and Therapeutic Advances in Hematologic Malignancies.* 1998. ISBN 0-7923-8206-4.
von Gunten, C.F. (ed.): *Palliative Care and Rehabilitation of Cancer Patients.* 1999. ISBN 0-7923-8525-X.
Burt, R.K., Brush, M.M. (eds): *Advances in Allogeneic Hematopoietic Stem Cell Transplantation.* 1999. ISBN 0-7923-7714-1.
Angelos, P. (ed.): *Ethical Issues in Cancer Patient Care* 2000. ISBN 0-7923-7726-5.
Gradishar, W.J., Wood, W.C. (eds): *Advances in Breast Cancer Management.* 2000. ISBN 0-7923-7890-3.
Sparano, Joseph, A. (ed.): *HIV & HTLV-I Associated Malignancies.* 2001. ISBN 0-7923-7220-4.
Ettinger, David, S. (ed.): *Thoracic Oncology.* 2001. ISBN 0-7923-7248-4.
Bergan, Raymond, C. (ed.): *Cancer Chemoprevention.* 2001. ISBN 0-7923-7259-X.
Raza, A., Mundle, S.D. (eds): *Myelodysplastic Syndromes & Secondary Acute Myelogenous Leukemia* 2001. ISBN: 0-7923-7396.
Talamonti, Marks, S. (ed.): *Liver Directed Therapy for Primary and Metastatic Liver Tumors.* 2001. ISBN 0-7923-7523-8.
Stack, M.S., Fishman, D.A. (eds): *Ovarian Cancer.* 2001. ISBN 0-7923-7530-0.
Bashey, A., Ball, E.D. (eds): *Non-Myeloablative Allogeneic Transplantation.* 2002. ISBN 0-7923-7646-3.
Leong, Stanley, P.L. (ed.): *Atlas of Selective Sentinel Lymphadenectomy for Melanoma, Breast Cancer and Colon Cancer.* 2002. ISBN 1-4020-7013-6.
Andersson, B., Murray, D. (eds): *Clinically Relevant Resistance in Cancer Chemotherapy.* 2002. ISBN 1-4020-7200-7.
Beam, C. (ed.): *Biostatistical Applications in Cancer Research.* 2002. ISBN 1-4020-7226-0.
Brockstein, B., Masters, G. (eds): *Head and Neck Cancer.* 2003. ISBN 1-4020-7336-4.
Frank, D.A. (ed.): *Signal Transduction in Cancer.* 2003. ISBN 1-4020-7340-2.
Figlin, Robert, A. (ed.): *Kidney Cancer.* 2003. ISBN 1-4020-7457-3.
Kirsch, Matthias, Black, Peter McL. (ed.): *Angiogenesis in Brain Tumors.* 2003. ISBN 1-4020-7704-1.
Keller, E.T., Chung, L.W.K. (eds): *The Biology of Skeletal Metastases.* 2004. ISBN 1-4020-7749-1.
Kumar, Rakesh (ed.): *Molecular Targeting and Signal Transduction.* 2004. ISBN 1-4020-7822-6.
Verweij, J., Pinedo, H.M. (eds): *Targeting Treatment of Soft Tissue Sarcomas.* 2004. ISBN 1-4020-7808-0.
Finn, W.G., Peterson, L.C. (eds): *Hematopathology in Oncology.* 2004. ISBN 1-4020-7919-2.
Farid, N. (ed.): *Molecular Basis of Thyroid Cancer.* 2004. ISBN 1-4020-8106-5.
Khleif, S. (ed.): *Tumor Immunology and Cancer Vaccines.* 2004. ISBN 1-4020-8119-7.
Balducci, L., Extermann, M. (eds): *Biological Basis of Geriatric Oncology.* 2004. ISBN
Abrey, L.E., Chamberlain, M.C., Engelhard, H.H. (eds): *Leptomeningeal Metastases.* 2005. ISBN 0-387-24198-1.
Platanias, L.C. (ed.): *Cytokines and Cancer.* 2005. ISBN 0-387-24360-7.
Leong, Stanley P.L., Kitagawa, Y., Kitajima, M. (eds): *Selective Sentinel Lymphadenectomy for Human Solid Cancer.* 2005. ISBN 0-387-23603-1.
Small, Jr. W., Woloschak, G. (eds): *Radiation Toxicity: A Practical Guide.* 2005. ISBN 1-4020-8053-0.

AIDS–ASSOCIATED VIRAL ONCOGENESIS

Edited by
CRAIG MEYERS, Ph.D.
Professor, Department of Microbiology and Immunology
The Pennsylvania State University College of Medicine
Hershey, Pennsylvania, USA

 Springer

Library of Congress Control Number: 2007921523

ISBN-10: 0-387-46804-8   e-ISBN-10: 0-387-46816-1
ISBN-13: 978-0-387-46804-4   e-ISBN-13: 978-0-387-46816-7

9 8 7 6 5 4 3 2 1

springer.com

CONTENTS

LIST OF CONTRIBUTORS

Charles Wood, Ph.D, Nebraska Center for Virology, and the School of Biological Sciences, University of Nebraska, Lincoln, NE

Anita Arora, MD, Center for Clinical Studies, Houston, TX

Elizabeth Chiao, MD, Baylor College of Medicine, Houston, TX

Stephen K. Tyring, MD, Ph.D, University of Texas Health Science Center, Houston, TX

Whitney Greene, Tumor Virology Program, Children's Cancer Research Institute, Departments of Pediatrics, Microbiology and Immunology, and Molecular Medicine, The University of Texas Health Science Center, San Antonio, TX

Kurt Kuhne, Tumor Virology Program, Children's Cancer Research Institute, Departments of Pediatrics, Microbiology and Immunology, and Molecular Medicine, The University of Texas Health Science Center, San Antonio, TX

Fengchun YE, Tumor Virology Program, Children's Cancer Research Institute, Departments of Pediatrics, Microbiology and Immunology, and Molecular Medicine, The University of Texas Health Science Center, San Antonio, TX

Jiguo Chen, Tumor Virology Program, Children's Cancer Research Institute, Departments of Pediatrics, Microbiology and Immunology, and Molecular Medicine, The University of Texas Health Science Center, San Antonio, TX

Fuchun Zhou, Tumor Virology Program, Children's Cancer Research Institute, Departments of Pediatrics, Microbiology and Immunology, and Molecular Medicine, The University of Texas Health Science Center, San Antonio, TX

Xiufeng Lei, Tumor Virology Program, Children's Cancer Research Institute, Departments of Pediatrics, Microbiology and Immunology, and Molecular Medicine, The University of Texas Health Science Center, San Antonio, TX

Shou-Jiang Gao, Tumor Virology Program, Children's Cancer Research Institute, Departments of Pediatrics, Microbiology and Immunology, and Molecular Medicine, The University of Texas Health Science Center, San Antonio, TX

Dirk P. Dittmer, Department of Microbiology & Immunology & Lineberger Comprehensive Cancer Center, University of North Carolina, Chapel Hill, NC

Blossom Damania, Department of Microbiology & Immunology & Lineberger Comprehensive Cancer Center, University of North Carolina, Chapel Hill, NC

Bharat G. Bajaj, Department of Microbiology and the Tumor Virology Program, Abramson Comprehensive Cancer Center, University of Pennsylvania Medical School, Philadelphia, PA

Masanao Murakami, Department of Microbiology and the Tumor Virology Program, Abramson Comprehensive Cancer Center, University of Pennsylvania Medical School, Philadelphia, PA

Erle S. Robertson, Department of Microbiology and the Tumor Virology Program, Abramson Comprehensive Cancer Center, University of Pennsylvania Medical School, Philadelphia, PA

Scott M. Long, Department of Biochemistry, St. Jude Children's Research Hospital, Memphis, TN

Clare E. Sample, Department of Biochemistry, St. Jude Children's Research Hospital, Memphis, TN

Jennifer E. Cameron, Tulane Health Sciences Center, Tulane Medical School, New Orleans, LA

Michael E. Hagensee, Department of Medicine, Louisiana State University Health Sciences Center, New Orleans, LA

Regis A. Vilchez, Department of Molecular Virology and Microbiology and Baylor–UTHouston Center for AIDS Research, Baylor College of Medicine, Houston, TX, Current affiliation: Schering Plough Research Institute, Kenilworth, New Jersey, USA

Janet S. Butel, Department of Molecular Virology and Microbiology and Baylor–UTHouston Center for AIDS Research, Baylor College of Medicine, Houston, TX

Jianming Hu, Department of Microbiology and Immunology –
The Pennsylvania State University College of Medicine, Hershey, PA

Laurie Ludgate, Department of Microbiology and Immunology –
The Pennsylvania State University College of Medicine, Hershey, PA

1. OVERVIEW

CHARLES WOOD, PH.D.

Nebraska Center for Virology and the School of Biological Sciences, University of Nebraska, Lincoln, NE

It has been 25 years since the acquired immunodeficiency syndrome (AIDS) was first described and over 23 years since the human immunodeficiency virus (HIV) associated with the disease was first discovered. In spite of the tremendous progress that was made in understanding both the disease and the virus, there are still millions of people infected, died, or living with the disease. As for the year 2005 alone, the Joint United Nations Programme on HIV/AIDS (http://www.UNAIDS.org) estimates that there are about 40 million people living with HIV/AIDS globally, and approximately 3 million people died from AIDS in the year. Globally, it is estimated that 25 million people have died of HIV/AIDS since 1981. The impact of the epidemic is enormous, with the greatest impact in sub-Saharan Africa where about two-third of those living with HIV/AIDS in the world reside. A number of countries in the region have infection rates to as high as 30–40% of the population. As for South and Southeast Asia, even though the adult prevalence is lower and estimated to be less than 1%, there are still 7.4 million people living with HIV/AIDS in this region, accounting for about 18% of those living with HIV/AIDS in the world.[104,114] North America, Latin America, and Eastern Europe and Central Asia, each have between 1.2 and 1.8 million people living with HIV/AIDS. Although the 300,000 people living with HIV/AIDS in the Caribbean constitute a small part of the global total, they are about 1.6% of adults in the region, making the Caribbean the only region other than sub-Saharan Africa to have an adult prevalence higher than 1%.[104,114] Essentially, no single country can escape from the impact of HIV/AIDS.

Since the beginning of the AIDS epidemic, it was soon realized that HIV-infected individuals that are immunosuppressed are affected not only by opportunistic infections but also by other AIDS-associated diseases, including malignancies. In fact, at the onset of the AIDS epidemic in 1981, the disease was first recognized through an increase of Kaposi's sarcoma (KS) in young adult homosexual male population, which is unusual as KS is a very rare form of malignancies which are found in certain ethic groups in the Mediterranean region and in Africa.[29,57] In 1992, the US Center for Disease Control and Prevention (CDC) developed the initial case definition for AIDS and included two AIDS-defining malignancies, KS and primary central nervous system lymphoma (PCNSL).[2] It was soon realized that patients infected by HIV are also at risk for developing both Hodgkin's and non–Hodgkin's lymphomas (NHL) when compared to normal uninfected individual, with a large number of homosexual male with AIDS developed NHL.[122] Subsequently, CDC has revised the definition of AIDS to include NHL in 1987[1] and invasive cervical cancer in 1992.[3] In addition to the common AIDS-defining malignancies it has been reported that AIDS and or immunosuppression may lead to an increase in other non–AIDS defining cancers, including multiple myeloma, Hodgkin's disease, leukemia, lung cancer, oral cavity cancers, and leiomyosarcoma in children.[44,47] These malignant tumors are generally characterized by a more aggressive behavior at diagnosis and a poorer outcome compared with the same tumors in the general population. A large-scale study by Frisch et al.[34] analyzing cancer registries from 11 sites in the US for over 300,000 HIV-infected adults showed that these individuals have a very high risk in developing AIDS-defining cancers, but in addition, non–AIDS defining cancers also showed a statistically significance increase as well. Those that showed a potential association with immunosuppression are Hodgkin's disease (HD), lung cancer, penile cancer, soft tissue malignancies, and testicular seminoma. However, most non–AIDS-defining cancers do not appear to be associated with HIV-associated immunosuppression and disease progression.[36,80] It is likely that other factors could be involved, including smoking and other viral coinfections.

A number of the common cancers associated with AIDS and immunosuppression were found to have infectious viral cofactors, including those classified as AIDS-defining disease, KS, NHL, and invasive cervical cancers. Kaposi's sarcoma and primary effusion non–Hodgkin's lymphomas have been linked to Kaposi's sarcoma-associated herpesvirus (KSHV) or human herpesvirus-8 (HHV-8); Hodgkin's disease and non–Hodgkin's lymphomas have been linked to EBV; and squamous cell carcinomas have been linked to human papilloma virus (HPV) (Table 1). These viral-associated AIDS-malignancies will be the major focus of this chapter.

AIDS-ASSOCIATED KAPOSI'S SARCOMA

KS is a very rare form of sarcoma that has become the most common neoplasm in HIV-infected individuals, since the onset of the AIDS epidemic. KS was first described by Moritz Kaposi in 1872 as an indolent tumor in elderly Mediterranean

Table 1. AIDS-associated malignancies with viral etiology

Malignancies	Virus
Kaposi's sarcoma	Human herpesvirus-8 (HHV-8) or Kaposi's sarcoma-associated herpesvirus (KSHV)
Non-Hodgkin's lymphomas	
Primary central nervous system lymphoma	Epstein-Barr Virus (EBV)
Primary effusion lymphoma	HHV-8
Burkitt's lymphoma	EBV
Plasmabalistic lymphoma	EBV
Hodgkin's disease	EBV
Squamous cell neoplasm	HPV
Genital dysplasia	

men with multifocal pigmented sarcoma.[62] KS is composed of a mixture of irregular shaped, round capillaries, and slitlike endothelium-lined vascular spaces and spindle-shaped cells with infiltrating mononuclear cells. It is not clear whether KS represents a clonal neoplastic process or a polyclonal inflammatory lesion. Studies have shown that varying monoclonality, oligoclonality, and polyclonality from lesions of various patients.[40] The origin of KS spindle cells is also not clear; it has been suggested that KS cells represent a heterogeneous population of cells, arising from pluripotent mesenchymal precursor cells, and may be of lymphatic endothelial cell origin.[28]

Since AIDS was first described in the early 1980s, the annual incidence of KS in the San Francisco Bay area showed an exponential increase with the incidence rate per age group followed a bimodal distribution that peaks from 30 to 36 years of age.[24] KS has become the most common neoplasm in HIV-infected individuals, and has reached epidemic proportions in the developing world where HIV is widespread, such as sub-Saharan Africa. Patients in this region with AIDS-associated KS have been shown to have a high tumor burden and rapid disease course, very different from those seen in the non-AIDS-related KS. There are four different epidemiological forms of KS that showed different clinical parameters. The first is the classical or sporadic form of KS. This is associated with elderly men in the Mediterranean countries like Italy and Israel.[57] This form of KS is usually nonaggressive and associated with lesions in the lower extremities. It is usually not associated with HIV infection. The older the patients are, the risk for disease progression is also greater, and disseminated KS in these individuals could occur if they are immunosuppressed. The second form of KS is endemic African KS. This form of KS was found in the African continent prior to the onset of the AIDS epidemic. It is found not only in men but in women and young children as well.[117] This form of KS tends to be more aggressive than the classical KS and can also involve the lymph nodes. With the spread of HIV since the 1980s, the prevalence of KS in the African continent has increased substantially. A study in Uganda has shown that prior to the 1970s, KS was diagnosed in no more than 7% of the male cancer population, and none in the female. However, by the early 1990s, it had risen to around

49% of the male cancer patients, and 18% of the female cancer patients,[117] and similar increases were reported in other African nations. In Zambia, in the late 1990s, KS was found to be one of the most common childhood cancers,[16] most likely due to HIV coinfection. The third type of KS is the Iatrogenic KS or transplant-associated KS. This normally occurs in patients after transplantation that were treated with immunosuppressive drugs, and withdrawal of therapy can lead to KS regression.[53] This form of KS tends to have a more rapid disease course as compared to the classical KS, but can also present as a chronic condition.[53] The fourth form of KS is the AIDS KS. This form of KS has been found to be associated with AIDS patients, it has increased dramatically since the onset of the AIDS epidemic and is one of the AIDS-defining illnesses.[43,45] This is the most aggressive form of KS and was first described in the early 1980s in the HIV-infected homosexual male population.[33] Unlike the classical KS, AIDS KS involves not only the lower extremities and skin, but also the upper body, the head regions, and the lymph nodes. It can also disseminate to other organs, such as the spleen, the lungs, the liver, and gastrointestinal track.[53]

Human Herpesvirus and KS

An infectious agent has long been suspected in the development of KS; herpesvirus-like particles were found in short-term KS tissue culture, and were subsequently identified as cytomegalovirus,[41] but the involvement of CMV in KS has not been confirmed. In 1994, a novel human herpesvirus was identified by Chang and Moore[14] using representational difference analyses. This virus is now known as KSHV or HHV-8 and is found to be necessary, but not sufficient for the development of all types of KS. It is clear that other cofactors, such as immunosuppression, are required for KS development. KSHV is found in all KS lesions, and is mainly located in the vascular endothelial cells and perivascular spindle-shaped cells[75] KSHV infection is not commonly found in low-risk population but found commonly in individuals at risk for KS.

KSHV belongs to the γ-herpesvirus family, which can further be divided into two subgroups, γ-1 or lymphocrytovirus and γ-2 or rhadinovirus. EBV is the prototype of γ-1 virus and the simian herpesvirus saimini is the prototype of γ-2 herpesvirus.[96] KSHV is classified as a γ-2 rhadinovirus and is the first human virus of this subfamily identified. Like other herpesviruses, HHV-8 is a double-stranded deoxyribonucleic acid (DNA) virus. Its genome is linear, it is about 165 kbp in length, and contains at least 87 viral genes. A feature of some DNA viruses, particularly of herpesviruses and KSHV, is the ability of these viruses to incorporate or pirate host genes into their genome: these genes can then play a role in the replication, survival, and transformation functions of the virus. KSHV was found to encode human homologous genes that regulate cell cycling like cyclin D, growth factors like interleukin 6, or genes that may prevent programmed cell death such as bcl-2. Deciphering the functions of these viral genes will lead to a better understanding of viral pathogenesis and oncogenesis.

Unlike most other herpesviruses, KSHV infection does not seem to be widely distributed in most populations. The detection of KSHV infection relies on the presence of antibodies against either lytic and/or latent antigens and varies among the different tests that were used in different seroprevalence studies. In general, the frequency of infection appears to be low in North America, certain Asian countries, and in Northern European nations such as the UK and Germany, with most studies reporting a seroprevalence rate in normal blood donors of less than 5%.[30,88,99,100] In these countries the seroprevalence of KSHV in different risk groups mirrors the incidence of AIDS KS, with a seroprevalence rate of 25–50% among homosexual men. In other countries, such as Italy, Greece, and Israel, especially Southern Italy, the infection rate seems to be much higher in the general population, and is more variable, ranging between 5 and 35%. In contrast to North America and Europe, KSHV infection is widespread in the African continent. High seroprevalence rates between 40 to 50% have been found in Central, West, and South Africa.[6,37,50,86] Therefore, KSHV seroprevalence tracks very closely with KS, with the highest infection rate in geographic areas where classic or endemic forms of KS are more common. KS has a particularly high incidence in Central African countries like the Republic of Congo, Uganda, and Zambia; these countries also have the highest KSHV infection rates in the world.[37] Very little is known about KSHV infection in China even though EBV infection was found to be ubiquitous. There were only two reported studies in China; Dilnur et al.[23] found that KSHV was associated with KS in China. The study by Du et al.[26] in the outskirt of China in the Xinjiang autonomous region, where there is a high incidence of HIV-1 infection, showed that there was a high KSHV infection rate. The study found KSHV infection varied among individuals of different racial origin, it was highest in the Khalkhas population at about 48% and lowest in the Kazak and Han population at over 12%, but the specimens were screened without dilution, the reproducibility of the assay was not determined, and the prevalence among different risk groups was not studied. Thus, there is a need to perform a systematic comparison of risk groups using established assays.

Impact of HAART on KS

Since the beginning of the AIDS epidemic in the early 1980s, AIDS KS has become one of the most common AIDS-associated malignancies with HIV-infected homosexual males at the highest risk, and those with AIDS had a 50% lifetime rate of developing KS early in the HIV epidemic.[63] However, the rate of AIDS KS has since steadily declined both in the US and Europe.[7] It has been suggested that the disease may have shifted from an early disease to a late manifestation during the HIV disease course. Since the introduction of highly active antiretroviral therapy, a further major decrease in AIDS KS was further observed,[71] and therapy has now made AIDS KS a relatively rare tumor in treated HIV-infected individuals.[9] Several studies have shown that there was a marked decrease in KS incidence since HAART was introduced, a decline of as high as 80-fold was observed. In addition, regression of KS following treatment has been reported.[48,58,91,92,112] Interestingly, the reduced KS risk was only observed with HAART, but not with double or single anti-HIV drugs.[59] Even though the incidence of KS in the treated HIV-infected individuals in the western world has decreased dramatically, in the setting where HAART is still not widely available, such as sub-Saharan Africa, AIDS KS still remains a major problem.

AIDS-ASSOCIATED NON-HODGKIN'S LYMPHOMA

ARL represents a heterogeneous group of tumors that are commonly found in HIV-infected, immunosuppressed individuals. The majority of the cancers are of B-cell in origin; the development of these cancers is characterized as an AIDS-defining illness.[68,74] ARL in general is a late event in the HIV disease course. The risk factors for the development of ARL include low CD4 and T cell counts, high HIV viral load, and increased age.[66] ARL in general is classified into three groups. The first are those that can also be found in immunocompetent individuals, like Burkitt's lymphoma (BL) and diffuse large B-cell lymphoma (DLBCL) which can further be classified into centroblastic, immunoblastic, and anaplastic subtypes. An example of DLBCL is the primary central nervous system lymphoma (PCNSL), which is a distinct extranodal presentation of DLBCL.[68] The second are those that are found only in HIV-infected immunosuppressed individuals. They include the primary effusion lymphoma (PEL) and plasmablastic lymphoma. The third are those that can be found also in immunosuppressed patients other than those due to HIV infection. These include post-transplant lymphoproliferative disorders. In general, the difference between ARL and NHL found in non-AIDS-associated patients is the presentation of advanced diseases, extranodal involvement, disease in soft tissues, and other locations, such as jaws and their association with viral infections, most prominently EBV and KSHV. In addition to viral coinfections, a large proportion of the ARL also have genetic abnormalities, especially among the large-cell lymphomas, Bcl-6 rearrangement occur in about 33% of the cases, c-myc rearrangement about 40%, and p53 mutations about 25% of the cases,[35] whether these genetic abnormalities are the causes or effect of malignant transformation is not clear.

The most commonly identified virus associated with AIDS-related lymphomas is EBV and there is a large body of published work on the oncogenic mechanisms of this agent.[12,121] The second most common is KSHV, which was found to be associated with PEL associated with HIV infection.[13] B-lymphocytes transformed by EBV (lymphoblastoid cell lines) in vitro express an array of virus-encoded proteins including six EBV nuclear antigens (EBNAs) and three LMPs. EBNAs are generated from differential splicing of a transcript that arises from one of two promoters (Cp or Wp).[121] This form of latency is termed Latency III, and is common in immunoblastic lymphomas.[12] A Type II form of latency where EBNA 1, LMP-1, and LMP-2a are expressed has been identified in some EBV-associated lymphomas. In Latency I (typical of Burkitt's lymphomas) only EBNA-1 (generated from the

Qp promoter) and EBV-encoded RNAs (EBERs) are expressed.[12,121] Recent studies have indicated that some heterogeneity in EBV gene expression and EBNA promoter usage exists among endemic BL.[65] It has been proposed that classical antigen driven B-cell proliferation may play a role in the ARL.[68] The hyperstimulation of B lymphocytes could be caused by HIV, EBV, and other infectious agents that may elicit the release of various growth factors and cytokines which promote the proliferation and transformation of B cells into ARC.

Among the ARL, the ones most commonly found in HIV-infected individuals are BL and DLBCL. They represent about 90% of all ARLs.[8] The others, such as PEL, represent less than 50% of the cases. Chemically immunosuppressed individuals, like those infected with HIV, have a much higher risk of developing lymphoproliferative disorders. The increased risk of developing ARC has been reported to range from 14-fold for low-grade lymphoma to 630-fold for high-grade immunoblastic lymphoma.[21] A study by Cote et al. has shown that the risk of HIV-infected individuals developing ARL within 3 years of being diagnosed with AIDS was 165-fold higher than those without AIDS. The same study also showed that the risk of developing BL is 261-fold higher.[20] In addition, it has been demonstrated that risk of AIDS patients developing PCNSL is 3,000 times higher and HD is ten times higher than HIV uninfected individuals.[108]

Burkitt's Lymphoma

AIDS-related BL is one of the most common ARLs found in HIV-infected individuals. Unlike other ARL, BL is unique in that presentation of the cancer can occur at relatively high CD4 T-cell count (>250 cells/mm^3).[90] AIDS BLs share features with endemic African BL in that both over express c-myc due to reciprocal translocations that bring the transactivator under the influence of potent promoter sequences within the immunoglobulin (Ig) genes loci. Inactivating mutations and deletions of p53 are also common as in all types of BL,[11] but Bcl-6 rearrangements are rarely observed. A distinguishing feature between AIDS-related and endemic BL is that the former is associated with EBV far less frequently than the latter.[11,72] EBV has been reported to present in a subset of between 33 and 67% of these AIDS-related BL, and the type III latency expressing viral EBNAs and LMP proteins was not consistently observed. With some expressing viral antigen patterns that are similar to those seen in Hodgkin's disease.[103] BL often carry genetic abnormities and chromosomal translocation, but there is no evident that EBV has been linked to these abnormalities, c-myc translocation is common in BL, with the c-myc gene transposed to the proximity of the immunoglobulin locus, and often led to the activation of c-myc expression. These tumors are incredibly aggressive with short doubling times. Flow cytometric analysis typically reveals that over 90% of the tumor cells are in S phase. Ongoing tumor lysis syndrome even in the absence of concomitant chemotherapy is often noted. AIDS-related BLs appear to carry a poor prognosis even when compared to AIDS-related DLBCL.[77]

Primary Central Nervous System Lymphoma

Primary central nervous system lymphoma (PCNSL) is a distinct extranodel form of DLBCL, which is usually of the immunoblastic type. It is a rapid and fatal disease with poor prognosis, associated with HIV infection and severe immunodeficiency, usually with CD4+ T cell counts of < 50 cells/mm^3.[85,107] HIV-infected persons with PCNSL usually have tumors confined to the craniospinal axis without systemic involvement. In contrast to PCNSL in HIV negative individuals, those found in the context of HIV infection is always associated with EBV in HIV-infected patients.[85] Detection of EBV sequences in the CSF by polymerase chain reaction in combination with thallium spectroscopy has been shown to be a useful diagnostic tool for the disease.[17] Since EBV is always present in PCNSL, it is likely that this virus plays a role in cancer development. The EBV genes that are expressed include the latent gene. EBNAs and LMP-1 and -2, typical of a type III latency seen in B cells, when they are transformed by EBV in vitro.[120] These EBV genes, such as LMP-2, are known to deregulate cellular replication and may play a role in the transformation of B cells. These gene products are also good targets of cytotoxic T lymphocytes, which could account for the reduction of PCNSL in patients that were undergoing antiretroviral therapy (ART) when their cytotoxic lymphocyte response was restored. However, it has been reported that abnormalities in the T cell functions in these patients can still be observed.[116] Thus, other factors, including the alternation of the EBV gene expression profile could have occurred upon ART. Treatment of patients with PCNSL with conventional chemotherapy in combination with radiation therapy has not been very effective. Only a 4-month survival has been observed. However, treatment with a high dose of zidovudine and ganciclovir has been shown to lead to long-term remission,[93,98] again suggesting that therapy targeting EVB may be effective in controlling PCNSL.

Plasmablastic Lymphoma

Plasmablastic lymphoma is a recently described variant form of ARL that only occurs in a low percentage of HIV-infected individuals. It was originally described by Delecluse et al. as a B-cell lymphoma stained by antibody specific for plasma cells, but these cells lack the classical B-cell antigen CD20.[22] These cancers occur only in a small percent of HIV-infected patients and are usually associated with EBV and KSHV infection.[11] Patients usually develop tumors in the jaw and oral cavity and the prognosis is poor. Patients with plasmablastic lymphoma normally do not respond to conventional chemotherapy and it has been suggested that therapy targeting EBV may be beneficial.[18] In fact, ART and intensive chemotherapy has been shown to be encouraging as five out of six HIV-infected and treated individuals were alive with a medium follow-up of 17 months.[113]

Primary Effusion Lymphoma

Primary effusion lymphoma (PEL) is a very rare subtype of NHL, predominantly associated with HIV-infected individuals. PEL was first identified as a subset of body-cavity-based lymphomas, which were subsequently called PELs.[13] PELs are unique as they were found to contain KSHV DNA and are most frequently found in men and in AIDS patients. This type of lymphoma is distinguished from others as having a distinctive morphology, bridging large cell immunoblastic lymphoma and anaplastic large cell lymphoma. PELs often present as lymphomatous effusions in the pleural, peritoneal, and/or pericardial cavity. These cells are usually CD20 negative, often express CD45 marker but lack B-cell-associated antigens. PELs have B cell origins with clonal immunoglobulin gene rearrangements. Gene expression profiling of PEL by Klein et al.[67] showed that PEL displayed a common gene expression profile that is distinct from ARL and NHL in HIV uninfected individuals. The profile also showed elements seen in EBV transformed lymphoblastoid cell lines, AIDS immunoblastic lymphoma, as well as multiple myeloma. Further confirming the notion that PEL is distinct and yet a subtype of NHL. Most PELs are coinfected with EBV and KSHV and lack e-myc gene rearrangements. PELs are extremely rare tumors, and estimated to be about 0.13% of all AIDS-related malignancies in AIDS patients in the US.[80] Thus, KSHV-associated lymphomas represent a rare, distinct pathobiologic category which often, but not always, associates with an effusion in AIDS patients. The role of KSHV in the development of these lymphomas is not clear since this type of malignancies is still rare even in the populations with high KSHV seroprevalence rate. However, KSHV has always been found in these lymphomas, suggesting that this virus is necessary, but other factors must be needed for the development of PELs. These factors could be EBV infection and/or immunosuppression. Recently, solid tumor variants with plasmablastic features have been reported and these tumors tend to be rapidly fatal although recent data suggest that some PEL lines are quite sensitive to inhibition of NF-kB,[64] suggesting that inhibitors of NF-kB potentially can be developed as therapy for patients with PEL.

ARL in the Era of Highly Active Antiretroviral Therapy

Since the introduction of HAART in the 1990s, the spectrum of ARL in the context of HIV infection has been substantially influenced. The epidemiology of ARL has been changed and the outcome of these tumors has been impacted. Prior to HAART, the prognosis of ARL was poor because the tumors tended to be more aggressive; there was an increase in hematological toxicity, and further complications occurred due to a high rate of opportunistic infection in the patients.[61,106] Risks for development of ARL increased as the HIV disease progressed and CD4+ T cells declined. With HAART, the prognosis of ARL has improved. There was a better tolerance to chemotherapy, a higher rate of complete remission, and the death rate has been reduced.[8,39] However, HAART appears to have differential effects on different subtypes of ARL. A recently completed study performed by the NCI sponsored AIDS malignancy consortium (AMC) demonstrated the feasibility of concomitant chemotherapy with HAART.[94] Probably the best reported results for chemotherapy in AIDS NHL were from Dr. Little's group at the National Cancer Institute. Using the EPOCH regimen, the group achieved remission in

22 of 24 patients with a progression free survival of 23 months. These patients had favorable prognostic factors (median CD4[+] lymphocyte count of 233 mm^3/ml).[119] Enhanced toxicity of rituximab and CHOP chemotherapy was recently noted in a large multicenter trial conducted by the AMC.[60] The addition of rituximab to standard-dose CHOP as compared to CHOP alone led to increased infectious complications and deaths attributable to sepsis. It is possible that delayed recovery of humoral immunity could contribute to this increased risk of life-threatening bacterial infections in HIV-infected patients. There have been several reports on the feasibility and efficacy of high-dose chemotherapy and autologous stem cell transplant for ARL.[69,95] It is reasonable to assume that patients with well-controlled HIV and good performance status should be considered candidates for this therapy. Newer approaches that may benefit patients with ARL include EBV specific cytotoxic T cells and agents that activate the lytic program of gamma herpesviruses, thereby sensitizing the tumors to antivirals.[70,97]

The effects of HAART on PCNSL are more dramatic than other systemic ARL. The standard treatment of PCNSL is whole brain irradiation, but the median survival time is still just 2.5 months or less. With the addition of HAART to radiotherapy, several studies have shown an improvement in survival.[56,85] In the meta analysis of 11 studies, it was estimated that the decline of PCNSL in the HAART era has declined by 58%. In addition, it was shown that HAART therapy alone has led to a regression of PCNSL.[81] It is possible that with HAART, the restoration of cytotoxic T cells against latently expressed EBV antigens may have played a role in the reduction in the incidence of PCNSL, further confirming the role of EBV in the development of this malignancy.

In contrary to PCNSL, the effects of HAART on systemic ARL seem to be less dramatic.[9,46] A number of studies, including a EuroSIDA study on 8599 HIV-infected individuals, showed that all types of lymphomas were reduced after the widespread use of HAART in the late 1990s, and demonstrated that there was a decline in systemic lymphomas ranging from twofold to sevenfold since HAART.[8,66,84] However, these studies also showed that the incidence of Burkitt's lymphomas appeared to be largely unaffected. In spite of the reduction of ARL in the HAART era, the risk of developing ARL in HAART-treated HIV-infected individuals is still about 20-fold higher than in uninfected individuals.[19] Whether this is due to the emergence of drug resistant viruses and immunosuppression, or due to prolonged mild immunosuppression and partial immune reconstitution, thus leading to an increase in developing cancers, is unknown. These risk factors remain to be elucidated.

HUMAN PAPILLOMA VIRUS-ASSOCIATED NEOPLASMS

Genital cancers have been a world-wide public health problem, especially in the context of HIV infection and immunosuppression. They are common malignancies found in AIDS patients. HIV infection has been shown to substantially enhance the development of cervical cancer and cervical cancer precursor lesions.[10,78] It is also the most common female malignancy in many developing countries.[118]

Genital cancers and cervical intraepithelial neoplasia (CIN) have been strongly implicated in association with HPV infection. The first report demonstrating the association of HPV and cervical cancer was published by Zur Hausen et al.,[123] showing that cervical cancer in the female genital tract has HPV-associated lesions. HPV infection of genital can either lead to asymptomatic infection or a wide range of genital lesions, ranging from genital warts to mild dysplasion to invasive carcinomas. Genital lesions are often referred to as cervical intraepithelial abnormalities or CIN, which is graded from I to III depending on the degree of epithelium abnormality. CIN I encompasses mild dysplasia or low-grade squamous intraepithelial lesions (SIL). CIN II represents high-grade SIL with moderate dysplasia, while severe dysplasia is referred to as CIN III.[79]

In 1993, the CDC has included invasive cervical cancer as one of the AIDS-defining illnesses. HIV infection has become an important risk factor for HPV infection and the development of genital cancers. HIV and HPV infections have a number of common features; both are sexually transmitted diseases (STD), and there is a high prevalence of HPV infection among HIV seropositive women, especially those that are immunosuppressed with low $CD4^+$ T cell counts.[5,111]

Up to 20% of the HIV and HPV coinfected individuals developed HPV-associated premalignant lesions of the uterine cervix within 3 years of HIV infection.[31] The progression of an untreated HPV-induced dysplasia could then lead to cervical cancer. A number of studies have shown that HIV positive women have 2–3 times more HPV DNA in cervicovaginal washings and 15 times more in anal swabs as compared to HIV negative individuals.[15,54,111] HIV-infected women were also shown to be much more likely to develop cervical intraepithelial abnormalities. The prevalence and severity of genital tract infection in these women are also more pronounced.[27,83] The rates of invasive cervical carcinoma are 15–18 times higher in women with AIDS compared to the general population.[32,102] Men fare no better – the incidence of anal cancer in men with history of anal intercourse is at 35 per 100,000 individuals, a number equivalent to that of cervical carcinoma before the advent of Pap-smear screening.[89,110] The mechanism by which HIV increases the risk of HPV infection and cervical dysplasia is likely due to immunodeficiency, resulting in the inability of the immune response to control HPV infection. Indeed, cervical dysplasia increased progressively as the patients immune function declined, as determined by $CD4^+$ T cell counts.[52] A large, long-term prospective cohort studying of over 1,800 HIV positive and over 500 HIV negative women was carried out to determine how HIV RNA levels and $CD4^+$ T cell counts are associated with the natural history of HPV infection.[109] The results demonstrated that in HIV-infected women, the HIV plasma viral load in combination with $CD4^+$ T cell counts has a strong correlation with the detection of HPV infection and reactivation. However, there was only a moderate correlation between HIV coinfection and HPV persistent infection. This partially explains why cervical cancer rates are not even higher than what was observed in HIV+ infected women.

The effects of HAART on HIV and the immune status of HIV-infected individuals are well established, but their effects on the course of HPV-related cervical lesions in HIV-infected women are still not well established. HAART has not been shown to affect HPV detection and its effects on the natural history of cervical intraepithelial neoplasia are unclear.[87] There were a number of studies examining the effects of HAART on the course of cervical lesions.[4,51] A study by the Women's Interagency HIV Study (WIHS) group has reported an association of the regression of cervical lesions with HAART[82] whereas others cannot.[76] In addition, a multicenter study of HAART on the repression of cervical lesions in over 700 women followed for over 5 years also did not show any correlation.[101] However, these studies primarily focused on evaluating the effects of HAART on prevalent cervical lesions and the date of onset of those lesions is not known, thus it is not surprising that the outcome of these studies was controversial. To address these concerns, a study by Ahdieh-Grant et al.[4] determined whether HAART alters the natural history of CIN among HIV-infected women that were regularly followed every 6 months for 7 years. This study provided evidence that HAART has a modest benefit for HIV-infected women that were at risk for cervical neoplasia, with women receiving HAART will survive longer and have better control over cervical HPV infection and low-grade squamous intraepithelial lesions. However, HAART does not seem to lead to a complete reconstitution of the immune system for it to control HPV, the rates of regression among HIV-infected women receiving HAART remained lower than HIV-infected women that were never on HAART or among HIV-uninfected women. Thus, there is a need to actively seek and treat CIN in HIV-infected women, including those that are responding to HAART.

HODGKIN'S DISEASE

Hodgkin's disease (HD) or Hodgkin's lymphoma has not been classified as an AIDS-defining illness but HIV-related HD is also increasing in the context of HIV infection, and is clearly related to the immunosuppression in the infected patients.[42,47] Almost all HIV-related HD are EBV positive. The clinical presentations of HD in HIV-infected individuals are unique as compared to uninfected individuals, patients are more likely to have more advanced disease stage, extranodal involvement, and often involve bone marrow.[25,105,115]

Prior to the HAART era, the treatment outcome for HIV HD is poor, with a medium survival rate between 1 and 2 years. A clinical trial on treatment using doxorubicin, bleomcin, vinblastine, and dacarbazine on AIDS patients showed severe hematologic toxicity and a poor survival rate of only 1.5 years.[73] More recently the outcome of treatment in combination with HAART appears to be improving, patients treated with HAART and responded to the treatment within 2 years of the development of HD have improved survival rates compared to those that are not responders.[38,55] The use of HAART in combination with chemotherapy has substantially improved both the response rate and the survival time. In a trial on using HAART and a combination of bleomycin, etoposide, doxorubicin, cyclophosphamide, vincristine, procarbazine, and prednisone (BEACOPP), 9 out of 12 treated patients have complete remission after a medium follow-up of 49 months.[49] These studies suggest that HAART in combination with conventional chemotherapy will be beneficial for patients with HD, whether the effect is due to the restoration of the immune response in HIV-infected individuals or due to the effects of EBV remained to be determined.

CONCLUSION

AIDS malignancies have been recognized as a major complication of the HIV disease course, and the high mortality rate in AIDS patients is in part due to this complication. A number of factors have been implicated to be playing a role in the increase incidence of malignancies in AIDS patients. These factors include immunosuppression and a deficient immune surveillance by T cells in eliminating the transformed cells; viral cofactors such as EBV and KSHV have also been associated with malignant transformation of the infected cells. The prognosis of ARL in HIV-infected individuals was extremely poor prior to the HAART era, but the advent of HAART, there was a dramatic improvement in the prognosis of ARL in these patients. The survival rate has improved, especially for those ARL that were found to be associated with herpesviruses such as EBV and KSHV. There were a substantial decrease in the number of cases of KS and NHL, in association with HAART, and the decrease in incidence appear to be independent of CD4 counts, substantiating the notion that coinfecting viruses may be playing a role in the disease. Most newly diagnosed cases of KS are patients that are either drug naïve or have virological failure upon treatment. In spite of the improvement in prognosis, HAART alone is inadequate for the majority of the antiviral naïve patients. A combination of HAART and conventional chemotherapy appears to be most effective. Moreover, the positive effects of HAART on some ARL and KS are not consistently observed for HPV-associated malignancies, including cervical and anal cancers.

Given the prolong survival rate of HIV-infected individuals with HAART, it is likely that ARL will continue to pose a challenge in the AIDS epidemic. HAART treatment, even though appears to be effective, still only leads to partial immune reconstitution. Prolonged immunosuppression is likely to lead to a resurgence of AIDS-associated cancers. In addition, HAART is still not widely available in parts of the world where AIDS has the greatest impact, such as the African continent. It is expected that AIDS-associated cancers will continue to pose a major challenge in the population for quite a while. Thus, more understanding on the role of oncogenic viral cofactor in the disease will be of great significance. ARL represents an intersection of virology, immunology, and tumor biology. A better understanding of the disease, the immune response, will provide useful information for the development of novel therapy and provide insights into cancer biology beyond the AIDS epidemic.

ACKNOWLEDGMENT

This publication was made possible, in part by support from the following NIH grants: PHS award CA76958, NCRR COBRE grant RR15635, and INBRE grant P20 RR016469. The author also wishes to acknowledge Ms. Dianna Wright for help with the preparation of this chapter.

REFERENCES

1. 1987. Revision of the CDC surveillance case definition for acquired immunodeficiency syndrome. Council of State and Territorial Epidemiologists; AIDS Program, Center for Infectious Diseases. MMWR Morb Mortal Wkly Rep **36**(Suppl 1):1S–15S.
2. 1982. Update on acquired immune deficiency syndrome (AIDS) – United States. MMWR Morb Mortal Wkly Rep **31**:507–8, 513–4.
3. 1992. Update: National Breast and Cervical Cancer Early Detection Program, July 1991–July 1992. MMWR Morb Mortal Wkly Rep **41**:739–43.
4. Ahdieh-Grant, L., R. Li, A. M. Levine, L. S. Massad, H. D. Strickler, H. Minkoff, M. Moxley, J. Palefsky, H. Sacks, R. D. Burk, and S. J. Gange. 2004. Highly active antiretroviral therapy and cervical squamous intraepithelial lesions in human immunodeficiency virus-positive women. J Natl Cancer Inst **96**:1070–6.
5. Ahdieh, L., R. S. Klein, R. Burk, S. Cu-Uvin, P. Schuman, A. Duerr, M. Safaeian, J. Astemborski, R. Daniel, and K. Shah. 2001. Prevalence, incidence, and type-specific persistence of human papillomavirus in human immunodeficiency virus (HIV)-positive and HIV-negative women. J Infect Dis **184**:682–90.
6. Ariyoshi, K., M. Schim van der Loeff, P. Cook, D. Whitby, T. Corrah, S. Jaffar, F. Cham, S. Sabally, D. O'Donovan, R. A. Weiss, T. F. Schulz, and H. Whittle. 1998. Kaposi's sarcoma in the Gambia, West Africa is less frequent in human immunodeficiency virus type 2 than in human immunodeficiency virus type 1 infection despite a high prevalence of human herpesvirus 8. J Hum Virol **1**:193–9.
7. Beral, V. 1991. Epidemiology of Kaposi's sarcoma. Cancer Surv **10**:5–22.
8. Besson, C., A. Goubar, J. Gabarre, W. Rozenbaum, G. Pialoux, F. P. Chatelet, C. Katlama, F. Charlotte, B. Dupont, N. Brousse, M. Huerre, J. Mikol, P. Camparo, K. Mokhtari, M. Tulliez, D. Salmon-Ceron, F. Boue, D. Costagliola, and M. Raphael. 2001. Changes in AIDS-related lymphoma since the era of highly active antiretroviral therapy. Blood **98**:2339–44.
9. Biggar, R. J. 2001. AIDS-related cancers in the era of highly active antiretroviral therapy. Oncology (Williston Park) **15**:439–48; discussion 448–9.
10. Bosch, F. X., A. Lorincz, N. Munoz, C. J. Meijer, and K. V. Shah. 2002. The causal relation between human papillomavirus and cervical cancer. J Clin Pathol **55**:244–65.
11. Carbone, A. 2003. Emerging pathways in the development of AIDS-related lymphomas. Lancet Oncol **4**:22–9.
12. Carter, K. L., E. Cahir-McFarland, and E. Kieff. 2002. Epstein-Barr virus-induced changes in B-lymphocyte gene expression. J Virol **76**:10427–36.
13. Cesarman, E., Y. Chang, P. S. Moore, J. W. Said, and D. M. Knowles. 1995. Kaposi's sarcoma-associated herpesvirus-like DNA sequences in AIDS-related body-cavity-based lymphomas. N Engl J Med **332**:1186–91.
14. Chang, Y., E. Cesarman, M. S. Pessin, F. Lee, J. Culpepper, D. M. Knowles, and P. S. Moore. 1994. Identification of herpesvirus-like DNA sequences in AIDS-associated Kaposi's sarcoma. Science **266**:1865–9.
15. Chiasson, M. A., T. V. Ellerbrock, T. J. Bush, X. W. Sun, and T. C. Wright, Jr. 1997. Increased prevalence of vulvovaginal condyloma and vulvar intraepithelial neoplasia in women infected with the human immunodeficiency virus. Obstet Gynecol **89**:690–4.
16. Chintu, C., U. H. Athale, and P. S. Patil. 1995. Childhood cancers in Zambia before and after the HIV epidemic. Arch Dis Child **73**:100–4; discussion 104–5.
17. Cingolani, A., A. De Luca, L. M. Larocca, A. Ammassari, M. Scerrati, A. Antinori, and L. Ortona. 1998. Minimally invasive diagnosis of acquired immunodeficiency syndrome-related primary central nervous system lymphoma. J Natl Cancer Inst **90**:364–9.
18. Cioc, A. M., C. Allen, J. R. Kalmar, S. Suster, R. Baiocchi, and G. J. Nuovo. 2004. Oral plasmablastic lymphomas in AIDS patients are associated with human herpesvirus 8. Am J Surg Pathol **28**:41–6.

19. Clifford, G. M., J. Polesel, M. Rickenbach, L. Dal Maso, O. Keiser, A. Kofler, E. Rapiti, F. Levi, G. Jundt, T. Fisch, A. Bordoni, D. De Weck, and S. Franceschi. 2005. Cancer risk in the Swiss HIV Cohort Study: associations with immunodeficiency, smoking, and highly active antiretroviral therapy. J Natl Cancer Inst **97**:425–32.

20. Cote, T. R., R. J. Biggar, P. S. Rosenberg, S. S. Devesa, C. Percy, F. J. Yellin, G. Lemp, C. Hardy, J. J. Goedert, and W. A. Blattner. 1997. Non-Hodgkin's lymphoma among people with AIDS: incidence, presentation and public health burden. AIDS/Cancer Study Group. Int J Cancer **73**:645–50.

21. Cote, T. R., A. Manns, C. R. Hardy, F. J. Yellin, and P. Hartge. 1996. Epidemiology of brain lymphoma among people with or without acquired immunodeficiency syndrome. AIDS/Cancer Study Group. J Natl Cancer Inst **88**:675–9.

22. Delecluse, H. J., I. Anagnostopoulos, F. Dallenbach, M. Hummel, T. Marafioti, U. Schneider, D. Huhn, A. Schmidt-Westhausen, P. A. Reichart, U. Gross, and H. Stein. 1997. Plasmablastic lymphomas of the oral cavity: a new entity associated with the human immunodeficiency virus infection. Blood **89**:1413–20.

23. Dilnur, P., H. Katano, Z. H. Wang, Y. Osakabe, M. Kudo, T. Sata, and Y. Ebihara. 2001. Classic type of Kaposi's sarcoma and human herpesvirus 8 infection in Xinjiang, China. Pathol Int **51**:845–52.

24. Dittmer, D. P., W. Vahrson, M. Staudt, C. Hilscher, and F. D. Fakhari. 2005. Kaposi's sarcoma in the era of HAART – an update on mechanisms, diagnostics and treatment. AIDS Rev **7**:56–61.

25. Doweiko, J., B. J. Dezube, and L. Pantanowitz. 2004. Unusual sites of Hodgkin's lymphoma: CASE 1. HIV-associated Hodgkin's lymphoma of the stomach. J Clin Oncol **22**:4227–8.

26. Du, W., G. Chen, and H. Sun. 2000. Antibody to human herpesvirus type-8 in the general populations of Xinjiang Autonomous Region(A.R.). Zhonghua Shi Yan He Lin Chuang Bing Du Xue Za Zhi **14**:44–6.

27. Duerr, A., M. F. Sierra, J. Feldman, L. M. Clarke, I. Ehrlich, and J. DeHovitz. 1997. Immune compromise and prevalence of Candida vulvovaginitis in human immunodeficiency virus-infected women. Obstet Gynecol **90**:252–6.

28. Dupin, N., C. Fisher, P. Kellam, S. Ariad, M. Tulliez, N. Franck, E. van Marck, D. Salmon, I. Gorin, J. P. Escande, R. A. Weiss, K. Alitalo, and C. Boshoff. 1999. Distribution of human herpesvirus-8 latently infected cells in Kaposi's sarcoma, multicentric Castleman's disease, and primary effusion lymphoma. Proc Natl Acad Sci USA **96**:4546–51.

29. Durack, D. T. 1981. Opportunistic infections and Kaposi's sarcoma in homosexual men. N Engl J Med **305**:1465–7.

30. Edelman, D. C. 2005. Human herpesvirus 8 – a novel human pathogen. Virol J **2**:78.

31. Ellerbrock, T. V., M. A. Chiasson, T. J. Bush, X. W. Sun, D. Sawo, K. Brudney, and T. C. Wright, Jr. 2000. Incidence of cervical squamous intraepithelial lesions in HIV-infected women. JAMA **283**:1031–7.

32. Franceschi, S., L. Dal Maso, S. Arniani, P. Crosignani, M. Vercelli, L. Simonato, F. Falcini, R. Zanetti, A. Barchielli, D. Serraino, and G. Rezza. 1998. Risk of cancer other than Kaposi's sarcoma and non-Hodgkin's lymphoma in persons with AIDS in Italy. Cancer and AIDS registry linkage study. Br J Cancer **78**:966–70.

33. Friedman-Kien, A. E. 1981. Disseminated Kaposi's sarcoma syndrome in young homosexual men. J Am Acad Dermatol **5**:468–71.

34. Frisch, M., R. J. Biggar, E. A. Engels, and J. J. Goedert. 2001. Association of cancer with AIDS-related immunosuppression in adults. JAMA **285**:1736–45.

35. Gaidano, G., F. Lo Coco, B. H. Ye, D. Shibata, A. M. Levine, D. M. Knowles, and R. Dalla-Favera. 1994. Rearrangements of the BCL-6 gene in acquired immunodeficiency syndrome–associated non-Hodgkin's lymphoma: association with diffuse large-cell subtype. Blood **84**:397–402.

36. Gallagher, B., Z. Wang, M. J. Schymura, A. Kahn, and E. J. Fordyce. 2001. Cancer incidence in New York State acquired immunodeficiency syndrome patients. Am J Epidemiol **154**:544–56.

37. Gao, S. J., L. Kingsley, M. Li, W. Zheng, C. Parravicini, J. Ziegler, R. Newton, C. R. Rinaldo, A. Saah, J. Phair, R. Detels, Y. Chang, and P. S. Moore. 1996. KSHV antibodies among Americans, Italians and Ugandans with and without Kaposi's sarcoma. Nat Med **2**:925–8.

38. Gerard, L., L. Galicier, E. Boulanger, L. Quint, M. G. Lebrette, E. Mortier, V. Meignin, and E. Oksenhendler. 2003. Improved survival in HIV-related Hodgkin's lymphoma since the introduction of highly active antiretroviral therapy. AIDS **17**:81–7.

39. Gerard, L., L. Galicier, A. Maillard, E. Boulanger, L. Quint, S. Matheron, B. Cardon, V. Meignin, and E. Oksenhendler. 2002. Systemic non-Hodgkin lymphoma in HIV-infected patients with effective suppression of HIV replication: persistent occurrence but improved survival. J Acquir Immune Defic Syndr **30**:478–84.

40. Gill, P. S., Y. C. Tsai, A. P. Rao, C. H. Spruck, III, T. Zheng, W. A. Harrington, Jr., T. Cheung, B. Nathwani, and P. A. Jones. 1998. Evidence for multiclonality in multicentric Kaposi's sarcoma. Proc Natl Acad Sci USA **95**:8257–61.

41. Giraldo, G., E. Beth, and E. S. Huang. 1980. Kaposi's sarcoma and its relationship to cytomegalovirus (CMNV). III. CMV DNA and CMV early antigens in Kaposi's sarcoma. Int J Cancer **26**:23–9.

42. Glaser, S. L., C. A. Clarke, M. L. Gulley, F. E. Craig, J. A. DiGiuseppe, R. F. Dorfman, R. B. Mann, and R. F. Ambinder. 2003. Population-based patterns of human immunodeficiency virus-related Hodgkin lymphoma in the Greater San Francisco Bay Area, 1988–1998. Cancer **98**:300–9.

43. Goedert, J. J. 2000. The epidemiology of acquired immunodeficiency syndrome malignancies. Semin Oncol **27**:390–401.

44. Goedert, J. J., T. R. Cote, P. Virgo, S. M. Scoppa, D. W. Kingma, M. H. Gail, E. S. Jaffe, and R. J. Biggar. 1998. Spectrum of AIDS-associated malignant disorders. Lancet **351**:1833–9.

45. Goedert, J. J., F. Vitale, C. Lauria, D. Serraino, M. Tamburini, M. Montella, A. Messina, E. E. Brown, G. Rezza, L. Gafa, and N. Romano. 2002. Risk factors for classical Kaposi's sarcoma. J Natl Cancer Inst **94**:1712–8.

46. Grulich, A. E. 1999. AIDS-associated non-Hodgkin's lymphoma in the era of highly active antiretroviral therapy. J Acquir Immune Defic Syndr **21**(Suppl 1):S27–30.

47. Grulich, A. E., Y. Li, A. McDonald, P. K. Correll, M. G. Law, and J. M. Kaldor. 2002. Rates of non-AIDS-defining cancers in people with HIV infection before and after AIDS diagnosis. AIDS **16**:1155–61.

48. Grulich, A. E., Y. Li, A. M. McDonald, P. K. Correll, M. G. Law, and J. M. Kaldor. 2001. Decreasing rates of Kaposi's sarcoma and non-Hodgkin's lymphoma in the era of potent combination anti-retroviral therapy. AIDS **15**:629–33.

49. Hartmann, P., U. Rehwald, B. Salzberger, C. Franzen, M. Sieber, A. Wohrmann, and V. Diehl. 2003. BEA-COPP therapeutic regimen for patients with Hodgkin's disease and HIV infection. Ann Oncol **14**:1562–9.

50. He, J., G. Bhat, C. Kankasa, C. Chintu, C. Mitchell, W. Duan, and C. Wood. 1998. Seroprevalence of human herpesvirus 8 among Zambian women of childbearing age without Kaposi's sarcoma (KS) and mother–child pairs with KS. J Infect Dis **178**:1787–90.

51. Heard, I., V. Schmitz, D. Costagliola, G. Orth, and M. D. Kazatchkine. 1998. Early regression of cervical lesions in HIV-seropositive women receiving highly active antiretroviral therapy. AIDS **12**:1459–64.

52. Heard, I., J. M. Tassie, V. Schmitz, L. Mandelbrot, M. D. Kazatchkine, and G. Orth. 2000. Increased risk of cervical disease among human immunodeficiency virus-infected women with severe immunosuppression and high human papillomavirus load(1). Obstet Gynecol **96**:403–9.

53. Hengge, U. R., T. Ruzicka, S. K. Tyring, M. Stuschke, M. Roggendorf, R. A. Schwartz, and S. Seeber. 2002. Update on Kaposi's sarcoma and other HHV8 associated diseases. Part 1: epidemiology, environmental predispositions, clinical manifestations, and therapy. Lancet Infect Dis **2**:281–92.

54. Hillemanns, P., T. V. Ellerbrock, S. McPhillips, P. Dole, S. Alperstein, D. Johnson, X. W. Sun, M. A. Chiasson, and T. C. Wright, Jr. 1996. Prevalence of anal human papillomavirus infection and anal cytologic abnormalities in HIV-seropositive women. AIDS **10**:1641–7.

55. Hoffmann, C., K. U. Chow, E. Wolf, G. Faetkenheuer, H. J. Stellbrink, J. van Lunzen, H. Jaeger, A. Stoehr, A. Plettenberg, J. C. Wasmuth, J. Rockstroh, F. Mosthaf, H. A. Horst, and H. R. Brodt. 2004. Strong impact of highly active antiretroviral therapy on survival in patients with human immunodeficiency virus-associated Hodgkin's disease. Br J Haematol **125**:455–62.

56. Hoffmann, C., S. Tabrizian, E. Wolf, C. Eggers, A. Stoehr, A. Plettenberg, T. Buhk, H. J. Stellbrink, H. A. Horst, H. Jager, and T. Rosenkranz. 2001. Survival of AIDS patients with primary central nervous system lymphoma is dramatically improved by HAART-induced immune recovery. AIDS **15**:2119–27.

57. Iscovich, J., P. Boffetta, S. Franceschi, E. Azizi, and R. Sarid. 2000. Classic Kaposi sarcoma: epidemiology and risk factors. Cancer **88**:500–17.

58. Jacobson, L. P., T. E. Yamashita, R. Detels, J. B. Margolick, J. S. Chmiel, L. A. Kingsley, S. Melnick, and A. Munoz. 1999. Impact of potent antiretroviral therapy on the incidence of Kaposi's sarcoma and non-Hodgkin's lymphomas among HIV-1-infected individuals. Multicenter AIDS Cohort Study. J Acquir Immune Defic Syndr **21**(Suppl 1):S34–41.

59. Jones, J. L., D. L. Hanson, M. S. Dworkin, and H. W. Jaffe. 2000. Incidence and trends in Kaposi's sarcoma in the era of effective antiretroviral therapy. J Acquir Immune Defic Syndr **24**:270–4.

60. Kaplan, L. D., J. Y. Lee, R. F. Ambinder, J. A. Sparano, E. Cesarman, A. Chadburn, A. M. Levine, and D. T. Scadden. 2005. Rituximab does not improve clinical outcome in a randomized phase 3 trial of CHOP with or without rituximab in patients with HIV-associated non-Hodgkin lymphoma: AIDS-Malignancies Consortium Trial 010. Blood **106**:1538–43.

61. Kaplan, L. D., D. J. Straus, M. A. Testa, J. Von Roenn, B. J. Dezube, T. P. Cooley, B. Herndier, D. W. Northfelt, J. Huang, A. Tulpule, and A. M. Levine. 1997. Low-dose compared with standard-dose m-BACOD chemotherapy for non-Hodgkin's lymphoma associated with human immunodeficiency

virus infection. National Institute of Allergy and Infectious Diseases AIDS Clinical Trials Group. N Engl J Med **336**:1641–8.

62. Kaposi, M. 1872. Arch Dermatol Syphilis **4**:265–73.

63. Katz, M. H., N. A. Hessol, S. P. Buchbinder, A. Hirozawa, P. O'Malley, and S. D. Holmberg. 1994. Temporal trends of opportunistic infections and malignancies in homosexual men with AIDS. J Infect Dis **170**:198–202.

64. Keller, S. A., E. J. Schattner, and E. Cesarman. 2000. Inhibition of NF-kappaB induces apoptosis of KSHV-infected primary effusion lymphoma cells. Blood **96**:2537–42.

65. Kelly, G. L., A. E. Milner, R. J. Tierney, D. S. Croom-Carter, M. Altmann, W. Hammerschmidt, A. I. Bell, and A. B. Rickinson. 2005. Epstein-Barr virus nuclear antigen 2 (EBNA2) gene deletion is consistently linked with EBNA3A, -3B, and -3C expression in Burkitt's lymphoma cells and with increased resistance to apoptosis. J Virol **79**:10709–17.

66. Kirk, O., C. Pedersen, A. Cozzi-Lepri, F. Antunes, V. Miller, J. M. Gatell, C. Katlama, A. Lazzarin, P. Skinhoj, and S. E. Barton. 2001. Non-Hodgkin lymphoma in HIV-infected patients in the era of highly active antiretroviral therapy. Blood **98**:3406–12.

67. Klein, U., A. Gloghini, G. Gaidano, A. Chadburn, E. Cesarman, R. Dalla-Favera, and A. Carbone. 2003. Gene expression profile analysis of AIDS-related primary effusion lymphoma (PEL) suggests a plasmablastic derivation and identifies PEL-specific transcripts. Blood **101**:4115–21.

68. Knowles, D. M. 2003. Etiology and pathogenesis of AIDS-related non-Hodgkin's lymphoma. Hematol Oncol Clin North Am **17**:785–820.

69. Krishnan, A., A. Molina, J. Zaia, D. Smith, D. Vasquez, N. Kogut, P. M. Falk, J. Rosenthal, J. Alvarnas, and S. J. Forman. 2005. Durable remissions with autologous stem cell transplantation for high-risk HIV-associated lymphomas. Blood **105**:874–8.

70. Kurokawa, M., S. K. Ghosh, J. C. Ramos, A. M. Mian, N. L. Toomey, L. Cabral, D. Whitby, G. N. Barber, D. P. Dittmer, and W. J. Harrington, Jr. 2005. Azidothymidine inhibits NF-kappaB and induces Epstein-Barr virus gene expression in Burkitt lymphoma. Blood **106**:235–40.

71. Lebbe, C., L. Blum, C. Pellet, G. Blanchard, O. Verola, P. Morel, O. Danne, and F. Calvo. 1998. Clinical and biological impact of antiretroviral therapy with protease inhibitors on HIV-related Kaposi's sarcoma. AIDS **12**:F45–9.

72. Levine, A. M. 2002. Challenges in the management of Burkitt's lymphoma. Clin Lymphoma **3**(Suppl 1): S19–25.

73. Levine, A. M., P. Li, T. Cheung, A. Tulpule, J. Von Roenn, B. N. Nathwani, and L. Ratner. 2000. Chemotherapy consisting of doxorubicin, bleomycin, vinblastine, and dacarbazine with granulocyte-colony-stimulating factor in HIV-infected patients with newly diagnosed Hodgkin's disease: a prospective, multi-institutional AIDS clinical trials group study (ACTG 149). J Acquir Immune Defic Syndr **24**:444–50.

74. Levine, A. M., L. Seneviratne, B. M. Espina, A. R. Wohl, A. Tulpule, B. N. Nathwani, and P. S. Gill. 2000. Evolving characteristics of AIDS-related lymphoma. Blood **96**:4084–90.

75. Li, J. J., Y. Q. Huang, C. J. Cockerell, and A. E. Friedman-Kien. 1996. Localization of human herpes-like virus type 8 in vascular endothelial cells and perivascular spindle-shaped cells of Kaposi's sarcoma lesions by in situ hybridization. Am J Pathol **148**:1741–8.

76. Lillo, F. B., D. Ferrari, F. Veglia, M. Origoni, M. A. Grasso, S. Lodini, E. Mastrorilli, G. Taccagni, A. Lazzarin, and C. Uberti-Foppa. 2001. Human papillomavirus infection and associated cervical disease in human immunodeficiency virus-infected women: effect of highly active antiretroviral therapy. J Infect Dis **184**:547–51.

77. Lim, S. T., R. Karim, B. N. Nathwani, A. Tulpule, B. Espina, and A. M. Levine. 2005. AIDS-related Burkitt's lymphoma versus diffuse large-cell lymphoma in the pre-highly active antiretroviral therapy (HAART) and HAART eras: significant differences in survival with standard chemotherapy. J Clin Oncol **23**:4430–8.

78. Massad, L. S., L. Ahdieh, L. Benning, H. Minkoff, R. M. Greenblatt, H. Watts, P. Miotti, K. Anastos, M. Moxley, L. I. Muderspach, and S. Melnick. 2001. Evolution of cervical abnormalities among women with HIV-1: evidence from surveillance cytology in the women's interagency HIV study. J Acquir Immune Defic Syndr **27**:432–42.

79. Massad, L. S., K. A. Riester, K. M. Anastos, R. G. Fruchter, J. M. Palefsky, R. D. Burk, D. Burns, R. M. Greenblatt, L. I. Muderspach, and P. Miotti. 1999. Prevalence and predictors of squamous cell abnormalities in Papanicolaou smears from women infected with HIV-1. Women's Interagency HIV Study Group. J Acquir Immune Defic Syndr **21**:33–41.

80. Mbulaiteye, S. M., R. J. Biggar, J. J. Goedert, and E. A. Engels. 2002. Pleural and peritoneal lymphoma among people with AIDS in the United States. J Acquir Immune Defic Syndr **29**:418–21.

81. McGowan, J. P., and S. Shah. 1998. Long-term remission of AIDS-related primary central nervous system lymphoma associated with highly active antiretroviral therapy. AIDS **12**:952–4.
82. Minkoff, H., L. Ahdieh, L. S. Massad, K. Anastos, D. H. Watts, S. Melnick, L. Muderspach, R. Burk, and J. Palefsky. 2001. The effect of highly active antiretroviral therapy on cervical cytologic changes associated with oncogenic HPV among HIV-infected women. AIDS **15**:2157–64.
83. Minkoff, H. L., D. Eisenberger-Matityahu, J. Feldman, R. Burk, and L. Clarke. 1999. Prevalence and incidence of gynecologic disorders among women infected with human immunodeficiency virus. Am J Obstet Gynecol **180**:824–36.
84. Mocroft, A., C. Katlama, A. M. Johnson, C. Pradier, F. Antunes, F. Mulcahy, A. Chiesi, A. N. Phillips, O. Kirk, and J. D. Lundgren. 2000. AIDS across Europe, 1994–98: the EuroSIDA study. Lancet **356**:291–6.
85. Newell, M. E., J. F. Hoy, S. G. Cooper, B. DeGraaff, A. E. Grulich, M. Bryant, J. L. Millar, B. J. Brew, and D. I. Quinn. 2004. Human immunodeficiency virus–related primary central nervous system lymphoma: factors influencing survival in 111 patients. Cancer **100**:2627–36.
86. Olsen, S. J., Y. Chang, P. S. Moore, R. J. Biggar, and M. Melbye. 1998. Increasing Kaposi's sarcoma-associated herpesvirus seroprevalence with age in a highly Kaposi's sarcoma endemic region, Zambia in 1985. AIDS **12**:1921–5.
87. Palefsky, J. M. 2003. Cervical human papillomavirus infection and cervical intraepithelial neoplasia in women positive for human immunodeficiency virus in the era of highly active antiretroviral therapy. Curr Opin Oncol **15**:382–8.
88. Pellett, P. E., D. J. Wright, E. A. Engels, D. V. Ablashi, S. C. Dollard, B. Forghani, S. A. Glynn, J. J. Goedert, F. J. Jenkins, T. H. Lee, F. Neipel, D. S. Todd, D. Whitby, G. J. Nemo, and M. P. Busch. 2003. Multicenter comparison of serologic assays and estimation of human herpesvirus 8 seroprevalence among US blood donors. Transfusion **43**:1260–8.
89. Piketty, C., T. M. Darragh, M. Da Costa, P. Bruneval, I. Heard, M. D. Kazatchkine, and J. M. Palefsky. 2003. High prevalence of anal human papillomavirus infection and anal cancer precursors among HIV-infected persons in the absence of anal intercourse. Ann Intern Med **138**:453–9.
90. Powles, T., G. Matthews, and M. Bower. 2000. AIDS related systemic non-Hodgkin's lymphoma. Sex Transm Infect **76**:335–41.
91. Rabkin, C. S. 2001. AIDS and cancer in the era of highly active antiretroviral therapy (HAART). Eur J Cancer **37**:1316–9.
92. Rabkin, C. S., M. A. Testa, J. Huang, and J. H. Von Roenn. 1999. Kaposi's sarcoma and non-Hodgkin's lymphoma incidence trends in AIDS Clinical Trial Group study participants. J Acquir Immune Defic Syndr **21**(Suppl 1):S31–3.
93. Raez, L., L. Cabral, J. P. Cai, H. Landy, G. Sfakianakis, G. E. Byrne, Jr., J. Hurley, E. Scerpella, D. Jayaweera, and W. J. Harrington, Jr. 1999. Treatment of AIDS-related primary central nervous system lymphoma with zidovudine, ganciclovir, and interleukin 2. AIDS Res Hum Retroviruses **15**:713–9.
94. Ratner, L., J. Lee, S. Tang, D. Redden, F. Hamzeh, B. Herndier, D. Scadden, L. Kaplan, R. Ambinder, A. Levine, W. Harrington, L. Grochow, C. Flexner, B. Tan, and D. Straus. 2001. Chemotherapy for human immunodeficiency virus-associated non-Hodgkin's lymphoma in combination with highly active antiretroviral therapy. J Clin Oncol **19**:2171–8.
95. Re, A., C. Cattaneo, M. Michieli, S. Casari, M. Spina, M. Rupolo, B. Allione, A. Nosari, C. Schiantarelli, M. Vigano, I. Izzi, P. Ferremi, A. Lanfranchi, M. Mazzuccato, G. Carosi, U. Tirelli, and G. Rossi. 2003. High-dose therapy and autologous peripheral-blood stem-cell transplantation as salvage treatment for HIV-associated lymphoma in patients receiving highly active antiretroviral therapy. J Clin Oncol **21**:4423–7.
96. Roizman, B., R. C. Desrosiers, B. Fleckenstein, C. Lopez, A. C. Minson, and M. J. Studdert. 1992. The family Herpesviridae: an update. Arch Virol **123**:425–49.
97. Rooney, C. M., M. A. Roskrow, C. A. Smith, M. K. Brenner, and H. E. Heslop. 1998. Immunotherapy for Epstein-Barr virus-associated cancers. J Natl Cancer Inst Monogr **23**:89–93.
98. Roychowdhury, S., R. Peng, R. A. Baiocchi, D. Bhatt, S. Vourganti, J. Grecula, N. Gupta, C. F. Eisenbeis, G. J. Nuovo, W. Yang, P. Schmalbrock, A. Ferketich, M. Moeschberger, P. Porcu, R. F. Barth, and M. A. Caligiuri. 2003. Experimental treatment of Epstein-Barr virus-associated primary central nervous system lymphoma. Cancer Res **63**:965–71.
99. Scadden, D. T. 2003. AIDS-related malignancies. Annu Rev Med **54**:285–303.
100. Schulz, T. F. 2000. KSHV (HHV8) infection. J Infect **41**:125–9.
101. Schuman, P., S. E. Ohmit, R. S. Klein, A. Duerr, S. Cu-Uvin, D. J. Jamieson, J. Anderson, and K. V. Shah. 2003. Longitudinal study of cervical squamous intraepithelial lesions in human immunodeficiency virus (HIV)-seropositive and at-risk HIV-seronegative women. J Infect Dis **188**:128–36.
102. Serraino, D. 1999. The spectrum of AIDS-associated cancers in Africa. AIDS **13**:2589–90.

103. Shibata, D., L. M. Weiss, A. M. Hernandez, B. N. Nathwani, L. Bernstein, and A. M. Levine. 1993. Epstein-Barr virus-associated non-Hodgkin's lymphoma in patients infected with the human immunodeficiency virus. Blood **81**:2102–9.
104. Simon, V., D. D. Ho, and Q. Abdool Karim. 2006. HIV/AIDS epidemiology, pathogenesis, prevention, and treatment. Lancet **368**:489–504.
105. Spina, M., M. Berretta, and U. Tirelli. 2003. Hodgkin's disease in HIV. Hematol Oncol Clin North Am **17**:843–58.
106. Spina, M., and U. Tirelli. 2004. HIV-related non-Hodgkin's lymphoma (HIV-NHL) in the era of highly active antiretroviral therapy (HAART): some still unanswered questions for clinical management. Ann Oncol **15**:993–5.
107. Stebbing, J., V. Marvin, and M. Bower. 2004. The evidence-based treatment of AIDS-related non-Hodgkin's lymphoma. Cancer Treat Rev **30**:249–53.
108. Straus, D. J. 2001. HIV-associated lymphomas. Curr Oncol Rep **3**:260–5.
109. Strickler, H. D., R. D. Burk, M. Fazzari, K. Anastos, H. Minkoff, L. S. Massad, C. Hall, M. Bacon, A. M. Levine, D. H. Watts, M. J. Silverberg, X. Xue, N. F. Schlecht, S. Melnick, and J. M. Palefsky. 2005. Natural history and possible reactivation of human papillomavirus in human immunodeficiency virus-positive women. J Natl Cancer Inst **97**:577–86.
110. Sun, X. W., J. P. Koulos, J. C. Felix, A. Ferenczy, R. M. Richart, T. W. Park, and T. O. Wright, Jr. 1995. Human papillomavirus testing in primary cervical screening. Lancet **346**:636.
111. Sun, X. W., L. Kuhn, T. V. Ellerbrock, M. A. Chiasson, T. J. Bush, and T. C. Wright, Jr. 1997. Human papillomavirus infection in women infected with the human immunodeficiency virus. N Engl J Med **337**:1343–9.
112. Tam, H. K., Z. F. Zhang, L. P. Jacobson, J. B. Margolick, J. S. Chmiel, C. Rinaldo, and R. Detels. 2002. Effect of highly active antiretroviral therapy on survival among HIV-infected men with Kaposi sarcoma or non-Hodgkin lymphoma. Int J Cancer **98**:916–22.
113. Teruya-Feldstein, J., E. Chiao, D. A. Filippa, O. Lin, R. Comenzo, M. Coleman, C. Portlock, and A. Noy. 2004. CD20-negative large-cell lymphoma with plasmablastic features: a clinically heterogenous spectrum in both HIV-positive and -negative patients. Ann Oncol **15**:1673–9.
114. UNAIDS. 2006. Report on the global AIDS epidemic: a UNAIDS 10th anniversary special edition. http://www.unaids.org/en/HIV_data/2006GlobalReport/default.asp.
115. Vaccher, E., M. Spina, and U. Tirelli. 2001. Clinical aspects and management of Hodgkin's disease and other tumours in HIV-infected individuals. Eur J Cancer **37**:1306–15.
116. van Baarle, D., E. Hovenkamp, M. F. Callan, K. C. Wolthers, S. Kostense, L. C. Tan, H. G. Niesters, A. D. Osterhaus, A. J. McMichael, M. H. van Oers, and F. Miedema. 2001. Dysfunctional Epstein-Barr virus (EBV)-specific CD8(+) T lymphocytes and increased EBV load in HIV-1 infected individuals progressing to AIDS-related non-Hodgkin lymphoma. Blood **98**:146–55.
117. Wabinga, H. R., D. M. Parkin, F. Wabwire-Mangen, and J. W. Mugerwa. 1993. Cancer in Kampala, Uganda, in 1989–91: changes in incidence in the era of AIDS. Int J Cancer **54**:26–36.
118. Waggoner, S. E. 2003. Cervical cancer. Lancet **361**:2217–25.
119. Yarchoan, R., G. Tosato, and R. F. Little. 2005. Therapy insight: AIDS-related malignancies – the influence of antiviral therapy on pathogenesis and management. Nat Clin Pract Oncol **2**:406–15; quiz 423.
120. Young, L., C. Alfieri, K. Hennessy, H. Evans, C. O'Hara, K. C. Anderson, J. Ritz, R. S. Shapiro, A. Rickinson, E. Kieff, et al. 1989. Expression of Epstein-Barr virus transformation-associated genes in tissues of patients with EBV lymphoproliferative disease. N Engl J Med **321**:1080–5.
121. Young, L. S., and A. B. Rickinson. 2004. Epstein-Barr virus: 40 years on. Nat Rev Cancer **4**:757–68.
122. Ziegler, J. L., J. A. Beckstead, P. A. Volberding, D. I. Abrams, A. M. Levine, R. J. Lukes, P. S. Gill, R. L. Burkes, P. R. Meyer, C. E. Metroka, et al. 1984. Non-Hodgkin's lymphoma in 90 homosexual men. Relation to generalized lymphadenopathy and the acquired immunodeficiency syndrome. N Engl J Med **311**:565–70.
123. zur Hausen, H., L. Gissmann, W. Steiner, W. Dippold, and I. Dreger. 1975. Human papilloma viruses and cancer. Bibl Haematol **43**:569–71.

2. AIDS MALIGNANCIES

**ANITA ARORA, MD*, ELIZABETH CHIAO, MD*,
AND STEPHEN K. TYRING, MD, PHD†**

*Center for Clinical Studies, Houston, TX
†University of Texas Health Science Center, Houston, TX

INTRODUCTION

Malignancies were one of the earliest recognized manifestations that lead to the eventual description of the acquired immune deficiency syndrome (AIDS) epidemic. Kaposi's sarcoma (KS), a rare skin cancer primarily seen in elderly men prior to the epidemic, became one of the first entities described in association with AIDS. Subsequently, non–Hodgkin's lymphoma (NHL) and cervical cancer also became defined by the Centers for Disease Control (CDC) as "AIDS-defining" illnesses. Although these cancers were not classic "Opportunistic Infections," it eventually became clear that these cancers were associated with coinfections with oncogenic viruses among HIV-infected individuals. In fact, the unique epidemiology of KS subsequently led to the identification of the novel human herpes virus, HHV-8. Other oncogenic viruses that are linked to HIV-associated malignancies include Epstein-Barr virus (EBV) and human papillomavirus (HPV). These three viruses have now been linked to other HIV-associated malignancies, including Hodgkin's lymphoma (HL), leiomyosarcoma, multicentric Castleman's disease (MCD), primary effusion lymphoma (PEL), and anal cancer.

The introduction of highly active antiretroviral therapy (HAART) in the 1990s has enormously impacted the outcomes of HIV infection. In addition to changing the natural history of HIV infection, in terms of survival and incidence of opportunistic diseases, it has also dramatically decreased the incidence of some virally mediated HIV-associated malignancies, such as KS, and primary CNS lymphoma

(PCNSL), but has had less of an impact on other virally mediated malignancies. This chapter will review available data demonstrating the effect of HAART on the epidemiology, presentation, treatment, and outcomes of HIV-associated malignancies mediated by EBV, HHV-8, and HPV.

EPSTEIN-BARR VIRUS-RELATED TUMORS

Epstein-Barr Virus

Epstein-Barr virus (EBV), a human herpesvirus of the *Lymphocrytovirus* genus,[1] was first discovered by electron microscopy in Burkitt's lymphoma (BL) cells in 1964.[2] Subsequently, the complete (172,282 bp) nucleotide sequence of EBV (B95-8 strain) was established in 1984.[3] Like all herpesviruses, EBV has latent and productive (lytic) phases in its life cycle, the former maintaining the virus long term in its host and the latter potentiating virus production and spread. Using a distinct set of latent genes EBV has oncogenic capability and the ability to induce immortalization of B lymphocytes in vitro.[4]

EBV has been implicated in a wide variety of diseases in individuals who are immunodeficient, including in the pathogenesis of a variety of malignancies in HIV/AIDS setting, such as in PCNSL, BL, Hodgkin's disease, diffuse large B-cell lymphoma (NHL), and leiomyosarcoma.[5-8] Individuals with AIDS have a significant defect of T-cell immunity to EBV and an elevated number of EBV-infected B cells in the circulation.[9] Specific T cells that target EBV-infected B cells are not decreased, but they lose their ability to respond to EBV antigens, including ones involved in virus reactivation and subsequent lymphomagenesis. The exact role of EBV in AIDS lymphomagenesis is unclear but it is hypothesized that chronic antigen exposure to HIV and EBV results in excessive B-cell stimulation, leading to proliferation of antigen-selected B-cell clones.[10] Each section on the EBV-related tumors further discusses EBV pathogenesis respective to each individual tumor type.

AIDS-RELATED PRIMARY CNS LYMPHOMA

PCNSL is a rare type of NHL, accounting for 1–2% of all NHLs and <5% of all primary brain tumors. In the last few decades the incidence has increased, approximately threefold.[11,12] Although immune deficiency is a risk factor for developing PCNSL, the increase in incidence is not completely explained by the human immunodeficiency (HIV) epidemic.[13]

EBV PATHOGENESIS IN PCNSL

In regards to PCNSL, the vast majority has been linked to EBV.[5] Macmahon and colleagues[14] described that EBV genes important for oncogenesis are abundant in patients with PCNSL, suggesting a pathogenic role of EBV in this setting. This association is suggestive that the pathogenesis of PCNSL might differ from systemic NHL, which only has a 40–50% association with EBV.[15-17]

ROLE OF EBV IN DIAGNOSIS OF PCNSL

EBV studies have been largely used as one of the least invasive ways to diagnose AIDS PCNSL. Specifically, using polymerase chain reaction (PCR) to detect EBV DNA in the CSF is useful for diagnosing AIDS-associated CNS lymphoma. PCR of EBV in PCNSL can detect most cases of AIDS-related PCNSL with a sensitivity of 80–100% and specificity for lymphoma of 93–100%.[18–21]

Brink and colleagues[22] demonstrated in a prospective study that the presence of EBV DNA in CSF was strongly linked with a diagnosis of CNS lymphoma, even when the clinical and radiological features of this disease were not present. In this study, nine of 96 patients had detectable EBV DNA in their CSF, two of them later developing lymphoma in the absence of clinical or radiological findings of PCNSL. This is suggestive that the presence of EBV DNA in CSF may predict later tumor development. For example, there have been four case reports noted where the appearance of EBV DNA in CSF was found to precede the diagnosis of AIDS PCNSL by 35 days.[23] Another report found that detectable EBV DNA was noted to precede the diagnosis of PCNSL by 17 months.[24]

EPIDEMIOLOGY OF PCNSL

In the era before HAART, the relative risk of PCNSL was approximately 1,000-fold and as high as 3,600-fold in individuals with AIDS compared to the general population.[25] With the introduction of HAART, the incidence of AIDS PCNSL has decreased.[26,27] In the Multicenter AIDS Cohort Study, incidence rate in 2,734 HIV-infected men went from 4.3 to 0.4.[27] Although incidence has decreased dramatically, survival rates have not significantly improved.[28]

CLINICAL PRESENTATION OF PCNSL

Clinical presentation of CNS lymphoma may be characterized by headaches, confusion, lethargy, memory loss, hemiparesis, aphasia, and seizures. Lesions are most common in the cerebrum, basal ganglia, and brainstem.[14,29] The lesions are usually large, few in number, and contrast enhanced on neuroimaging. In the pre-HAART era, most patients with PCNSL were near the late stages of AIDS, with the majority dying within 4–6 weeks of diagnosis with supportive care alone.[30]

TREATMENT AND SURVIVAL OF NON-AIDS-ASSOCIATED PCNSL

Treatment for PCNSL was first evaluated in immunocompetent patients. Without treatment most patients with PCNSL are likely to die within three months, even in the absence of AIDS. Treatment modalities for non–AIDS-related PCNSL include glucocorticoids, radiation therapy, chemotherapy, autologous stem cell transplant, and immunotherapy.[31]

In non–AIDS-related PCNSL, the conventional chemotherapy for systemic lymphoma consisting of cyclophosphamide, doxorubicin, vincristine, and pred-nisone (CHOP) was ineffective for PCNSL.[32,33] These findings may be due to the compounds in CHOP penetrating poorly through the blood–brain barrier and also due to molecular differences between systemic lymphoma and PCNSL.[31]

Methotrexate administered intravenously is able to penetrate the blood–brain barrier and achieve levels high enough to be cytotoxic. In non-AIDS-related PCNSL, using methotrexate with radiation therapy (RT) and other chemotherapy regimens demonstrated a significant survival prolongation compared with RT alone. Medial survival was 42 months and 5-year survival was 22.3 months, compared to 3–5% rate of 5-year survival in individuals treated with RT only.[34]

TREATMENT AND SURVIVAL OF AIDS-ASSOCIATED PCNSL

Radiation

The standard treatment of patients with AIDS-related CNS lymphoma is palliative whole-brain radiation. Radiation can improve symptoms and extend median survival to between 2 and 5.5 months.[35] Survival was often limited and patients often develop opportunistic infections in the pre-HAART era.[36] Survival appears to be more affected by an individual's baseline functional status than on the dose of radiation received.[37]

Chemotherapy (Table 1)[38]

In AIDS-related PCNSL, there is no optimum regimen of chemotherapy. In the largest combined chemotherapy and radiation regimen trial, a single cycle of CHOD (cyclophosphamide–doxorubicin–vincristine–dexamethasone) followed by RT, the median survival rate of 2.4 months was similar to that of radiation therapy alone.[39] Chamberlain and colleagues[40] studied whole-brain radiation therapy plus hydroxyurea radiosensitization, followed by procarbazine–lomustine–vincristine (PCV), and while the study did show a longer than average survival, approximately 13 months, the individuals represented a highly selected population of patients, all that had a much higher pretreatment functional status in regards to their disease.

In AIDS-related PCNSL, an uncontrolled pilot study using intravenous methotrexate resulted in median survival of 2.4 months in the ten patients that had histologically confirmed lymphoma. Two of ten patients survived beyond a year, and half of the patients did not respond to treatment. Analysis in this study was complicated because not all patients had histological confirmation of lymphoma.[41] In addition, steroids were also administered and individuals ultimately received antiretrovirals and a protease inhibitor.[41]

Zidovudine, Ganciclovir, Interleukin-2

The results of a study using intravenous zidovudine, ganciclovir, and low–dose interleukin-2 (IL-2) provided long-term remission, but further studies are necessary as the study was small and not all patients were treated similarly.[42]

Role of HAART

HAART should be initiated in all individuals with AIDS-related PCNSL that are undertaking treatment. In the pre-HAART era, RT prolonged survival for

Table 1. Chemotherapy Regimens for AIDS–Related PCNSL

Study	No. of patients	Histologic diagnosis	Median age	Median CD4 count (µL)	Regimen	Overall survival in months	HAART vs no HAART
Ambinder et al.[39]	34	All	36 [22–54]	10.5	CHOD, followed by RT	2.4	No HAART
Chamberlain and Kormanik[40]	7	All	Not reported	>200	RT with hydroxyurea, followed by PCV	13	No HAART
Jacomet et al.[41]	10	All	Not reported	7	Methotrexate ± steroids	2.4	Most responders received antiretrovirals including protease inhibitor
Raez et al.[42]	5	3	33 [18–40]	<50	Zidovudine IV, ganciclovir, IL–2; oral lamivudine; Indinavir	0.5, 4, 7, 22+, 13+*	3 patients received no HAART and 2 received AZT for <3 months

*One patient received radiation, one responded but required another treatment (surviving 7 months), and one patient was lost to follow-up after treatment. (Adapted from reference 38)

2–5.5 months compared to palliative care.[35] McGowan and Shah[43] were the first to describe a case of remission maintained for over 2 years in an individual who had PCNSL after treatment with HAART alone. Hoffmann and colleagues,[44] in a retrospective analysis, showed that survival time of patients receiving HAART in addition to RT differed significantly from those receiving RT or palliative care alone. Four of the six patients receiving HAART survived for more than 1.5 years. Skiest and Crosby[45] in a retrospective analysis demonstrated a prolonged median survival of 667 days in those individuals who did receive HAART. These findings are suggestive that immune recovery is contributing to longer remission in patients studied.

AIDS-RELATED NON-HODGKIN'S LYMPHOMA

AIDS-related lymphomas (ARL) include a heterogeneous group of tumors.[46] The first cases of AIDS-related NHL were described in 1982[47,48] and NHL was included as an AIDS-defining malignancy in 1985.[49] NHL is now known as the second most common neoplasm occurring among HIV-infected individuals.[50,51] Diffuse large B-cell lymphoma (DLBCL) and BL are the most common of ARL, representing approximately 90%.[52,53] The World Health Organization (WHO) has divided ARL into three categories (Table 2): (1) lymphomas also occurring in immunocompetent patients such as BL and DLBCL; (2) lymphomas occurring more specifically in HIV-infected patients such as PEL and plasmablastic lymphoma; and (3) lymphomas also occurring in other immunodeficiency states such as polymorphic or post-transplant lymphoproliferative disorder-like B-cell lymphoma.[54] AIDS-related NHLs are grouped into three distinct histologic categories, including small non-cleaved cell lymphoma, diffuse large-cell lymphoma, and anaplastic large-cell lymphoma. BL is defined as an aggressive, high-grade, small non-cleaved cell lymphoma, and DLBCL as an intermediate-grade lymphoma diffuse large cell lymphoma.[55,56] BL and DLBCL will be discussed separately in the following sections.

BURKITT'S LYMPHOMA

In the WHO classification, three clinical variants of BL are described: endemic, sporadic, and immunodeficiency-associated types.[56] Immunodeficiency-associated BL occurs mainly in HIV-infected individuals, but also occurs in transplant patients and individuals with congenital immunodeficiency.[57,58] AIDS-related BL are further discussed in this chapter.

Table 2. AIDS-related lymphomas (WHO classification)[54]

Lymphomas also occurring in immunocompetent patients
Burkitt's lymphoma
DLBCL: centroblastic, immunoblastic, and anaplastic variants
Lymphomas occurring more specifically in HIV-infected patients
Primary effusion lymphoma
Plasmablastic lymphoma
Lymphomas also occurring in other immunodeficiency states
Polymorphic or post-transplant lymphoproliferative disorder-like B-cell lymphoma

EPIDEMIOLOGY OF AIDS-RELATED BL

BL was estimated to be 1,000 times more common in HIV-infected individuals as compared to the general population.[50] BL accounts for 30–40% of NHL in HIV-infected individuals.[59–61] Neoplastic cells in BL have been reported as EBV[+] in 15–30% of cases or fewer in some series.[62]

CLINICAL PRESENTATION OF AIDS-RELATED BL

Several cases of BL were described in homosexual men in the beginning of the AIDS epidemic and these were the first descriptions of NHL in association with HIV.[47,48] In BL, 30% of patients present with limited-stage disease (Stage I or II according to Ann Arbor staging system[56]), while 70% present with widespread disease.[63] Patients often present with bulky disease, frequently with an elevated lactate dehydrogenase level (LDH).[56] BL commonly involves the lymph nodes, bone marrow (positive in 30–38%), and the CNS (involved in 13–17% of cases) in HIV-infected adults.[59,64] BL patients may be younger and have higher mean CD4 counts than patients with AIDS-related DLBCL.[56]

TREATMENT OF AIDS-RELATED BL

In regards to treatment, current studies have uniformly targeted intermediate-grade DLBCL and high-grade BL in HIV-infected individuals. In non-HIV-infected individuals, multiagent, dose intensive chemotherapy regimens are used instead of standard regimens (CHOP) for the treatment of BL, but these regimens have included very few patients infected with HIV.[49] Lim and colleagues[65] reviewed 363 individuals with ARL (including DLBCL and BL) from 1982 to 2003, including 262 in the pre-HAART era, and 101 in the HAART era. Although overall median survival was similar for both DLBCL and BL (8.3 months and 6.4 months, respectively), survival was significantly worse in individuals with HIV-BL in the HAART era (5.7 months in HIV-BL vs 43.2 months in DLBCL).

It is now recommended that similar regimens used for both DLBCL and BL in HIV-infected individuals be re-evaluated with prospective trials.[65] Cortes and colleagues used HyperCVAD (hyperfractionated cyclophosphamide, vincristine, doxorubicin, and dexamethasone alternating with high-dose methotrexate and cytarabine) regimen in 13 patients with HIV-related BL, nine of whom also received HAART. Complete remission (CR) was 92% although median survival was only 12 months.[66] While Wang and colleagues, in a retrospective study, compared 14 HIV-infected individuals with BL with 24 non-HIV-infected individuals with BL who received CODOX-M/IVAC (cyclophosphamide, doxorubicin, high-dose methotrexate/ifosfamide, etoposide, and high-dose cytarabine) or less intensive regimens. Patients receiving CODOX-M/IVAC, independent of HIV status, had a better 2-year event-free survival compared with those receiving less intensive regimens (80% vs 38%, respectively, $p = 0.05$).[67]

EBV PATHOGENESIS IN AIDS-RELATED NHL

The pathogenesis of HIV-NHL is most likely multifactorial, involving HIV, immune dysfunction, cytokine dysregulation, and other viral antigens.[49] In AIDS-related systemic NHL, only approximately 40–50% contain EBV.[15–17] Of AIDS-related NHLs, EBV occurs more frequently in the immunoblastic lymphoma (IBL) variant of DLBCL than in BL.[16] Ballerini et al.[16] reported 100% EBV infection in the IBL variant of DLBCL. Neoplastic cells were EBV[+] in 77% of cases in another study.[68] In a study of plasmablastic lymphomas, Teruya-Feldstein and colleagues reported EBV[+] cells in 73% of cases.[69]

The expression of the latent EBV transforming proteins EBNA-2 and LMP-1 are known to play central roles in the initiation and maintenance of EBV-induced B-cell growth and proliferation.[70,71] Both EBNA-2 and LMP-1 can serve as targets for cytotoxic T cells, and thus their expression induces T-cell immunosurveillance and regulates the lymphomagenesis in individuals who are immunocompetent. With immunodeficiency states such as in the late stages of HIV, EBNA-2, and LMP-1 expression may become unregulated, and subsequently lead to uncontrolled proliferation of EBV-infected cells.[71]

EPIDEMIOLOGY OF AIDS-RELATED NHL

The CDC has calculated the risk of NHL in the United States in the pre-HAART era to be 60 times greater in individuals who have AIDS than in non-HIV-infected individuals and the incidence rate of AIDS-related NHL was reported as 2.9%. These data are based on 2,824 NHLs occurring among 97,258 HIV-infected individuals between 1981 and mid-1989.[50] The risk also varies by histologic subtype, up to 600-fold excess risk for IBL.[72]

IMPACT OF HAART ON INCIDENCE OF AIDS-RELATED NHL

In a meta-analysis by Appleby et al.,[73] including 47,936 HIV-infected individuals with NHL (inclusive of PCNSL), the incidence fell from 6.2 cases/1,000 person-years (p-y) in the pre-HAART era to 3.6 cases/1,000 p-y in the post-HAART era ($p < 0.0001$). The EuroSIDA group[74] reported a fall in incidence rates from its database of 26,764 p-y of observation in ARLs with antiretrovirals, 1.99 cases/100 p-y before HAART to 0.30 cases/100 p-y after HAART ($p < 0.001$). Among patients who started HAART, the incidence of NHL decreased from 0.88/100 p-y within the first 12 months after starting HAART to 0.45 cases/100 p-y after more than 24 months ($p = 0.004$). Besson and colleagues[53] reported in a large population of HIV-infected patients in the French Hospital Database a fall in incidence of systemic ARL from 86.0 cases/10,000 p-y in 1993–1994 (pre-HAART) compared to 42.9/10,000 p-y in 1997–1998 (post-HAART) ($p < 0.001$). Grulich et al.,[75] used a national registry in Australia and found the incidence to decrease from 7–7.5 cases/1,000 p-y pre-HAART to 4.3 cases/1,000 p-y in the post-HAART era ($p < 0.012$).

In addition, The Swiss cohort study[76] with 6,636 HIV-infected individuals found the incidence of systemic NHL post-HAART to be 4 cases/1,000 p-y.

However, other studies found nonsignificant increases in AIDS-related NHL incidence. In the San Francisco clinic cohort study,[77] including 622 patients, the incidence of systemic NHL increased from 14 cases/1,000 p-y (pre-HAART) to 18 cases/1,000 p-y post HAART ($p = 0.2$). In the adult/adolescent spectrum of HIV disease project (CDC)[78] of 19,684 individuals, including those with PCNSL and systemic NHL, incidence rates increased from 11.4 cases/1,000 p-y pre-HAART to 12.5 cases/1,000 p-y post-HAART (p-value not significant).

PRESENTATION OF AIDS-RELATED NHL

Approximately 2/3 of ARLs are accounted by DLBCL.[79] It is more frequent for DLBCL to arise extranodally, for instance in the GI tract and CNS and it rarely involves the bone marrow.[64] With DLBCL, particularly with the immunoblastic variant, it is common for the patient to have a lower CD4 T cell count (<50/mm^3) at diagnosis, be older in age, and carry a prior diagnosis of AIDS.[50,80,81] Plasmablastic lymphoma is a rare AIDS-related NHL which typically involves the jaw and oral cavity of HIV-infected individuals,[82] and also has been more recently documented in other sites such as the anorectum, nasal and paranasal regions, skin, testes, bones, and lymph nodes.[83,84] Plasmablastic lymphomas are morphologically different from other NHLs by lacking CD20 antigen, but containing antigens such as CD138, P63, and variable CD79A.[69,85]

The clinical prognostic factors in the international prognostic index (IPI), initially described in the pre-HAART era, have been used in risk stratification for patients with DLBCL for more than a decade (Table 3).[86,87] The IPI includes clinical features that reflect the growth and invasive potential of the tumor (tumor stage, serum LDH level, and number of extranodal disease sites), the patient's response to the tumor (performance status), and the patient's ability to tolerate

Table 3. 5-Year relapse-free and overall survival rates according to the International Prognostic Index (IPI) and age-adjusted IPI (adapted from references[86])[87]

Risk group	Number of adverse factors*	5-Year relapse survival (%)	5-Year overall survival (%)
International prognostic index			
Low	0 or 1	70	73
Low-intermediate	2	50	51
High-intermediate	3	49	43
High	4 or 5	40	26
Age-adjusted international prognostic index			
Low	0	86	83
Low-intermediate	1	66	69
High-intermediate	2	53	46
High	3	58	32

*Adverse risk factors for IPL are: stage III or IV disease, age > 60 years, elevated lactate dehydrogenase (LDH), and performance status ≥ 2, ≥2 extranodal sites. Adverse risk factors for age-adjusted IPI are: stage III or IV disease, elevated LDH, performance status ≥2.

intensive therapy (age and performance status). The simplified model for younger patients (the age-adjusted IPI) uses a subgroup of these clinical features (tumor stage, LDH level, and performance status).

Lim et al.[88] compared the prognostic factors for survival and the use of the IPI in pre- and post-HAART HIV-infected individuals with diffuse-large cell lymphoma. In groups with low-, low-intermediate-, and high-intermediate-risk IPI disease, 3-year overall survival rates were 20%, 22%, and 5% in the pre-HAART era and 64%, 64%, and 50% in the HAART era, respectively.

TREATMENT OF AIDS-RELATED NHL

The treatment for ARLs is similar to that of non–HIV-infected individuals, with some exceptions. Intrathecal chemotherapy prophylaxis is necessary as patients with HIV are at an increased risk for CNS involvement.[89] The use of hematopoetic stimulants such as granulocyte-colony stimulating factor (G-CSF) may aid in reducing chemotherapy-induced cytopenic complications. For those that are receiving concomitant HAART with chemotherapy, zidovudine should not be used due to increased risk of myelosuppression, and one should be cautious in using didanosine, stavudine, and zalcitabine, which may potentiate vincristine-induced neuropathy.[90]

Chemotherapy in the Pre-HAART Era

In the pre-HAART era, HIV-infected individuals with NHL had a poor prognosis were managed on low-dose chemotherapy regimens because of concern of toxicity, and a median survival of 5–8 months was common.[91–93]

Chemotherapy in the HAART Era

In the HAART era, more recent standard chemotherapy regimens have been reported without excessive toxicity due to restored immunity.[90] The AIDS Malignancy Consortium (AMC) reported using, in 65 patients, reduced doses of cyclophosphamide and doxorubicin, combined with vincristine and prednisone (modified CHOP, mCHOP) and full doses of CHOP combined with G-CSF with concomitant HAART. Complete response rates were 30% and 48% in the reduced- and full-dose groups, respectively.[94] No long-term outcomes were reported in this study. Other studies of CHOP-based chemotherapy and concurrent HAART have resulted in median survival of 2 years.[53,95]

Risk-adaptive chemotherapy has also been studied comparing the post- to pre-HAART era. 485 HIV-infected individuals were assigned randomly to chemotherapy after risk stratification based on an HIV score (comprising performance status, prior AIDS, and CD4-positive cell counts < 100/mm^3). Two hundred eighteen good risk patients (HIV score 0) received doxorubicin, cyclophosphamide, vindesine, bleomycin, and prednisone (ACVBP) or CHOP, 177 intermediate risk patients (HIV score 1), CHOP or low-dose CHP, and 90 poor risk patients (HIV score 2 or 3), low-dose CHOP or vincristine and steroid. Five-year overall survival in the good risk group was 51% for ACVBP versus 47% for CHOP ($p = 0.85$), in the intermediate risk group, 28% for CHOP versus 24% for low-dose CHOP ($p = 0.19$), and in the poor risk group 11% for low-dose CHOP versus 3% for vincristine and steroid ($p = 0.14$). Only significant factors in this study for overall survival proved to be HAART (relative risk (RR) 1.6, $p = 0.0002$), HIV score (RR 1.7, $p = 0.0001$), and the IPI score (RR 1.5, $p = 0.0012$), but not the intensity of chemotherapy.[96]

An infusional regimen of cyclophosphamide, doxorubicin, and etoposide (CDE) with and without HAART (only didanosine) resulted in complete response rate of 45% and median overall survival was 12.8 months. At the time of the analysis, 30% in the pre-HAART group were alive compared with 47% in the HAART group. Further, patients in the HAART group experienced less nonhematologic toxicity (22% vs 42%), thrombocytopenia (31% vs 52%), and anemia (9% vs 27%).[97]

Chemotherapy Without HAART in the HAART Era

Chemotherapy without antiretrovirals has been studied due to concerns of drug interactions with chemotherapy and noncompliance with HAART resulting in increased resistance.[81] Further, protease inhibitors (in HAART regimens) have been associated with neutropenia with concomitant chemotherapy.[98]

The National Cancer Institute used a dose-adjusted regimen of etoposide, vincristine, and doxorubicin (4 days), and daily oral prednisone (5 days), followed by cyclophosphamide. The study consisted of 39 patients and antiretrovirals were not given until after the final cycle of chemotherapy. A CR rate of 74% was achieved.[99]

Rituximab

Kaplan and colleagues[100] reported a randomized trial in the HAART era using CHOP versus CHOP and rituximab (anti-CD20 antibody) given with each cycle and with an additional three monthly doses after complete response was attained. Median event-free survival, approximately a year, was similar between both groups. The group treated with rituximab and CHOP, however, had an increased risk of death from infection (14% vs 2%, $p = 0.027$). Up to 60% of deaths were in patients with a CD4 count of <50/mm³, and 40% occurred during the maintenance phase of rituximab.

In non-HIV-infected individuals, rituximab benefit is limited to lymphomas that overexpress blc-2, and this overexpression is found less commonly in ARL, which may explain the decreased response to immunotherapy in this study.[99]

HEMATOPOIETIC-CELL TRANSPLANTION FOR AIDS-RELATED NHL

Poor outcomes were observed in using allogeneic or syngeneic bone marrow transplantation in the setting of HIV disease in 1980s.[101–104] More recently, several authors have reported on small series of patients treated at individual institutions for high-first risk remission, relapsed, or refractory ARL. No definitive conclusions regarding efficacy can be made due to the small and varied group of patients.

Overall survival has been reported at 50% at 9 months,[105] 85% at 32 months,[106] 55% at 9 months,[107] 71% at 21 months in these series.[108] There were a few opportunistic infections reported in patients receiving transplants, and success rates have ranged from 80 to 100%.[105,107–109]

There has been a sufficient number of HIV-infected individuals that have undergone autologous hematopoietic-cell transplantion (HCT) to say it is safe and feasible approach for ARL patients that meet criteria for transplantation.[79]

All studies except the French series by Diez-Martin[108] and colleagues have required HIV disease to be under control for HCT, either by low to undetectable HIV viral loads or CD4 counts >100/mm³.

AIDS-RELATED HODGKIN'S LYMPHOMA

Lymphomas are categorized in two divisions: NHL and HL. HL is characterized by the presence of the Reed-Sternberg cell, an atypical, large cell.[89] HIV-related HL is the most common non-AIDS defining tumor in HIV-infected individuals.[110] The World Health Organization (WHO) classifies HL into two distinct entities: the more commonly diagnosed classical HL and the rare nodular lymphocyte-predominant HL. Classical HL is further subdivided into nodular sclerosis, mixed cellularity, lymphocyte depleted, and lymphocyte rich.[89,111] EBV association with HL in non-HIV-infected individuals was shown to be lower than 50%.[110] In contrast, nearly all of AIDS-associated HL are EBV positive.[112]

EPIDEMIOLOGY OF HL

While the incidence of non–AIDS-associated HL has decreased over the past 30 years, the majority of recent studies show an increase in incidence of HL in HIV-infected individuals[113] and a clear relationship between the incidence of disease and immunodeficiency.[114] Glaser et al.[115] reported in a decade in the Greater San Francisco Bay area that incidence rates by race of HIV-associated HL were overall higher for whites (11%), blacks (22%) and Hispanics (14%) compared to those with non-HIV-associated HL. Further, data from the AIDS cohort and registry matching studies depicted a relative risk of HL in the HIV setting ranging from 2.5 to 8.5.[116–118]

CLINICAL PRESENTATION OF HL

HL occurring in the HIV population exhibits clinical features that are distinct from HL in the general population. HIV-infected individuals are more likely to present with systemic symptoms (B symptoms), including fever, night sweats, and/or weight loss, advanced disease stage, and extranodal disease.[115,119–121] In the HIV population, the subgroups often seen are the ones associated with worse prognosis, including mixed cellularity and lymphocyte-depleted variant.[115,122,123] Rubio and colleagues in a cooperative study in Spain reported that the individuals with AIDS-associated HL had the histologic subtypes of mixed cellularity (41.3%), lymphoid depletion (21.7%), nodular sclerosis (21.7%), and lymphocytic predominance (4.3%).[124]

In the pre-HAART era, approximately 75% of patients have been shown to have advanced disease stages (stage III–IV according to Ann Arbor staging classification, Table 4), and bone marrow involvement occurs in 40–50% of individuals, and may be the first sign of HL in 20% of cases.[122,124–127]

TREATMENT OF HL

Optimal regimen for HL in HIV-infected individuals has not yet been defined. Antineoplastic treatment of AIDS-associated HL has many problems considering the underlying immunodeficiency caused by HIV itself that is further compromised by additional chemotherapy. Further, CD4[+] counts in these individuals may drop significantly during treatment increasing the risk of opportunistic infections (OI).[110]

Prior to HAART, HIV-infected individuals with HL had a limited median survival of 1–2 years.[110] Gerard and colleagues[129] in a retrospective study over 15 years estimated the 2-year survival probability was 45% in the pre-HAART period, and 62% in the post-HAART period. HIV-infected individuals who receive and respond to HAART within 2 years of their HL diagnosis were shown to have overall survival of 89% at 24 months (median survival was not reached) and median survival time in patients without HAART response was 18.6 months.[123]

Pre-HAART Treatment

A trial of doxorubicin, bleomycin, vinblastine, and dacarbazine (ABVD), the standard treatment for HL in non-HIV-infected individuals,[130] used with no antiretroviral therapy (only initiated at the end) was associated with significant hematologic toxicity and a poor median survival of 1.5 years.[131]

Chemotherapy and HAART

Several prospective trials to date have now been done using antiretrovirals concomitantly with different chemotherapy regimens. The regimen bleomycin, etoposide, doxorubicin, cyclophosphamide, vincristine, procarbazine, and prednisone

Table 4. Modified Ann Arbor staging system

Stage	Area of involvement
I	Single lymph node group
II	Multiple lymph node groups on same side of diaphragm
III	Multiple lymph node groups on both sides of diaphragm
IV	Multiple extranodal sites or lymph nodes and extranodal disease
Modifiers	
X	Bulk > 10 cm
E	Extranodal extension or single isolated site of extranodal disease that can be encompassed in a single radiation port
A/B	B symptoms: weight loss > 10%, fever, drenching night sweats

Adapted from Lister and colleagues.[128]

(BEACOPP) was used in 12 patients and five of the 12 patients received concurrent HAART. CR was achieved in all patients. Of 12 patients, nine patients remained in CR for their individual follow-up period, median of 49 months. The most common observed toxicity was bone marrow suppression with grade 3/4 leukopenia in 75% of all cases.[132]

A prospective, nonrandomized trial with epirubicin, bleomycin, vinblastine, and prednisone regimen concomitantly with HAART and G-CSF was used on 35 previously untreated patients. The median survival was 16 months, with a survival rate of 32% and a disease-free survival of 53% at 36 months. Toxicity was moderate with grade 3/4 leukopenia observed in 32% of patients and thrombocytopenia in 10% of patients. Of the 23 patients who died, 48% died of HL and 9% of patients died of OI.[133]

Spina and colleagues[134] used the Stanford V regimen (doxorubicin, vinblastine, mecloretamine, etoposide, vincristine, bleomycine, and prednisone) and radiotherapy in a phase II prospective study plus HAART and G-CSF on 59 HIV-infected individuals. Sixty-nine percent of patients completed treatment with no dose reduction or delayed chemotherapy administration. Bone marrow toxicity and neurotoxicity were the most notable dose-limiting adverse effects. CR was seen in 81% of patients, and 56% of patients were alive and disease-free at median follow-up of 17 months.

AIDS-RELATED LEIOMYOSARCOMA

The range of AIDS-associated pediatric cancers resembles that seen in HIV-infected adults, with the addition of leiomyosarcoma. Leiomyosarcoma, a tumor of smooth muscle origin, is a rare tumor in childhood, occurring in the general population at a rate of one case per one million children. In HIV-positive children 14 years of age and younger, leiomyosarcoma was found to be the third most common cause of cancer after NHL and Kaposi's sarcoma (KS).[135]

ROLE OF EPSTEIN-BARR VIRUS

Leiomyosarcomas in HIV-infected children differ from cases occurring in the general population by EBV detected by in situ hybridization and PCR in nearly all cases.[6–8] Jenson and colleagues[8] reported several EBV antigens, including latent antigen EBNA-1, expressed from cells cultured from leiomyosarcoma. Human umbilical cord lymphocytes that were transformed with the virus that was isolated from the tumor expressed EBV antigens similar to EBV-transformed lymphocyte cell lines. These results confirmed that EBV may have a contributory role in the oncogenesis of leiomyosarcomas.

EPIDEMIOLOGY

In HIV-infected children, leiomyosarcoma incidence has been accounted at around one case per 5,000 children.[6,136,137] Biggar and colleagues[135] used an AIDS-cancer registry and found that leiomyosarcomas appear to occur several years after AIDS onset, with three of the four cases occurring 33–76 months after AIDS diagnosis.

The relative risk of leiomyosarcoma during the time period 2–5 years after AIDS diagnosis was 1,915 (95% CI, 232–6,915).

Granovsky and colleagues,[138] in a retrospective analysis of cancer cases from the Children's Cancer Group and National Cancer Institute, found that 11 among 64 children had leiomyosarcomas, representing 17% of all tumors and second behind NHL. Median survival for leiomyosarcoma after diagnosis was 12 months. The monthly death rate after leiomyosarcoma diagnosis increased from 5% in the first 6 months to about 20% thereafter.

CLINICAL PRESENTATION

In immunocompetent children, leiomyosarcoma generally presents with indolent disease, but in HIV-infected children, leiomyosarcomas are usually fairly aggressive.[139] Unusual localizations, such as spleen, pleural space, adrenal glands, and lungs have been reported as the site of the leiomyosarcoma, although they present most commonly in the gastrointestinal tract. Several cases of intracranial or dural leiomyosarcomas have also been described.[140]

TREATMENT IN THE PRE-HAART ERA

Smooth muscle tumors are in general not responsive to chemotherapy; thus local excision or radiotherapy, if feasible, is the first line of therapy. Intensive and prolonged chemotherapy as used in the general population is not usually well tolerated in children.[141] There are no studies evaluating the effects of HAART on treatment outcomes.

HUMAN HERPES VIRUS-8-RELATED TUMORS

Human Herpes Virus-8

With the HIV epidemic in the early 1980s, homosexual men were found to have a 20-fold higher risk than other risk groups in developing Kaposi's sarcoma (KS). The uneven distribution among HIV risk groups lead to the search for an infectious transmissible agent, which eventually led to the discovery of the KS herpes virus or human herpes virus-8 in 1994. HHV-8, a gammaherpesvirus$_2$ (genus *Rhadinovirus*) is a large (165 kb) double-stranded DNA virus that has an extensive group of regulatory genes obtained from the host genome.[142] The DNA sequences of HHV-8 on discovery were found to be homologous to, yet distinct from genes of gammaherpesvirinae, herpesvirus saimiri, and EBV.[143]

HHV-8 has been associated in all forms of KS and it is now known to be the etiologic agent of KS.[143–146] Aside from KS, HHV-8 has been found in approximately half of the cases of multicentric Castleman's disease (MCD) occurring in non-HIV-infected individuals and nearly in all cases of MCD infected with HIV.[147] HHV-8 sequences have also been detected in peripheral blood of individuals with MCD, and exacerbation of symptoms has been associated with large increases in HHV-8 viral load.[147–149] In addition to KS and MCD, a link between primary effusion lymphoma (PEL) and HHV-8 was first reported in 1995.[150]

Diseases caused by HHV-8 are greatly dependent on the host's underlying immune status and it has been suggested that most individuals attain a high-risk of developing KS, and other HHV-8-associated diseases such as MCD and PEL with HIV infection.[142] For further HHV-8 pathogenesis in KS, MCD, and PEL, refer to each section, respectively.

AIDS-RELATED KAPOSI'S SARCOMA

KS before the AIDS epidemic was known as a rare, chronic skin sarcoma in the United States.[144] Four types of KS have been described: classic, endemic or African, posttransplant or iatrogenic, and AIDS-associated KS. Classic KS, characterized mostly by cutaneous involvement, is typically seen in elderly men of Mediterranean and Eastern European background.[151] The endemic form, found primarily in sub-Saharan Africa often is more aggressive than the classic form, with visceral involvement being more common.[152,153] The third form, posttransplant KS, was described after the initiation of organ transplantation and both cutaneous and visceral disease are common.[154,155] The fourth form of KS was first described in homosexual men in 1981, at the advent of the HIV epidemic.[156,157] Now, KS is the most common neoplasm occurring in HIV-infected individuals.[158] Both cutaneous lesions and visceral involvement are seen in individuals with AIDS-associated KS, and visceral involvement often portends a poor prognosis.[159-161]

HHV-8 PATHOGENESIS IN KS

Multiple factors contribute to the development of KS. Among individuals with HIV, immunosuppression confers the greatest risk and is predictive of development of KS.[162,163] Since the discovery of HHV-8, there have been a large number of studies evaluating its underlying role in KS. Engels et al.[164] demonstrated HHV-8 viremia to be an early marker of KS and Newton et al.[165] found that the risk of developing disease increased with HHV-8 antibody titers.

The role of HHV-8 in oncogenesis is likely associated with its viral proteins, those that are homologous to interleukin-6 (IL-6), chemokines of the macrophage inflammatory protein family, cell cycle regulators, and antiapoptosis molecules.[166-169] HHV-8 has been shown to have certain genes to be homologous to cyclin D1 and G-protein coupled receptors, which are known to contribute to oncogenesis. The presence of these genes suggests that HHV-8 can contribute to the development of KS.[150,170,171]

HIV PATHOGENESIS IN KS

The HIV virus itself may be an oncogene in KS development. A regulatory protein of HIV released by infected cells, called the Tat protein, guards KS cells from apoptosis,[172] stimulates growth and angiogenesis,[173,174] and also increases the production and release of matrix metalloproteinases (MMPs) from endothelial and inflammatory cells. MMPs contribute to the angiogenesis found in KS lesions.[175,176]

EPIDEMIOLOGY OF KS

Incidence of KS has dramatically declined with widespread use of HAART. Mocroft et al.,[177] in a pan-European study from 1994 to 2003 found that the incidence in 2003 to be less than 10% of the incidence reported in 1994. In a large multicenter HIV/AIDS surveillance study from 1990 to 1998, individuals with KS had a 50% reduction in incidence when initiated on triple antiretroviral drug therapy.[178]

Prior to HAART, the median survival for HIV-infected persons with KS visceral involvement was 15 months, and those with cutaneous involvement the median survival was 27 months.[179] Further, in the pre-HAART era, HIV-infected male homosexuals were estimated to have a 20-fold higher risk of KS development in comparison with other HIV transmission risk groups.[180,181]

In areas of Southern Africa where KS is endemic, KS has reached epidemic proportions due to lack of antiretroviral therapy.[182,183] For instance, in Zimbabwe, KS is reported as to represent 40% of all cancers is men.[184] Individuals there with AIDS-associated KS have high tumor burdens and aggressive disease progression and expected mortality is less than 6 months.[182]

CLINICAL PRESENTATION OF KS

KS is an angioproliferative disease varying from an indolent to fulminant disease with significant morbidity and mortality. The disease usually presents with disseminated and pigmented skin lesions, which can evolve from patches to plaques and eventually ulcerating tumors. Lesions are often associated with edema, lymph node, and visceral involvement.[185]

In the pre-HAART era, the AIDS Clinical Trials Group (ACTG) developed a staging system based on tumor extent (T), CD4 cell count (L), and presence of systemic illness (S). Two different risk categories were noted based on this staging system. A good risk is defined as (T0L0S0) and is seen in disease involving the skin, and/or presence of minimal oral disease, CD4 > 150 μL, no opportunistic infectious (OI)/B symptoms and performance status (PS) > 70. A poor risk is defined as (T1L1S1) and describes disease with edema or ulcerations or extensive oral KS and visceral involvement, CD4 < 150 μL, OI and/or B symptoms and PS < 70.[179] Nasti et al.[186] concluded from epidemiological, clinical, staging, and survival data from 211 patients in two Italian prospective cohort studies that in the era of HAART, a refinement of the ACTG staging system is needed. CD4 level in this study did not seem to provide prognostic information; only the combination of poor tumor stage (T1) and poor systemic disease (S1) identified patients with unfavorable prognosis. The 3-year survival rate for patients with T1S1 was 53%, which was significantly lower compared with the 3-year survival rates of patients with T0S0, T1S0, and T0S1, which were 88%, 80%, and 81%, respectively.

Extracutaneous disease is common in individuals with HIV, with oral cavity involvement found in 33% of cases, and can be the initial site of diagnosis 15% of the time.[187] Gastrointestinal KS can occur and has been reported in 40% of cases at initial diagnosis[187] with any segment involved, and individuals may have symptoms of weight loss, abdominal pain, nausea, vomiting, or bleeding.[188] Pulmonary KS is also common and patients may be symptomatic with shortness of breath or cough or hemoptysis or may present with an asymptomatic radiographic finding.[189] Chest X-ray findings can consist of nodular, interstitial, or alveolar infiltrates, isolated pulmonary nodules, pleural effusions, and hilar or mediastinal lymphadenopathy.[190]

TREATMENT AND SURVIVAL OF KS

Treatment is guided by the extent of symptomatic and extracutaneous KS, immune system conditions, and concurrent complications of HIV infection since KS is not considered curable by standard therapies.

Local Treatment

Localized KS lesions can be treated with alitretinoin gel, intralesional chemotherapy, radiation therapy, laser therapy, cryotherapy, and surgical excision.[90,110] With intralesional therapy, vinblastine is likely the most used and yields a response rate of 70%.[191] Radiotherapy, whether given as whole-body electron beam therapy, fractionated focal radiation therapy, and single treatments have produced CRs in 50–80% of patients.[192–194]

Chemotherapy

Systemic chemotherapy is used with individuals with progressive, widespread disease, and in particular with visceral involvement. Large randomized studies have established liposomal anthracyclines (doxorubicin, daunorubicin) as first line single-agent chemotherapy agents with promising results compared with combination chemotherapy treatment.[195–197] Paclitaxel has also been recently used as a cytotoxic agent against KS. Compared with other regimens, the median duration of response ranged from 7.4 to 10.4 months, among the longest observed.[198–201] Although well tolerated, paclitaxel has many adverse effects, including neutropenia and toxic interactions with HAART due to both being metabolized by the cytochrome P-450 pathways.[202,203]

Interferon

For HIV-infected individuals who have appropriate immune reconstitution with antiretrovirals, but still have disseminated cutaneous KS, interferon-alpha may be a favorable option. Response rates from 20 to 40% have been observed in patients.[168] Adverse side effects from interferon-alpha encompass fever, chills, neutropenia, hepatotoxicity, and cognitive impairment.

Anti-HHV-8 Therapy

The discovery of HHV-8 in all forms of KS has raised the possibility of using antiviral agents to target this virus. HIV-infected individuals that used foscarnet instead of ganciclovir to treat cytomegalovirus disease had a considerably longer time of their existing KS progressing to a more severe form.[204] In another trial, the use of ganciclovir for treatment of cytomegalovirus disease reduced the risk of developing KS.[205] A pilot study using cidofovir for the treatment of AIDS-associated and classic KS showed no effect on disease progression, no decrease in viral load of HHV-8 among seven patients, and no change in expression of early lytic and latent gene expressions from cutaneous KS lesions. Only one of seven patients demonstrated decreased production of a late lytic gene.[206]

Most KS tumor cells are latently infected with HHV-8 and thus more resistant to antiherpesvirus drugs that rely on lytic replication. This resistance can be overcome by inducing HHV-8 to reenter the lytic cycle during antiviral therapy, leading to destruction of virally infected cells. Another way would be to disrupt the maintenance and replication of HHV-8 during the latency, via glycyrrhizic acid or hydroxyurea, and thus interrupt the role of the virus in tumor formation.[207]

Other Potential Targets

Potential targets based on KS pathogenesis are the focus of many current trials. Angiogenesis inhibitors including fumagillin, thalidomide, MMP inhibitor COL-3, and Imatinib mesylate are all agents being tested on individuals with AIDS-associated KS.[187,208–212] Further, platelet-derived growth factor[213] and c-kit[214] have a role in active tumor formation. In one small study, the use of Imatinib mesylate, a c-kit and platelet-derived growth factor (PDGF) receptor inhibitor, resulted in regression of AIDS-related KS lesions.[211]

Impact of HAART

It is essential for tumor control for the majority of individuals now with HIV and concurrent KS to be treated with HAART.[187,215] The inhibition of HIV replication, decreased production of the Tat protein, restored immunity to HHV-8 and the direct antiangiogenic activity of some protease inhibitors are among the benefits of antiretroviral therapy.[216,217]

HAART has been associated with a lengthening of time to treatment failure with either local or systemic therapy for KS. Bower and colleagues[218] reported in a retrospective study a median time of 1.7 years from the initiation of HAART versus 0.5 years with no antiretroviral therapy to detect treatment failure among HIV-infected individuals with KS.

It has also been shown that antiretroviral therapy alone is linked with regression in size and number of existing KS lesions.[219–221] Nasti et al.[222] demonstrated that HIV-infected individuals who were already receiving HAART at KS diagnosis had a less aggressive presentation versus individuals who were naïve to HAART at time of KS diagnosis.

Tam et al.,[223] in a retrospective analysis from 1990 to 1999, demonstrated improved survival, an 81% risk reduction of death among HIV-infected individuals, with the initiation of HAART after the diagnosis of KS.

Thus, HAART taken along with chemotherapy extends the time to treatment failure of anti-KS therapies[218] and patients are able to sustain extended course of chemotherapy without relapse at its discontinuation.[224]

AIDS-ASSOCIATED MULTICENTRIC CASTLEMAN'S DISEASE

Castleman's disease, originally described by Benjamin Castleman in 1956 as localized lymph node hyperplasia resembling a thymoma, is a rare lymphoproliferative disorder.[225] Two clinical forms have been described, one in which the disease is localized and the other generalized (multicentric). The localized form presents as lymph node hyperplasia in a single node, usually in the abdomen or mediastinum, and is asymptomatic in 51% of the cases and can resolve with resection.[226] MCD, described by Leibetseder and Thurner[227] in 1973 and by Gaba et al.[228] in 1978, is always symptomatic and is characterized by generalized lymphadenopathy with systemic symptoms of fever, fatigue, and weight loss and the potential for malignant transformation.[229,230]

In non-human immunodeficiency virus (HIV)-infected individuals, MCD can be associated with other diseases such as cutaneous KS, B-cell lymphoma, Hodgkin's disease, and polyneuropathy, organomegaly, endocrinopathies, monoclonal gammopathy and skin changes (POEMS) syndrome, and autoimmune diseases.[147,229] MCD was diagnosed in two homosexual men with AIDS in 1985.[231] In individuals with HIV, MCD is linked with malignant transformation to NHL at 15-fold higher rate than those not diagnosed with MCD.[232]

HHV-8 PATHOGENESIS IN MULTICENTRIC CASTLEMAN'S DISEASE

HHV-8 encodes for viral proteins involved in signal transduction, cell cycle regulation, and/or inhibition of apoptosis.[233] One of them is a viral homologue of human interleukin-6 (IL-6), called viral IL-6. In vitro, viral IL-6 acts similarly to human IL-6[234] and possibly serves as an autocrine growth factor in MCD.[235]

MCD associated with HHV-8 is also known as the plasmablastic variant because of the presence of plasmablastic cells harboring HHV-8 in the mantle zones of the follicles.[236] HHV-8 is considered to infect naïve B lymphocytes, transforming these cells into polyclonal plasmablasts, with the potential of developing into plasmablastic lymphoma.[237,238]

EPIDEMIOLOGY OF MCD

The precise incidence of MCD is not known, but it has been found to be more common in HIV-infected individuals. Further, HIV-infected individuals with MCD have a 15-fold higher risk of NHL than the general HIV-infected population.[232]

Pre-HAART, the median survival with chemotherapy was less than 7 months.[237] In the post-HAART era, prognosis of AIDS-associated HIV remains poor with a median survival of 48 months with the same chemotherapy regimen used in pre-HAART era study.[239]

CLINICAL PRESENTATION OF MCD

Patients with MCD often have constitutional symptoms of fever and fatigue. On physical examination, diffuse lymphadenopathy, hepatosplenomegaly, and peripheral edema may be encountered. Further laboratory examination may reveal cytopenia, hypergammaglobulinemia, hypoalbuminemia, and raised C-reactive protein. Diagnosis of MCD is clinicopathological and thus based on lymph node biopsy with the clinical signs of a lymphoproliferative disorder and multisystem involvement.[230,240] It has been suggested that HHV-8 load in the peripheral blood is the most accurate marker for monitoring MCD.[149]

TREATMENT OF MCD

In the localized form of Castleman's disease, 5-year survival rate after surgical resection has been reported at 100%, with follow-up recommended due to some cases of recurrence.[241] Surgery is less frequently used with MCD due to systemic involvement, with the exception of splenectomy when indicated.[230,242,243] Less invasive thoracoscopic surgery has also been shown to be a safe and effective alternative.[244] Radiotherapy has been successfully used in localized disease, and two of four patients treated with radiotherapy were symptom free at a follow-up of 40 months, and the other two died of causes unrelated to Castleman's disease.[245]

Chemotherapy and Steroids

In HIV-infected individuals, several chemotherapy regimens have been tried, but generally low-dose alkylating agents have been used.[246] In a pre-HAART study by Oksenhendler et al.,[237] 12 of 20 patients were treated with chemotherapy. Of the 12 patients, nine received vinblastine with resulting partial response with loss of constitutional symptoms and regression of lymphadenopathy. Four of them remained stable, but required maintenance therapy every 2 weeks, and five of these individuals relapsed and required combination chemotherapy or splenectomy. In another study, two HIV-infected individuals not on HAART and with aggressive MCD were treated with oral etoposide. Remission, documented by computerized tomography (CT) was achieved at a follow-up of 1.5 and 6 months, with minimal side effects. It was concluded by Scott et al.[247] in this study that oral etoposide may be a safe, tolerable, and active agent in MCD.

In a retrospective study by Loi and colleagues,[242] 11 patients, ten of whom were already on HAART prior to chemotherapy, were treated mostly with cyclophosphamide and chorambucil, combined with prednisolone. Seven of the 11 patients achieved a response to treatment and the median duration to response was 16.2 months. Mortality from MCD and its resulting complications was 45% in this series.

Steroids are generally used along with chemotherapy, thus results are limited in treatment for MCD. Short (5 days) and long courses have been used with varying results. They may have an antitumor effect, but it is not maintained once the steroids are terminated.[226]

Antiviral Therapy

Inconsistent results have been reported in various studies from the use of antiviral therapy targeting HHV-8. For instance, remission of HIV-associated MCD, along with a decline in plasma HHV-8 DNA levels, was attained in three patients treated with intravenous ganciclovir therapy. Two patients, only one on HAART, achieved a reduction in the frequency of recurrent flares of MCD and detectable HHV-8 DNA with intravenous or oral ganciclovir. The third patient, not on HAART, recovered from an acute episode of renal and respiratory failure with intravenous ganciclovir therapy, but subsequently died of a fungal infection.[248] In a prior study, antiviral therapy has not been effective in reducing HHV-8 levels, and clinical improvement was attained after the use of corticosteroids along with chemotherapy.[249]

Rituximab

Plasmablasts located in the mantle zone of follicles and harboring HHV-8 in HHV-8-related MCD express variable levels of CD20 surface antigen.[236] The anti–CD20 monoclonal antibody rituximab was tested therapeutically in a small group of HIV-infected individuals, with CR of clinical symptoms and HHV-8 viremia in three patients for up to 14 months.[250,251]

Interferon–alpha

Interferon–alpha has also been used effectively in two case reports on HIV-infected patients. On HAART, an HIV-positive patient remained in remission with IFN–alpha for 24 months, despite failing two HAART regiments.[252] Twenty-four months after initiation of IFN–alpha, the patient was receiving IFN–alpha treatment and was in remission, with a CD4$^+$ cell count of 123 cells/mm^3 and a viral load of 1,763 copies/mm^3. The second case report also describes a one year remission of MCD with a tapering schedule of IFN–alpha after multiple relapses with chemotherapy, splenectomy, and antiviral therapy.[253] IFN-alpha has been used as adjunct therapy in three HIV-infected individuals in combination with vinblastine and splenectomy. Two of the three patients tolerated it well, and had a CR.[237]

Interleukin-6

The dysregulated production of human IL-6 is thought to have a role in the pathogenesis of MCD.[254] Using IL-6 antibody to block IL-6 signaling has achieved good results, but all reports are in HIV-negative patients.[255,256]

Thalidomide

Thalidomide, an agent that disrupts cytokines, has been successfully used in one HIV-infected patient on HAART with MCD. Although various chemotherapeutic agents were used, the individual continued to relapse with fever and thrombocytopenia without chronic etoposide administration. Thalidomide was subsequently started along with etoposide, and after 4 weeks the etoposide was held. At 38 weeks follow-up, the patient remained without fevers and platelet counts were stable, bone marrow biopsy revealed no evidence of MCD, and initial hypergammaglobulinemia had normalized.[257]

Role of HAART in MCD

With HAART, both cases have been described: ones in which reduction of HHV-8 DNA are achieved in the peripheral blood along with remission of clinical manifestations[219,258] and cases in which antiretroviral therapy failed to decrease the HHV-8 load in the peripheral blood even with immune reconstitution.[242,259]

Two cases have been described where HAART alone led to recovery from MCD, with a follow-up of 12 months.[260] Another study described no effect of a protease inhibitor-based regimen, even with immune reconstitution.[261]

Aaron and colleagues[239] reported case histories of seven HIV-infected individuals with MCD on HAART and chemotherapy. Four had KS along with their MCD. Although the KS resolved with immune reconstitution on HAART, their MCD relapsed when chemotherapy was terminated. This study suggests that MCD is not affected by immune reconstitution. The main causes of death in the pre-HAART studies of MCD were associated with immune deficiency from HIV disease itself, such as opportunistic infections. Therefore, the receipt of antiretrovirals may itself account for the longer survival rates (median survival of 48.4 months) with MCD seen in the post-HAART era.[237,239]

AIDS-ASSOCIATED PRIMARY EFFUSION LYMPHOMA

Knowles et al.[262] first described a lymphoma syndrome in HIV-infected individuals characterized by malignant effusions in the absence of nodal disease in 1989, which was subsequently named body-cavity-based lymphoma by Cesarman and colleagues in 1995.[150] Now it is called primary effusion lymphoma (PEL), a name more commonly used in the literature to describe this syndrome.

HHV-8 AND EBV ETIOLOGY IN PEL

Cesarman and colleagues[150] found HHV-8 sequences among eight HIV-infected individuals with PEL, and quantitatively, these lymphomas contained 60–80 copies of HHV-8 per cells, compared with one copy in the KS lesion.

As with HHV-8, EBV expression is also present in nearly all cases of PEL.[262,263] Horenstein and colleagues[263] analyzed the pattern of EBV latent (persistent) gene expression in five patients with PEL confirmed to be coinfected with EBV and HHV-8. A significant finding was that PELs express EBNA1 mRNA, which has been shown to induce B-cell lymphomas when expressed in B cells as a transgene in mice.[264] However, overall it was found that PELs exhibit a restricted pattern of EBV latency, lacking significant expression of most of the major EBV growth transforming factors. It is likely, with this background, that HHV-8 has a significant role in the pathogenesis of PELs. Furthermore, there are a few cases containing HHV-8, but lacking EBV, suggesting a secondary role for EBV in PELs.[263]

EPIDEMIOLOGY OF PEL

PEL, a rare lymphoma with poor prognosis, presents similarly in terms of epidemiology to that of KS, occurring mostly in male homosexual AIDS patients and rarely in elderly HIV-negative individuals.[150,265,266] Boulanger and colleagues[267] reported 28 cases in a 11-year period from six centers in France, and a median survival rate of 6.2 months among HIV-infected individuals. Prior reported median survival was shorter than 6 months, and a poor performance status and absence of HAART before PEL diagnosis was found to be independently associated with a weaker clinical outcome in the French study.

FEATURES OF PEL

PEL is a distinct, high grade B-cell NHL characterized by pleural effusion, pericardial effusion, and/or ascites, generally in an HIV-infected individual, but has been described in elderly individuals with chronic heart or liver disease. In addition, there have also been case reports of PEL associated with KS in HIV-negative individuals, one individual with prior KS and the other after cardiac transplantation.[268,269]

Clinically, it presents as malignant effusions with a lack of nodal or extranodal solid tumors.[270] Cytologic examination is necessary for diagnosis and fluid can be attained through thoracentesis or paracentesis.[271]

TREATMENT OF PEL

Simonelli and colleagues,[272] in a retrospective study with 277 patients with HIV infection and systemic lymphoma, had 11 individuals in the group that were classified in the PEL category. Eight of the 11 patients received CHOP-like regimen and CR was reached in 42% of the patients, with a median survival of 6 months. One of the three patients receiving no cytotoxic therapy also had CR, which may have occurred as result of HAART. Only two patients survived beyond a year, and both had a CR after cytotoxic therapy or occurring along with antiretroviral therapy. Initiation of HAART alone has been reported to cause spontaneous regression of PEL.[273]

Currently, no definite treatment recommendations can be made due to lack of sufficient data. Since administering HAART has proven effective in some cases, a course of antiretroviral therapy with deferment of chemotherapy should be tried for all individuals.[271]

HUMAN PAPILLOMAVIRUS

HPV is a virus in the papovavirus family. It is a double-stranded DNA virus and generally infects stratified squamous epithelium, such as mucosal and squamous cells.[274] The genome is approximately 8,000 bp in size, and based on genome sequence comparisons, over 100 HPV types have been identified to date. Approximately 30 of the known HPV types infect genital epithelium, and HPV is one of the most common sexually transmitted infections.

Among HIV-infected individuals, human HPV infection has a well-established relationship with the elevated risk of invasive cervical cancer, anal cancer, and precursor lesions of both types of cancer, cervical intraepithelial neoplasia (CIN) and anal intraepithelial neoplasia (AIN).[275–278] In addition, other HPV-associated intraepithelial neoplastic lesions and malignancies appear to be more prevalent among HIV-infected individuals. Petry et al.[279] reported an increased prevalence among HIV-infected women of high-risk HPV infection associated vulval intraep-ithelial neoplasia (VIN). Frisch et al.[280] reported an increase in both invasive and in situ forms of not only cervical and anal cancer, but also of vulvar/vaginal and penile cancers among HIV-infected individuals. In Africa, several studies have found evidence of increasing conjuctival carcinoma as well as an association between HPV infection and the risk of squamous cell carcinoma of the conjunctiva.[281,282] Moubayed et al.[283] found that HPV 6/11, 16 or 18 were found in the majority of 14 cases of conjunctival carcinoma in Tanzania using a highly sensitive in situ hybridization technique. Nine of these 14 cases were HIV positive.

HIV AND HPV INTERACTION

The increased prevalence of HPV disease associated with HIV infection may be mediated by impaired T-cell and antigen-presenting cell function, however local effects of HIV infection may also upregulate HPV replication and oncogenesis. The HPV viral oncogenes E6 and E7 are able to immortalize primary keratinocytes and can transform cells in culture.[284,285] In an animal mode of estrogen-stimulated HPV-induced cervical cancer, expression of E7 alone resulted in precancers and cancer, while the expression of E6 and E7 together resulted in larger cancers.[286] Although the exact mechanisms of HIV-related immunosuppression and HPV coinfection have not been determined, several in vitro studies have shown that HIV tat protein can drive the replication of HPV 16 and HPV 18 through the overex-pression of E7 and other genes in the early region.[287,288]

HIV-ASSOCIATED SQUAMOUS CELL CANCER OF THE CERVIX

Epidemiology of HIV-Associated Cervical Cancer

Since 1993, invasive cervical cancer has been listed by the CDC guidelines as an "AIDS-defining" condition. However, quantifying the contribution of HIV infec-tion in the development of cervical cancer among HIV-infected women remains unclear. Although immunosuppression after organ transplantation has been shown to enhance the onset and progression of HPV-related tumors,[289] a 1996 study by the International Agency for Research on Cancer (IARC) which evaluated the relationship between HIV and cervical cancer found no conclusive evidence that HIV per se increased the risk of cervical cancer among HIV-infected women.[289] Subsequent studies continue to yield conflicting results. In developed countries, with access to HAART, there have been several studies that have shown an increased risk of cervical cancer. Studies conducted in Southern European countries have found that the increased risk of invasive cervical cancer ranged from 3.1 to 15.5 among HIV-infected women.[116,290,291] In addition, using a national

AIDS-cancer linked registry database of cases through 1998, Frisch et al.[280] found a relative risk of 5.4 for invasive cervical cancer among HIV-positive women compared with the general population. Data from South Africa have also shown an increased risk of cervical cancer.[292] However, no increase in cervical cancer has been noted among case–control studies from multiple African countries where endemic rates of cervical cancer are higher and women have shorter survival, such as Rwanda, Cote d'Ivoire, Zambia, or Kenya.[293–296]

There appears to have been little change in the incidence of invasive cervical cancer in recent years associated with HAART. A study conducted by the International Collaboration on HIV and Cancer 2000 compiled 23 studies from North America, Europe, and Australia, evaluating the incidence of cervical cancer among HIV-infected women from 1992 to 1999. No significant increase in the incidence rate of cervical cancer was noted during the study period, particularly since the widespread use of HAART.[297] However, longer follow-up time is necessary before definitive conclusions are drawn regarding the effect of HAART on invasive cervical cancer incidence.

Epidemiology of CIN in HIV-Infected Women

The relationship between HIV-infection and increased prevalence of CIN has been shown in many studies. Mandelblatt et al.[298] performed a meta-analysis of 15 cross-sectional studies published between 1986 and 1998 that evaluated prevalence of cervical neoplasia, HPV infection and HIV infection among women. They found that among women infected with HPV, HIV-infected women were significantly more likely to develop cervical neoplasia, and that this effect was related to degree of immunodeficiency. Several other recent studies have also shown that HIV-infected women are at higher risk for CIN. A cross-sectional study conducted by the Women's Interagency Health Study (WIHS) group[299] found that 38% of HIV-positive women had abnormal cytologic findings compared with 16% of HIV-negative women. In another cross-sectional multicenter study, Duerr et al.[300] also found that HIV-positive women were more likely than HIV-negative women to present with atypical cervical cytology. Finally, Ahdieh et al.[301] found that 13% of HIV-positive women versus 2% of HIV-negative women had abnormal cytological findings. They also found that HIV-positive women had a much lower rate of HPV clearance on follow-up exams, and that in a multivariate model, the increased rate of CIN among HIV-positive women was fully accounted for by HPV persistence.

Screening Guidelines for HIV-Infected Women

Current United States Public Health Service (USPHS) and Infectious Diseases Society of America (IDSA) guidelines[302] recommend that HIV-positive women undergo a complete history, and pelvic exam as part of cervical cytologic assessment at the time of initial evaluation. A pap smear every 6 months is recommended for the first year after HIV diagnosis, and if both are normal, a yearly pap smear may be obtained thereafter. Women with abnormal pap smears should undergo colposcopy and can be subsequently managed according to published universal guidelines.[303] In addition, Tate et al.[304] found that the recurrence rates of CIN are high among HIV-infected women, but a separate study found that HAART may decrease the recurrence of CIN among HIV-infected women after therapy.[305]

Effect of HAART on Cervical Cancer and CIN

Although HAART has significantly improved the survival of HIV-infected individuals through immune reconstitution and has decreased the incidence of opportunistic infections, the effects of HAART on HPV infection and CIN among HIV-infected women remain unclear. While three small retrospective studies did not find a significant reduction in risk of cervical dysplasia among women on HAART,[306–308] several other studies have found a reduction in cervical dysplasia risk related to HAART. In a cross-sectional study, Heard et al.[309] did not find any change in prevalence of HPV infection 5 months after HAART initiation in 73 HIV-positive women, but the prevalence of CIN decreased significantly. Greater increases in CD4 cell counts after initiating HAART were also associated with CIN regression. The largest retrospective analysis performed by the WIHS group found that among 741 HIV-positive women, women on HAART were 40% (95% CI, 4–81) more likely to exhibit a regression of cervical lesions and were also significantly less likely to have progression of CIN (OR 0.68).[310] Heard et al.[311] conducted another prospective analysis evaluating the effect on HAART on time to regression of CIN in 168 women and found that the hazard ratio for the regression of CIN was 1.93 (95% CI, 1.14–3.20) for women on HAART compared with those on less effective or no therapy.

The conflicting observations regarding the effect of HAART on cervical squamous intraepithelial lesions (SIL) or CIN from these reports are likely in part due to the design differences, differences in inclusion criteria, differences in end points, and different statistical methodologies employed by the authors. However, taken together, it appears that HAART has a modest beneficial effect on progression of HPV-related cervical disease.

Treatment of Cervical Cancer in HIV-Positive Women

Women with AIDS and cervical cancer have been shown to have a significantly higher mortality than those without AIDS. In the pre-HAART era, data from the National Center for Health Certificates from 1990 through 1995 found that relative risk of cervical cancer for women aged 25–44, adjusted for race, was 5.5 for in HIV-infected women compared with HIV-negative women.[312] In the post-HAART era (1996–2000), a recent study linking the New York state cancer registry and the New York City HIV/AIDS registry found that the 24 month survival for women with AIDS was 64% compared with 74% among women without AIDS. The adjusted hazard ratio of death at 24 months for women with AIDS was 1.8 (95% CI 1.1–3.2) compared with women without AIDS.

Invasive cervical cancer diagnosed in women with HIV infection is treated using the same criteria and protocols as those women without HIV infection, as long as no other contraindications for treatment exist. In addition, close post-treatment surveillance is also recommended.

HIV-ASSOCIATED SQUAMOUS CELL CANCER OF THE ANUS

Epidemiology of HIV-Associated Anal Cancer

In the 1960s, the annual incidence of squamous cell cancer of the anus (SCCA) among men in the United States was relatively low and stable, with approximately 0.5 cases per 100,000 persons.[313,314] The disease was primarily found among older persons, with a 2:1 female to male ratio.[313,314] However, a recent US population based analysis of the Surveillance, Epidemiology and End Results (SEER) program data found that the incidence of SCCA in the US among men increased from 1.06 per 100,000 in 1973–1979 to 2.04 per 100,000 in 1996–2004.[315] Other recent studies have shown a steady increase in the incidence of SCCA during the past three decades,[313,314] with significant increases among never married men in the San Francisco Bay Area.[314,316]

In addition, large retrospective AIDS-Cancer Registry Match studies and prospective studies of HIV-infected cohorts have shown an increased risk of SCCA among HIV-infected individuals. Compared to the general population, the relative risk for SCCA in the national US AIDS Cancer Registry match was 37.9 (33.0–43.4) for men and 6.8 (2.7–14) for women.[158] The incidence of SCCA among HIV-infected populations reported from prospective cohort studies ranged from 3.9 to 92 per 100,000 and the relative risks reported ranged from 33.4 to 115. In addition, the widespread introduction of HAART in industrialized nations does not appear to have changed the incidence of SCCA in these studies.[317–319]

Epidemiology of HIV-Associated Anal Intraepithelial Neoplasia

There have been many studies that have reported the prevalence of AIN among HIV-infected men and women. Sixteen studies demonstrated that between 41 and 97% of HIV-infected men are found to have anal dysplasia on anal pap smear screening.[320–333] Palefsky et al.[334] also found that the relative risk (RR) of developing high grade squamous intraepithelial lesions (HSIL) was 3.7; (95% CI 2.6–5.7) for HIV-positive men who have sex with men (MSM) compared to HIV-negative MSMs. In addition, five studies have shown that between 14 and 28% of HIV-infected women are found to have anal dysplasia.[276,335–338] In a study evaluating the prevalence of HPV DNA in anal swabs of HIV-positive and HIV-negative women, 76% of HIV-infected and 42% of HIV-negative women had HPV detected.[336]

Screening for Anal Cancer

As shown in the previous section, HIV-infected individuals are at an increased risk for SCCA and AIN. In addition, SCCA shares many biologic similarities with cervical cancer, including detectable dysplastic precursor lesions and high-risk HPV infection. Thus, the institution of annual anal pap screening for HIV-infected patients has been recommended.[339,340] Anal pap smears are obtained by randomly obtaining squamous cells from the anal canal using a Dacron swab. They are then fixed in liquid fixative medium. Similar to cervical cytology protocols, abnormal anal cytologic findings are confirmed by high-resolution anoscopy (HRA) directed biopsy of visualized lesions. The 2001 revised Bethesda System of cytologic classi-fication includes a basic primer on anal cytology and uses the system of cervical cytologic classification as an accepted means for classifying anal cytology.[341] Anal pap smears have a similar sensitivity and specificity to cervical pap smears.[342–346] Although there are no definitive clinical studies showing that anal pap smears decrease SCCA-related morbidity and mortality among HIV-infected individuals, a recent cost-effectiveness analysis found that the incremental cost-effectiveness ratio per quality-adjusted life year saved was $16,000, which is similar to other widely accepted screening procedures.[347]

Effect of HAART on Anal Cancer and Anal Dysplasia

Only eight small case series (ranging from 4 to 26 patients) describe outcomes of HIV-associated SCCA, with 5-year survival ranging from 47 to 60%.[348–355] The majority of cases were diagnosed at either stage 1 or 2 (localized tumor without lymph node invasion) and received combined chemoradiotherapy. Most series reported some toxicity associated with therapy and with one reporting up to 50% of patients unable to complete planned therapy.[356–362] Among three studies that specifically compared survival among patients with SCCA in the pre-HAART versus HAART eras, there was a nonsignificant trend toward improved survival, better tolerability of chemoradiotherapy, and improved local tumor control in the HAART era.[319,356,357,362] However, a recent study linking the New York state cancer registry with the New York city HIV/AIDS registry found that in the HAART era (1990–1996) the 24-month survival was 76% for patients with AIDS, compared to 78% for patients without AIDS, suggesting that in the HAART era, HIV-infected patients with SCCA have equivalent survival with HIV-negative patients.[368]

Like studies evaluating the effect of HAART on cervical dysplasia, studies eval-uating the effect of HAART on anal dysplasia also report conflicting results, likely related to the significant design and methodological differences among these stud-ies. Palefsky et al.[320] compared the rates of progression and regression of anal dys-plasia after 6 months of HAART. They found that the likelihood of lesion progression or regression was not affected by HAART initiation, but they noted that among the patients starting HAART at higher CD4 counts, HAART demon-strated a nonsignificant benefit on anal dysplasia lesions. In a subsequent study, Palefsky et al.[363] performed a cross-sectional analysis on the prevalence of AIN 3 among a cohort of 433 HIV-positive men. They found that men on HAART had an increased risk of 12.6 (95% CI, 2.4–64) for AIN 2 or 3 after adjustment for CD4 count. In contrast, Wilkin et al.[323] conducted a cross-sectional study evaluating anal HPV infection and anal dysplasia in 98 HIV-positive men. In a multivariate analysis they found that HAART and higher nadir CD4 count were significantly protective for anal dysplasia by histology, but were not protective of anal HPV infection. Therefore, it remains unclear if HAART initiation influences the natural history of AIN in HIV-infected individuals.

Treatment outcomes for anal dysplasia have only been reported for small case series. Current treatment options are similar for HIV-positive and HIV-negative individuals. These treatments include surgical ablation, infrared coagulation, imiquimod, and topical 5-FU.[364–367]

SUMMARY

Among individuals with HIV-infection, coinfection with oncogenic viruses including EBV, HHV-8, and HPV cause significant cancer-related morbidity and mortality. It is clear that these viruses interact with HIV in unique ways that predispose HIV-infected individuals to malignant diseases. In general, treatment directed specifically against these viruses does not appear to change the natural history of the malignant disease, and once the malignancy develops, if their health permits, HIV-infected patients should be treated using similar treatment protocols to HIV-negative patients. However, for the less frequent HIV-related malignancies, such as PEL, or MCD, optimal treatments are still emerging. For certain AIDS-defining malignancies, it is clear that the widespread access to HAART has significantly decreased the incidence, and improved outcomes. However, for other cancers, such as the HPV-related tumors, the role of HAART is much less clear. Further research into prevention and treatment of these oncogenic virally mediated AIDS-related malignancies is necessary.

REFERENCES

1. Ruf, I. K., A. Moghaddam, F. Wang, and J. Sample. 1999. Mechanisms that regulate Epstein-Barr virus EBNA-1 gene transcription during restricted latency are conserved among lymphocryptoviruses of Old World primates. J Virol **73**(3):1980–9.
2. Epstein, M. A., B. G. Achong, and Y. M. Barr. 1964. Virus particles in cultured lymphoblasts from Burkitt's lymphoma. Lancet **15**:702–3.
3. Baer, R., A. T. Bankier, M. D. Biggin, P. L. Deininger, P. J. Farrell, T. J. Gibson, G. Hatfull, G. S. Hudson, S. C. Satchwell, C. Seguin, et al. 1984. DNA sequence and expression of the B95-8 Epstein-Barr virus genome. Nature **310**(5974):207–11.
4. Macsween, K. F., and D. H. Crawford. 2003. Epstein-Barr virus-recent advances. Lancet Infect Dis **3**(3):131–40.
5. Ambinder, R. F. 2001. Epstein-Barr virus associated lymphoproliferations in the AIDS setting. Eur J Cancer **37**(10):1209–16.
6. McClain, K. L., C. T. Leach, H. B. Jenson, V. V. Joshi, B. H. Pollock, R. T. Parmley, F. J. DiCarlo, E. G. Chadwick, and S. B. Murphy. 1995. Association of Epstein-Barr virus with leiomyosarcomas in children with AIDS. N Engl J Med **332**(1):12–8.
7. Jenson, H. B., C. T. Leach, K. L. McClain, V. V. Joshi, B. H. Pollock, R. T. Parmley, E. G. Chadwick, and S. B. Murphy. 1997. Benign and malignant smooth muscle tumors containing Epstein-Barr virus in children with AIDS. Leuk Lymphoma **27**(3–4):303–14.
8. Jenson, H. B., E. A. Montalvo, K. L. McClain, Y. Ench, P. Heard, B. A. Christy, P. J. Dewalt-Hagan, and M. P. Moyer. 1999. Characterization of natural Epstein-Barr virus infection and replication in smooth muscle cells from a leiomyosarcoma. J Med Virol **57**(1):36–46.
9. Birx, D. L., R. R. Redfield, and G. Tosato. 1986. Defective regulation of Epstein-Barr virus infection in patients with acquired immunodeficiency syndrome (AIDS) or AIDS-related disorders. N Engl J Med **314**(14):874–9.
10. Knowles, D. M. 1996. Etiology and pathogenesis of AIDS-related non-Hodgkin's lymphoma. Hematol Oncol Clin North Am **10**(5):1081–109.

11. Olson, J. E., C. A. Janney, R. D. Rao, J. R. Cerhan, P. J. Kurtin, D. Schiff, R. S. Kaplan, and B. P. O'Neill. 2002. The continuing increase in the incidence of primary central nervous system non-Hodgkin lymphoma: a surveillance, epidemiology, and end results analysis. Cancer **95**(7):1504–10.

12. Corn, B. W., S. M. Marcus, A. Topham, W. Hauck, and W. J. Curran, Jr. 1997. Will primary central nervous system lymphoma be the most frequent brain tumor diagnosed in the year 2000? Cancer **79**(12):2409–13.

13. Levine, A. M. 1992. Acquired immunodeficiency syndrome-related lymphoma. Blood **80**(1):8–20.

14. MacMahon, E. M., J. D. Glass, S. D. Hayward, R. B. Mann, P. S. Becker, P. Charache, J. C. McArthur, and R. F. Ambinder. 1991. Epstein-Barr virus in AIDS-related primary central nervous system lymphoma. Lancet **338**(8773):969–73.

15. Subar, M., A. Neri, G. Inghirami, D. M. Knowles, and R. Dalla-Favera. 1988. Frequent c-myc oncogene activation and infrequent presence of Epstein-Barr virus genome in AIDS-associated lymphoma. Blood **72**(2):667–71.

16. Ballerini, P., G. Gaidano, J. Z. Gong, V. Tassi, G. Saglio, D. M. Knowles, and R. Dalla-Favera. 1993. Multiple genetic lesions in acquired immunodeficiency syndrome-related non-Hodgkin's lymphoma. Blood **81**(1):166–76.

17. Hamilton-Dutoit, S. J., G. Pallesen, J. Karkov, P. Skinhoj, M. B. Franzmann, and C. Pedersen. 1989. Identification of EBV-DNA in tumour cells of AIDS-related lymphomas by in-situ hybridisation. Lancet **1**(8637):554–62.

18. Arribas, J. R., D. B. Clifford, C. J. Fichtenbaum, R. L. Roberts, W. G. Powderly, and G. A. Storch. 1995. Detection of Epstein-Barr virus DNA in cerebrospinal fluid for diagnosis of AIDS-related central nervous system lymphoma. J Clin Microbiol **33**(6):1580–3.

19. Cingolani, A., A. De Luca, L. M. Larocca, A. Ammassari, M. Scerrati, A. Antinori, and L. Ortona. 1998. Minimally invasive diagnosis of acquired immunodeficiency syndrome-related primary central nervous system lymphoma. J Natl Cancer Inst **90**(5):364–9.

20. Cinque, P., L. Vago, H. Dahl, M. Brytting, M. R. Terreni, C. Fornara, S. Racca, A. Castagna, A. D. Monforte, B. Wahren, A. Lazzarin, and A. Linde. 1996. Polymerase chain reaction on cerebrospinal fluid for diagnosis of virus-associated opportunistic diseases of the central nervous system in HIV-infected patients. AIDS **10**(9):951–8.

21. Bossolasco, S., P. Cinque, M. Ponzoni, M. G. Vigano, A. Lazzarin, A. Linde, and K. I. Falk. 2002. Epstein-Barr virus DNA load in cerebrospinal fluid and plasma of patients with AIDS-related lymphoma. J Neurovirol **8**(5):432–8.

22. Brink, N. S., Y. Sharvell, M. R. Howard, J. D. Fox, M. J. Harrison, and R. F. Miller. Detection of Epstein-Barr virus and Kaposi's sarcoma-associated herpesvirus DNA in CSF from persons infected with HIV who had neurological disease. J Neurol Neurosurg Psychiatry **65**(2):191–5.

23. Cingolani, A., R. Gastaldi, L. Fassone, F. Pierconti, M. L. Giancola, M. Martini, A. De Luca, A. Ammassari, C. Mazzone, E. Pescarmona, G. Gaidano, L. M. Larocca, and A. Antinori. 2000. Epstein-Barr virus infection is predictive of CNS involvement in systemic AIDS-related non-Hodgkin's lymphomas. J Clin Oncol **18**(19):3325–30.

24. al-Shahi, R., M. Bower, M. R. Nelson, and B. G. Gazzard. 2000. Cerebrospinal fluid Epstein-Barr virus detection preceding HIV-associated primary central nervous system lymphoma by 17 months. J Neurol **247**(6):471–2.

25. Cote, T. R., A. Manns, C. R. Hardy, F. J. Yellin, and P. Hartge. 1996. Epidemiology of brain lymphoma among people with or without acquired immunodeficiency syndrome. AIDS/Cancer Study Group. J Natl Cancer Inst **88**(10):675–9.****

26. Wolf, T., H. R. Brodt, S. Fichtlscherer, K. Mantzsch, D. Hoelzer, E. B. Helm, P. S. Mitrou, and K. U. Chow. 2005. Changing incidence and prognostic factors of survival in AIDS-related non-Hodgkin's lymphoma in the era of highly active antiretroviral therapy (HAART). Leuk Lymphoma **46**(2):207–15.

27. Sacktor, N., R. H. Lyles, R. Skolasky, C. Kleeberger, O. A. Selnes, E. N. Miller, J. T. Becker, B. Cohen, and J. C. McArthur. 2001. HIV-associated neurologic disease incidence changes: multicenter AIDS Cohort Study, 1990–1998. Neurology **56**(2):257–60.

28. Conti, S., M. Masocco, P. Pezzotti, V. Toccaceli, M. Vichi, S. Boros, R. Urciuoli, C. Valdarchi, and G. Rezza. 2000. Differential impact of combined antiretroviral therapy on the survival of Italian patients with specific AIDS-defining illnesses. J Acquir Immune Defic Syndr **25**(5):451–8.

29. Loureiro, C., P. S. Gill, P. R. Meyer, R. Rhodes, M. U. Rarick, and A. M. Levine. 1988. Autopsy findings in AIDS-related lymphoma. Cancer **62**(4):735–9.

30. Fine, H. A., and R. J. Mayer. 1993. Primary central nervous system lymphoma. Ann Intern Med **119**(11):1093–104.

31. Rubenstein, J., and M. Berger. 2005. Management of primary central nervous system lymphoma and primary intraocular lymphoma. In: Volberding, P. A., J. M. Palefsky, editors. Viral and Immunological Malignancies. 1st ed. Hamilton: BC Decker; 2005, pp. 166–79.

32. Schultz, C., C. Scott, W. Sherman, B. Donahue, J. Fields, K. Murray, B. Fisher, R. Abrams, and J. Meis-Kindblom. 1996. Preirradiation chemotherapy with cyclophosphamide, doxorubicin, vincristine, and dexamethasone for primary CNS lymphomas: initial report of radiation therapy oncology group protocol 88-06. J Clin Oncol **14**(2):556–64.

33. O'Neill, B. P., J. R. O'Fallon, J. D. Earle, J. P. Colgan, L. D. Brown, and R. L. Krigel. 1995. Primary central nervous system non-Hodgkin's lymphoma: survival advantages with combined initial therapy? Int J Radiat Oncol Biol Phys **33**(3):663–73.

34. Abrey, L. E., L. M. DeAngelis, and J. Yahalom. 1998. Long-term survival in primary CNS lymphoma. J Clin Oncol **16**(3):859–63.

35. Donahue, B. R., J. W. Sullivan, and J. S. Cooper. 1995. Additional experience with empiric radiotherapy for presumed human immunodeficiency virus-associated primary central nervous system lymphoma. Cancer **76**(2):328–32.

36. Baumgartner, J. E., J. R. Rachlin, J. H. Beckstead, T. C. Meeker, R. M. Levy, W. M. Wara, and M. L. Rosenblum. 1990. Primary central nervous system lymphomas: natural history and response to radiation therapy in 55 patients with acquired immunodeficiency syndrome. J Neurosurg **73**(2):206–11.

37. Goldstein, J. D., D. W. Dickson, F. G. Moser, A. D. Hirschfeld, K. Freeman, J. F. Llena, B. Kaplan, and L. Davis. 1991. Primary central nervous system lymphoma in acquired immune deficiency syndrome. A clinical and pathologic study with results of treatment with radiation. Cancer **67**(11):2756–65.

38. Kasamon, Y. L., and R. F. Ambinder. 2005. AIDS-related primary central nervous system lymphoma. Hematol Oncol Clin North Am **19**(4):665–87, vi–vii.

39. Ambinder, R. F., S. Lee, W. J. Curran, Jr., et al. 2003. Phase II intergroup trial of sequential chemotherapy and radiotherapy for AIDS-related primary central nervous system lymphoma. Cancer Ther **1**:215–21.

40. Chamberlain, M. C., and P. A. Kormanik. 1999. AIDS-related central nervous system lymphomas. J Neurooncol **43**(3):269–76.

41. Jacomet, C., P. M. Girard, M. G. Lebrette, V. L. Farese, L. Monfort, and W. Rozenbaum. 1997. Intravenous methotrexate for primary central nervous system non-Hodgkin's lymphoma in AIDS. AIDS **11**(14):1725–30.

42. Raez, L., L. Cabral, J. P. Cai, H. Landy, G. Sfakianakis, G. E. Byrne, Jr., J. Hurley, E. Scerpella, D. Jayaweera, and W. J. Harrington, Jr. 1999. Treatment of AIDS-related primary central nervous system lymphoma with zidovudine, ganciclovir, and interleukin 2. AIDS Res Hum Retroviruses **15**(8):713–9.

43. McGowan, J. P., and S. Shah. 1998. Long-term remission of AIDS-related primary central nervous system lymphoma associated with highly active antiretroviral therapy. AIDS **12**(8):952–4.

44. Hoffmann, C., S. Tabrizian, E. Wolf, C. Eggers, A. Stoehr, A. Plettenberg, T. Buhk, H. J. Stellbrink, H. A. Horst, H. Jager, and T. Rosenkranz. 2001. Survival of AIDS patients with primary central nervous system lymphoma is dramatically improved by HAART-induced immune recovery. AIDS **15**(16):2119–27.

45. Skiest, D. J., and C. Crosby. 2003. Survival is prolonged by highly active antiretroviral therapy in AIDS patients with primary central nervous system lymphoma. AIDS **17**(12):1787–93.

46. Carbone, A., and A. Gloghini. 2005. AIDS-related lymphomas: from pathogenesis to pathology. Br J Haematol 130(5):662–70.

47. Doll, D. C., and A. F. List. 1982. Burkitt's lymphoma in a homosexual. Lancet 1(8279):1026–7.

48. Ziegler, J. L., W. L. Drew, R. C. Miner, L. Mintz, E. Rosenbaum, J. Gershow, E. T. Lennette, J. Greenspan, E. Shillitoe, J. Beckstead, C. Casavant, and K. Yamamoto. 1982. Outbreak of Burkitt's-like lymphoma in homosexual men. Lancet **2**(8299):631–3.

49. Gates, A. E., and L. D. Kaplan. 2003. Biology and management of AIDS-associated non-Hodgkin's lymphoma. Hematol Oncol Clin North Am **17**(3):821–41.

50. Beral, V., T. Peterman, R. Berkelman, and H. Jaffe. 1991. AIDS-associated non-Hodgkin lymphoma. Lancet **337**(8745):805–9.

51. Knowles, D. M., and A. Chadburn. 2001. Lymphadenopathy and the lymphoid neoplasms associated with the acquired immune deficiency syndrome. In: Knowles, D. M., editor. Neoplastic Hematopathology. Philadelphia: Lippincott-Williams; 2001, pp. 987–1089.

52. Levine, A. M., L. Seneviratne, B. M. Espina, A. R. Wohl, A. Tulpule, B. N. Nathwani, and P. S. Gill. 2000. Evolving characteristics of AIDS-related lymphoma. Blood **96**(13):4084–90.

53. Besson, C., A. Goubar, J. Gabarre, W. Rozenbaum, G. Pialoux, F. P. Chatelet, C. Katlama, F. Charlotte, B. Dupont, N. Brousse, M. Huerre, J. Mikol, P. Camparo, K. Mokhtari, M. Tulliez, D. Salmon-Ceron, F. Boue, D. Costagliola, and M. Raphael. 2001. Changes in AIDS-related lymphoma since the era of highly active antiretroviral therapy. Blood **98**(8):2339–44.

54. Raphael, M., B. Borisch, and E. Jaffe. 2001. Lymphomas associated with infection by the human immunodeficiency virus (HIV). In: Jaffe, E. S., N. L. Harris, and H. Stein, editors. World Health

Organization Classification of Tumours: Pathology and Genetics Tumours of Haematopoietic and Lymphoid Tissues. Washington, DC: IARC Press; 2001.

55. Parekh, S., H. Ratech, and J. A. Sparano. 2003. Human immunodeficiency virus-associated lymphoma. Clin Adv Hematol Oncol **1**(5):295–301.

56. Ferry, J. A. 2006. Burkitt's Lymphoma: clinicopathologic features and differential diagnosis. Oncologist **11**(4):375–83.

57. Gong, J. Z., T. T. Stenzel, E. R. Bennett, A. S. Lagoo, C. H. Dunphy, J. O. Moore, D. A. Rizzieri, J. H. Tepperberg, P. Papenhausen, and P. J. Buckley. 2003. Burkitt lymphoma arising in organ transplant recipients: a clinicopathologic study of five cases. Am J Surg Pathol **27**(6):818–27.

58. Xicoy, B., J. M. Ribera, J. Esteve, S. Brunet, M. A. Sanz, P. Fernandez-Abellan, and E. Feliu. 2003. Post-transplant Burkitt's leukemia or lymphoma. Study of five cases treated with specific intensive therapy (PETHEMA ALL-3/97 trial). Leuk Lymphoma **44**(9):1541–3.

59. Blum, K. A., G. Lozanski, and J. C. Byrd. 2004. Adult Burkitt leukemia and lymphoma. Blood **104**(10):3009–20.

60. Kasamon, Y. L., and L. J. Swinnen. 2004. Treatment advances in adult Burkitt lymphoma and leukemia. Curr Opin Oncol **16**(5):429–35.

61. Knowles, D. M. 2003. Etiology and pathogenesis of AIDS-related non-Hodgkin's lymphoma. Hematol Oncol Clin North Am **17**(3):785–820.

62. Burmeister, T., S. Schwartz, H. A. Horst, H. Rieder, N. Gokbuget, D. Hoelzer, and E. Thiel. 2005. Molecular heterogeneity of sporadic adult Burkitt-type leukemia/lymphoma as revealed by PCR and cytogenetics: correlation with morphology, immunology and clinical features. Leukemia **19**(8):1391–8.

63. Diebold, J. Burkitt lymphoma. In: Jaffe, E., N. Harris, H. Stein, et al., editors. Pathology and Genetics of Tumours of Haematopoietic and Lymphoid Tissues. Washington, DC: IARC Press; 2001, pp. 181–4.

64. Knowles, D. M., G. A. Chamulak, M. Subar, J. S. Burke, M. Dugan, J. Wernz, C. Slywotzky, G. Pelicci, R. Dalla-Favera, and B. Raphael. 1988. Lymphoid neoplasia associated with the acquired immunodeficiency syndrome (AIDS). The New York University Medical Center experience with 105 patients (1981–1986). Ann Intern Med **108**(5):744–53.

65. Lim, S. T., R. Karim, B. N. Nathwani, A. Tulpule, B. Espina, and A. M. Levine. 2005. AIDS-related Burkitt's lymphoma versus diffuse large-cell lymphoma in the pre-highly active antiretroviral therapy (HAART) and HAART eras: significant differences in survival with standard chemotherapy. J Clin Oncol **23**(19):4430–8.

66. Cortes, J., D. Thomas, A. Rios, C. Koller, S. O'Brien, S. Jeha, S. Faderl, and H. Kantarjian. 2002. Hyperfractionated cyclophosphamide, vincristine, doxorubicin, and dexamethasone and highly active antiretroviral therapy for patients with acquired immunodeficiency syndrome-related Burkitt lymphoma/leukemia. Cancer **94**(5):1492–9.

67. Wang, E. S., D. J. Straus, J. Teruya-Feldstein, J. Qin, C. Portlock, C. Moskowitz, A. Goy, E. Hedrick, A. D. Zelenetz, and A. Noy. 2003. Intensive chemotherapy with cyclophosphamide, doxorubicin, high-dose methotrexate/ifosfamide, etoposide, and high-dose cytarabine (CODOX-M/IVAC) for human immunodeficiency virus-associated Burkitt lymphoma. Cancer **98**(6):1196–205.

68. Hamilton-Dutoit, S. J., M. Raphael, J. Audouin, J. Diebold, I. Lisse, C. Pedersen, E. Oksenhendler, L. Marelle, and G. Pallesen. 1993. In situ demonstration of Epstein-Barr virus small RNAs (EBER 1) in acquired immunodeficiency syndrome-related lymphomas: correlation with tumor morphology and primary site. Blood **82**(2):619–24.

69. Teruya-Feldstein, J., E. Chiao, D. A. Filippa, O. Lin, R. Comenzo, M. Coleman, C. Portlock, and A. Noy. 2004. CD20-negative large-cell lymphoma with plasmablastic features: a clinically heterogenous spectrum in both HIV-positive and -negative patients. Ann Oncol **15**(11):1673–9.

70. Liebowitz, D., and E. Kieff. 1989. Epstein-Barr virus latent membrane protein: induction of B-cell activation antigens and membrane patch formation does not require vimentin. J Virol **63**(9):4051–4.

71. Gaidano, G., and R. Dalla-Favera. Molecular pathogenesis of AIDS-related lymphomas. Adv Cancer Res **67**:113–53.

72. Cote, T. R., R. J. Biggar, P. S. Rosenberg, S. S. Devesa, C. Percy, F. J. Yellin, G. Lemp, C. Hardy, J. J. Geodert, and W. A. Blattner. 1997. Non-Hodgkin's lymphoma among people with AIDS: incidence, presentation and public health burden. AIDS/Cancer Study Group. Int J Cancer **73**(5):645–50.

73. Appleby, P., V. Beral, R. Newton, and G. Reeves. 2000. Highly active antiretroviral therapy and incidence of cancer in human immunodeficiency virus-infected adults. J Natl Cancer Inst **92**:1823–30.

74. Kirk, O., C. Pedersen, A. Cozzi-Lepri, F. Antunes, V. Miller, J. M. Gatell, C. Katlama, A. Lazzarin, P. Skinhoj, S. E. Barton. 2001. Non-Hodgkin lymphoma in HIV-infected patients in the era of highly active antiretroviral therapy. Blood **98**(12):3406–12.

75. Grulich, A. E., Y. Li, A. M. McDonald, P. K. Correll, M. G. Law, and J. M. Kaldor. 2001. Decreasing rates of Kaposi's sarcoma and non-Hodgkin's lymphoma in the era of potent combination anti-retroviral therapy. AIDS **15**(5):629–33.

76. Ledergerber, B., A. Telenti, and M. Egger. 1999. Risk of HIV related Kaposi's sarcoma and non-Hodgkin's lymphoma with potent antiretroviral therapy: prospective cohort study. Swiss HIV Cohort Study. BMJ **319**(7201):23–4.

77. Buchbinder, S. P., S. D. Holmberg, S. Scheer, G. Colfax, P. O'Malley, and E. Vittinghoff.1992. Combination antiretroviral therapy and incidence of AIDS-related malignancies. J Acquir Immune Defic Syndr **21**(Suppl 1):S23–6.

78. Jones, J., D. Hanson, M. Dworkin, J. Ward, and H. W. Jaffe. 1992. Effect on antiretroviral therapy on recent trends in selected cancers among HIV-infected persons. Adult/Adolescent Spectrum of HIV Disease Project Group. J Acquir Immune Defic Syndr **21**(Suppl 1):S11–7.

79. Navarro, W. H., and L. D. Kaplan. AIDS-related lymphoproliferative disease. Blood **107**(1):13–20.

80. Roithmann, S., J. M. Tourani, and J. M. Andrieu. 1991. AIDS-associated non-Hodgkin lymphoma. Lancet **338**(8771):884–5.

81. Powles, T., G. Matthews, and M. Bower. 2000. AIDS related systemic non-Hodgkin's lymphoma. Sex Transm Infect **76**(5):335–41.

82. Colomo, L., F. Loong, and S. Rives. 2004. Plasmablastic lymphomas (PBL): diverse lymphomas associated with immunodeficiency and HIV. Paper presented at the Eighth International Conference of Malignancies in AIDS and Other Immunodeficiencies, Abstract 41; 2004.

83. Chetty, R., N. Hlatswayo, R. Muc, R. Sabaratnam, and K. Gatter. 2003. Plasmablastic lymphoma in HIV+ patients: an expanding spectrum. Histopathology **42**(6):605–9.

84. Schichman, S. A., R. McClure, R. F. Schaefer, and P. Mehta. 2004. HIV and plasmablastic lymphoma manifesting in sinus, testicles, and bones: a further expansion of the disease spectrum. Am J Hematol **77**(3):291–5.

85. Lin, O., R. Gerhard, M. C. Zerbini, and J. Teruya-Feldstein. 2005. Cytologic features of plasmablastic lymphoma. Cancer **105**(3):139–44.

86. 1993. A predictive model for aggressive non-Hodgkin's lymphoma. The International Non-Hodgkin's Lymphoma Prognostic Factors Project. N Engl J Med **329**(14):987–94.

87. Sweetenham, J. W. 2005. Diffuse large B-cell lymphoma: risk stratification and management of relapsed disease. Hematology (Am Soc Hematol Educ Program) 252–9.

88. Lim, S. T., R. Karim, A. Tulpule, B. N. Nathwani, and A. M. Levine. 2005. Prognostic factors in HIV-related diffuse large-cell lymphoma: before versus after highly active antiretroviral therapy. J Clin Oncol **23**(33):8477–82.

89. Chari, A., and L. Kaplan. 2005. Diagnosis and management of non-Hodgkin's lymphoma and Hodgkin's lymphoma. In: Volberding, P. A., J. M. Palefsky, editors. Viral and Immunological Malignancies. 1st ed. Hamilton: BC Decker; 2005, pp. 180–205.

90. Cheung, M. C., L. Pantanowitz, and B. J. Dezube. 2005. AIDS-related malignancies: emerging challenges in the era of highly active antiretroviral therapy. Oncologist **10**(6):412–26.

91. Lyter, D. W., J. Bryant, R. Thackeray, C. R. Rinaldo, and L. A. Kingsley. 1995. Incidence of human immunodeficiency virus-related and nonrelated malignancies in a large cohort of homosexual men. J Clin Oncol **13**(10):2540–6.

92. Kaplan, L. D., D. J. Straus, M. A. Testa, J. Von Roenn, B. J. Dezube, T. P. Cooley, B. Herndier, D. W. Northfelt, J. Huang, A. Tulpule, and A. M. Levine. 1997. Low-dose compared with standard-dose m-BACOD chemotherapy for non-Hodgkin's lymphoma associated with human immunodeficiency virus infection. National Institute of Allergy and Infectious Diseases AIDS Clinical Trials Group. N Engl J Med **336**(23):1641–8.

93. Sandler, A. S., and L. D. Kaplan. 1996. Diagnosis and management of systemic non-Hodgkin's lymphoma in HIV disease. Hematol Oncol Clin North Am **10**(5):1111–24.

94. Ratner, L., J. Lee, S. Tang, D. Redden, F. Hamzeh, B. Herndier, D. Scadden, L. Kaplan, R. Ambinder, A. Levine, W. Harrington, L. Grochow, C. Flexner, B. Tan, and D. Straus. 2001. Chemotherapy for human immunodeficiency virus-associated non-Hodgkin's lymphoma in combination with highly active anti-retroviral therapy. J Clin Oncol **19**(8):2171–8.

95. Vaccher, E., M. Spina, G. di Gennaro, R. Talamini, G. Nasti, O. Schioppa, G. Vultaggio, and U. Tirelli. 2001. Concomitant cyclophosphamide, doxorubicin, vincristine, and prednisone chemotherapy plus highly active antiretroviral therapy in patients with human immunodeficiency virus-related, non-Hodgkin lymphoma. Cancer **91**(1):155–63.

96. Mounier, N., M. Spina, J. Gabarre, et al. AIDS-related non-Hodgkin's lymphoma: final analysis of 485 patients treated with risk-adapted intensive chemotherapy. Blood 2006; **107**(10):3832–40.

97. Sparano, J. A., S. Lee, M. G. Chen, T. Nazeer, A. Einzig, R. F. Ambinder, D. H. Henry, J. Manalo, T. Li, and J. H. Von Roenn. 2004. Phase II trial of infusional cyclophosphamide, doxorubicin, and etoposide in patients with HIV-associated non-Hodgkin's lymphoma: an Eastern Cooperative Oncology Group Trial (E1494). J Clin Oncol **22**(8):1491–500.

98. Bower, M., N. McCall-Peat, N. Ryan, L. Davies, A. M. Young, S. Gupta, M. Nelson, B. Gazzard, and J. Stebbing. 2004. Protease inhibitors potentiate chemotherapy-induced neutropenia. Blood **104**(9):2943–6.

99. Little, R. F., S. Pittaluga, N. Grant, S. M. Steinberg, M. F. Kavlick, H. Mitsuya, G. Franchini, M. Gutierrez, M. Raffeld, E. S. Jaffe, G. Shearer, R. Yarchoan, and W. H. Wilson. 2003. Highly effective treatment of acquired immunodeficiency syndrome-related lymphoma with dose-adjusted EPOCH: impact of anti-retroviral therapy suspension and tumor biology. Blood **101**(12):4653–9.

100. Kaplan, L., J. Lee, and D. Scadden. 2003. No benefit from rituximab in a randomized phase III trial of CHOP with or without rituximab for patients with HIV-associated non-Hodgkin's lymphoma: updated data from AIDS Malignancies Consortium Study 010. Blood **23**:1002.

101. Vilmer, E., A. Rhodes-Feuillette, C. Rabian, M. Benbunan, J. Meletis, A. Devergie, J. H. Bourrhis, J. C. Gluckman, J. C. Chermann, and E. Gluckman E. 1987. Clinical and immunological restoration in patients with AIDS after marrow transplantation, using lymphocyte transfusions from the marrow donor. Transplantation **44**(1):25–9.

102. Davis, K. C., A. Hayward, G. Ozturk, and P. F. Kohler. 1983. Lymphocyte transfusion in case of acquired immunodeficiency syndrome. Lancet **1**(8324):599–600.

103. Hassett, J. M., C. G. Zaroulis, M. L. Greenberg, and F. P. Siegal. 1983. Bone marrow transplantation in AIDS. N Engl J Med **309**(11):665.

104. Holland, H. K., R. Saral, J. J. Rossi, A. D. Donnenberg, W. H. Burns, W. E. Beschorner, H. Farzadegan, R. J. Jones, G. V. Quinnan, G. B. Vogelsang, et al. Allogeneic bone marrow transplantation, zidovudine, and human immunodeficiency virus type 1 (HIV-1) infection. Studies in a patient with non-Hodgkin lymphoma. Ann Intern Med **111**(12):973–81.

105. Gabarre, J., N. Azar, B. Autran, C. Katlama, and V. Leblond. 2000. High-dose therapy and autologous haematopoietic stem-cell transplantation for HIV-1-associated lymphoma. Lancet **355**(9209):1071–2.

106. Krishnan, A., A. Molina, J. Zaia, D. Smith, D. Vasquez, N. Kogut, P. M. Falk, J. Rosenthal, J. Alvarnas, and S. J. Forman. 2005. Durable remissions with autologous stem cell transplantation for high-risk HIV-associated lymphomas. Blood **105**(2):874–8.

107. Re, A., C. Cattaneo, M. Michieli, S. Casari, M. Spina, M. Rupolo, B. Allione, A. Nosari, C. Schiantarelli, M. Vigano, I. Izzi, P. Ferremi, A. Lanfranchi, M. Mazzuccato, G. Carosi, U. Tirelli, and G. Rossi. 2003. High-dose therapy and autologous peripheral-blood stem-cell transplantation as salvage treatment for HIV-associated lymphoma in patients receiving highly active antiretroviral therapy. J Clin Oncol **21**(23):4423–7.

108. Diez-Martin, J., P. Balsalobre, R. Carrion, et al. 2003. Long term survival after autologous stem cell transplant (ASCT) in AIDS related lymphoma patients [abstract 868]. Blood **102**:247a.

109. Molina, A., J. Zaia, and A. Krishnan. 2003. Treatment of human immunodeficiency virus-related lymphoma with haematopoietic stem cell transplantation. Blood Rev **17**(4):249–58.

110. Berretta, M., R. Cinelli, F. Martellotta, M. Spina, E. Vaccher, and U. Tirelli. 2003. Therapeutic approaches to AIDS-related malignancies. Oncogene **22**(42):6646–59.

111. Ansell, S. M., and J. O. Armitage. 2006. Management of Hodgkin lymphoma. Mayo Clin Proc **81**(3):419–26.

112. Uccini, S., F. Monardo, L. P. Ruco, C. D. Baroni, A. Faggioni, A. M. Agliano, A. Gradilone, V. Manzari, L. Vago, G. Costanzi, et al. 1989. High frequency of Epstein-Barr virus genome in HIV-positive patients with Hodgkin's disease. Lancet **1**(8652):1458.

113. Grulich, A. E., Y. Li, A. McDonald, P. K. Correll, M. G. Law, and J. M. Kaldor. 2002. Rates of non-AIDS-defining cancers in people with HIV infection before and after AIDS diagnosis. AIDS **16**(8):1155–61.

114. IARC Working Group on the Evaluation of Carcinogenic Risks to Humans. 1996. Human Immunodeficiency Viruses and Human T-Cell Lymphotropic Viruses, Lyon, France, 1–18 June 1996. IARC Monogr Eval Carcinog Risks Hum 67:1–424.

115. Glaser, S. L., C. A. Clarke, M. L. Gulley, F. E. Craig, J. A. DiGiuseppe, R. F. Dorfman, R. B. Mann, and R. F. Ambinder. 2003. Population-based patterns of human immunodeficiency virus-related Hodgkin lymphoma in the Greater San Francisco Bay Area, 1988–1998. Cancer **98**(2):300–9.

116. Franceschi, S., L. Dal Maso, S. Arniani, P. Crosignani, M. Vercelli, L. Simonato, F. Falcini, R. Zanetti, A. Barchielli, D. Serraino, and G. Rezza. 1998. Risk of cancer other than Kaposi's sarcoma and non-Hodgkin's lymphoma in persons with AIDS in Italy. Cancer and AIDS Registry Linkage Study. Br J Cancer **78**(7):966–70.

117. Spina, M., S. Sandri, and U. Tirelli. 1999. Hodgkin's disease in HIV-infected individuals. Curr Opin Oncol **11**(6):522–6.

118. Spina, M., E. Vaccher, A. Carbone, and U. Tirelli. Neoplastic complications of HIV infection. Ann Oncol **10**(11):1271–86.

119. Doweiko, J., B. J. Dezube, and L. Pantanowitz. 2004. Unusual sites of Hodgkin's lymphoma: CASE 1. HIV-associated Hodgkin's lymphoma of the stomach. J Clin Oncol **22**(20):4227–8.

120. Vaccher, E., M. Spina, and U. Tirelli. 2001. Clinical aspects and management of Hodgkin's disease and other tumours in HIV-infected individuals. Eur J Cancer **37**(10):1306–15.
121. Spina, M., M. Berretta, and U. Tirelli. 2003. Hodgkin's disease in HIV. Hematol Oncol Clin North Am **17**(3):843–58.
122. Tirelli, U., D. Errante, R. Dolcetti, A. Gloghini, D. Serraino, E. Vaccher, S. Franceschi, M. Boiocchi, and A. Carbone. 1995. Hodgkin's disease and human immunodeficiency virus infection: clinicopathologic and virologic features of 114 patients from the Italian Cooperative Group on AIDS and Tumors. J Clin Oncol **13**(7):1758–67.
123. Hoffmann, C., K. U. Chow, E. Wolf, G. Faetkenheuer, H. J. Stellbrink, J. van Lunzen, H. Jaeger, A. Stoehr, A. Plettenberg, J. C. Wasmuth, J. Rockstroh, F. Mosthaf, H. A. Horst, and H.. R. Brodt. 2004. Strong impact of highly active antiretroviral therapy on survival in patients with human immunodeficiency virus-associated Hodgkin's disease. Br J Haematol **125**(4):455–62.
124. Rubio, R. 1994. Hodgkin's disease associated with human immunodeficiency virus infection. A clinical study of 46 cases. Cooperative Study Group of Malignancies Associated with HIV Infection of Madrid. Cancer **73**(9):2400–7.
125. Ree, H. J., J. A. Strauchen, A. A. Khan, J. E. Gold, J. P. Crowley, H. Kahn, and R. Zalusky. 1991. Human immunodeficiency virus-associated Hodgkin's disease. Clinicopathologic studies of 24 cases and preponderance of mixed cellularity type characterized by the occurrence of fibrohistiocytoid stromal cells. Cancer **67**(6):1614–21.
126. Errante, D., U. Tirelli, R. Gastaldi, D. Milo, A. M. Nosari, G. Rossi, G. Fiorentini, A. Carbone, E. Vaccher, and S. Monfardini. 1994. Combined antineoplastic and antiretroviral therapy for patients with Hodgkin's disease and human immunodeficiency virus infection. A prospective study of 17 patients. The Italian Cooperative Group on AIDS and Tumors (GICAT). Cancer **73**(2):437–44.
127. Tirelli, U., E. Vaccher, V. Zagonel, R. Talamini, D. Bernardi, M. Tavio, A. Gloghini, M. C. Merola, S. Monfardini, and A. Carbone. 1995. CD30 (Ki-1)-positive anaplastic large-cell lymphomas in 13 patients with and 27 patients without human immunodeficiency virus infection: the first comparative clinicopathologic study from a single institution that also includes 80 patients with other human immunodeficiency virus-related systemic lymphomas. J Clin Oncol **13**(2):373–80.
128. Lister, T. A., D. Crowther, S. B. Sutcliffe, E. Glatstein, G. P. Canellos, R. C. Young, S. A. Rosenberg, C. A. Coltman, M. Tubiana. 1989. Report of a committee convened to discuss the evaluation and staging of patients with Hodgkin's disease: Cotswolds meeting. J Clin Oncol **7**(11):1630–6.
129. Gerard, L., L. Galicier, E. Boulanger, L. Quint, M. G. Lebrette, E. Mortier, V. Meignin, and E. Oksenhendler. 2003. Improved survival in HIV-related Hodgkin's lymphoma since the introduction of highly active antiretroviral therapy. AIDS **17**(1):81–7.
130. Johnson, P. W., J. A. Radford, M. H. Cullen, M. R. Sydes, J. Walewski, A. S. Jack, K. A. MacLennan, S. P. Stenning, S. Clawson, P. Smith, D. Ryder, and B. W. Hancock. 2005. Comparison of ABVD and alternating or hybrid multidrug regimens for the treatment of advanced Hodgkin's lymphoma: results of the United Kingdom Lymphoma Group LY09 Trial (ISRCTN97144519). J Clin Oncol **23**(36):9208–18.
131. Levine, A. M., P. Li, T. Cheung, A. Tulpule, J. Von Roenn, B. N. Nathwani, and L. Ratner. 2000. Chemotherapy consisting of doxorubicin, bleomycin, vinblastine, and dacarbazine with granulocyte-colony-stimulating factor in HIV-infected patients with newly diagnosed Hodgkin's disease: a prospective, multi-institutional AIDS clinical trials group study (ACTG 149). J Acquir Immune Defic Syndr **24**(5):444–50.
132. Hartmann, P., U. Rehwald, B. Salzberger, C. Franzen, M. Sieber, A. Wohrmann, and V. Diehl. 2003. BEACOPP therapeutic regimen for patients with Hodgkin's disease and HIV infection. Ann Oncol **14**(10):1562–9.
133. Errante, D., J. Gabarre, A. L. Ridolfo, G. Rossi, A. M. Nosari, C. Gisselbrecht, Y. Kerneis, F. Mazzetti, E. Vaccher, R. Talamini, A. Carbone, and U. Tirelli. 1999. Hodgkin's disease in 35 patients with HIV infection: an experience with epirubicin, bleomycin, vinblastine and prednisone chemotherapy in combination with antiretroviral therapy and primary use of G-CSF. Ann Oncol **10**(2):189–95.
134. Spina, M., J. Gabarre, G. Rossi, M. Fasan, C. Schiantarelli, E. Nigra, M. Mena, A. Antinori, A. Ammassari, R. Talamini, E. Vaccher, G. di Gennaro, and U. Tirelli. 2002. Stanford V regimen and concomitant HAART in 59 patients with Hodgkin disease and HIV infection. Blood **100**(6):1984–8.
135. Biggar, R. J., M. Frisch, and J. J. Goedert. 2000. Risk of cancer in children with AIDS. AIDS-Cancer Match Registry Study Group. JAMA **284**(2):205–9.
136. Chadwick, E. G., E. L. Connor, I. C. Hanson, V. V. Joshi, H. Abu-Farsakh, R. Yogev, G. McSherry, K. McClain, and S. B. Murphy. 1990. Tumors of smooth-muscle origin in HIV-infected children. JAMA **263**(23):3182–4.
137. Levin, T. L., H. M. Adam, K. H. van Hoeven, and H. S. Goldman. 1994. Hepatic spindle cell tumors in HIV positive children. Pediatr Radiol **24**(1):78–9.

138. Granovsky, M. O., B. U. Mueller, H. S. Nicholson, P. S. Rosenberg, and C. S. Rabkin. 1998. Cancer in human immunodeficiency virus-infected children: a case series from the Children's Cancer Group and the National Cancer Institute. J Clin Oncol **16**(5):1729–35.

139. Chiao, E. Y., and S. E. Krown. 2003. Update on non-acquired immunodeficiency syndrome-defining malignancies. Curr Opin Oncol **15**(5):389–97.

140. Mueller, B. U. 1999. Cancers in children infected with the human immunodeficiency virus. Oncologist **4**(4):309–17.

141. Miser, J., T. Triche, T. Kinsella, et al. 1997. Other soft tissue sarcomas of childhood. In: Pizzo, P. A., D. G. Poplack, editors. Principles and Practice of Pediatric Oncology. Philadelphia: Lippincott-Raven Publishers; 1997, pp. 865–88.

142. Moore, P. S., and Y. Chang. 2001. Molecular virology of Kaposi's sarcoma-associated herpesvirus. Philos Trans R Soc Lond B Biol Sci **356**(1408):499–516.

143. Chang, Y., E. Cesarman, M. S. Pessin, F. Lee, J. Culpepper, D. M. Knowles, and P. S. Moore. Identification of herpesvirus-like DNA sequences in AIDS-associated Kaposi's sarcoma. Science **266**(5192):1865–9.

144. Aoki, Y., and G. Tosato. 2003. Pathogenesis and manifestations of human herpesvirus-8-associated disorders. Semin Hematol **40**(2):143–53.

145. Schwartz, E. J., R. F. Dorfman, and S. Kohler. 2003. Human herpesvirus-8 latent nuclear antigen-1 expression in endemic Kaposi sarcoma: an immunohistochemical study of 16 cases. Am J Surg Pathol **27**(12):1546–50.

146. Aoki, Y., and G. Tosato. 2003. Targeted inhibition of angiogenic factors in AIDS-related disorders. Curr Drug Targets Infect Disord **3**(2):115–28.

147. Soulier, J., L. Grollet, E. Oksenhendler, P. Cacoub, D. Cazals-Hatem, P. Babinet, M. F. d'Agay, J. P. Clauvel, M. Raphael, L. Degos, et al. 1995. Kaposi's sarcoma-associated herpesvirus-like DNA sequences in multicentric Castleman's disease. Blood **86**(4):1276–80.

148. Chadburn, A., E. Cesarman, R. G. Nador, Y. F. Liu, and D. M. Knowles. 1997. Kaposi's sarcoma-associated herpesvirus sequences in benign lymphoid proliferations not associated with human immunodeficiency virus. Cancer **80**(4):788–97.

149. Oksenhendler, E., G. Carcelain, Y. Aoki, E. Boulanger, A. Maillard, J. P. Clauvel, and F. Agbalika. 2000. High levels of human herpesvirus 8 viral load, human interleukin-6, interleukin-10, and C reactive protein correlate with exacerbation of multicentric castleman disease in HIV-infected patients. Blood **96**(6):2069–73.

150. Cesarman, E., Y. Chang, P. S. Moore, J. W. Said, and D. M. Knowles. 1995. Kaposi's sarcoma-associated herpesvirus-like DNA sequences in AIDS-related body-cavity-based lymphomas. N Engl J Med **332**(18):1186–91.

151. Iscovich, J., P. Boffetta, S. Franceschi, E. Azizi, and R. Sarid. 2000. Classic Kaposi sarcoma: epidemiology and risk factors. Cancer **88**(3):500–17.

152. Cook-Mozaffari, P., R. Newton, V. Beral, and D. P. Burkitt. 1998. The geographical distribution of Kaposi's sarcoma and of lymphomas in Africa before the AIDS epidemic. Br J Cancer **78**(11):1521–8.

153. Friedman-Kien, A. E., and B. R. Saltzman. 1990. Clinical manifestations of classical, endemic African, and epidemic AIDS-associated Kaposi's sarcoma. J Am Acad Dermatol **22**(6 Pt 2):1237–50.

154. Siegel, J. H., R. Janis, J. C. Alper, H. Schutte, L. Robbins, and M. D. Blaufox. 1969. Disseminated visceral Kaposi's sarcoma. Appearance after human renal homograft operation. JAMA **207**(8):1493–6.

155. Penn, I. 1979. Kaposi's sarcoma in organ transplant recipients: report of 20 cases. Transplantation **27**(1):8–11.

156. Friedman-Kien, A. E. 1981. Disseminated Kaposi's sarcoma syndrome in young homosexual men. J Am Acad Dermatol **5**(4):468–71.

157. Hengge, U. R., T. Ruzicka, S. K. Tyring, M. Stuschke, M. Roggendorf, R. A. Schwartz, and S. Seeber. 2002. Update on Kaposi's sarcoma and other HHV8 associated diseases. Part 1: epidemiology, environmental predispositions, clinical manifestations, and therapy. Lancet Infect Dis **2**(5):281–92.

158. Frisch, M., R. J. Biggar, E. A. Engels, and J. J. Goedert. 2001. Association of cancer with AIDS-related immunosuppression in adults. JAMA **285**(13):1736–45.

159. Kaplan, L. D., and D. W. Northfelt. 1997. Malignancies associated with AIDS. In: Sande, M. A., P. A. Volberding, editors. The Medical Management of AIDS. Philadelphia: WB Saunders; 1997, pp. 413–22.

160. Friedman, S. L., T. L. Wright, and D. F. Altman. 1985. Gastrointestinal Kaposi's sarcoma in patients with acquired immunodeficiency syndrome. Endoscopic and autopsy findings. Gastroenterology **89**(1):102–8.

161. Mitchell, D. M., M. McCarty, J. Fleming, and F. M. Moss. 1992. Bronchopulmonary Kaposi's sarcoma in patients with AIDS. Thorax **47**(9):726–9.

162. Renwick, N., T. Halaby, G. J. Weverling, N. H. Dukers, G. R. Simpson, R. A. Coutinho, J. M. Lange, T. F. Schulz, and J. Goudsmit. 1998. Seroconversion for human herpesvirus 8 during HIV infection is highly predictive of Kaposi's sarcoma. AIDS **12**(18):2481–8.

163. Jacobson, L. P., F. J. Jenkins, G. Springer, A. Munoz, K. V. Shah, J. Phair, Z. Zhang, and H. Armenian. 2000. Interaction of human immunodeficiency virus type 1 and human herpesvirus type 8 infections on the incidence of Kaposi's sarcoma. J Infect Dis **181**(6):1940–9.
164. Engels, E. A., R. J. Biggar, V. A. Marshall, M. A. Walters, C. J. Gamache, D. Whitby, and J. J. Goedert. 2003. Detection and quantification of Kaposi's sarcoma-associated herpesvirus to predict AIDS-associated Kaposi's sarcoma. AIDS **17**(12):1847–51.
165. Newton, R., J. Ziegler, D. Bourboulia, D. Casabonne, V. Beral, E. Mbidde, L. Carpenter, D. M. Parkin, H. Wabinga, S. Mbulaiteye, H. Jaffe, R. Weiss, and C. Boshoff. Infection with Kaposi's sarcoma-associated herpesvirus (KSHV) and human immunodeficiency virus (HIV) in relation to the risk and clinical presentation of Kaposi's sarcoma in Uganda. Br J Cancer **89**(3):502–4.
166. Verma, S. C., and E. S. Robertson. 2003. Molecular biology and pathogenesis of Kaposi sarcoma-associated herpesvirus. FEMS Microbiol Lett **222**(2):155–63.
167. Dourmishev, L. A., A. L. Dourmishev, D. Palmeri, R. A. Schwartz, D. M. Lukac. 2003. Molecular genetics of Kaposi's sarcoma-associated herpesvirus (human herpesvirus-8) epidemiology and pathogenesis. Microbiol Mol Biol Rev **67**(2):175–212, table of contents.
168. Tirelli, U., D. Bernardi, M. Spina, and E. Vaccher. 1997. AIDS-related tumors: integrating antiviral and anticancer therapy. Crit Rev Oncol Hematol **41**(3):299–315.
169. Nicholas, J., V. R. Ruvolo, W. H. Burns, G. Sandford, X. Wan, D. Ciufo, S. B. Hendrickson, H. G. Guo, G. S. Hayward, and M. S. Reitz. 1997. Kaposi's sarcoma-associated human herpesvirus-8 encodes homologues of macrophage inflammatory protein-1 and interleukin-6. Nat Med **3**(3):287–92.
170. Bais, C., B. Santomasso, O. Coso, L. Arvanitakis, E. G. Raaka, J. S. Gutkind, A. S. Asch, E. Cesarman, M. C. Gershengorn, and E. A. Mesri. 1998. G-protein-coupled receptor of Kaposi's sarcoma-associated herpesvirus is a viral oncogene and angiogenesis activator. Nature **391**(6662):86–9.
171. Uccini, S., S. Scarpino, F. Ballarini, A. Soriani, M. Chilosi, M. A. Montesu, M. V. Masala, F. Cottoni, and L. Ruco. 2003. In situ study of chemokine and chemokine-receptor expression in Kaposi sarcoma. Am J Dermatopathol **25**(5):377–83.
172. Deregibus, M. C., V. Cantaluppi, S. Doublier, M. F. Brizzi, I. Deambrosis, A. Albini, and G. Camussi. 2002. HIV-1-Tat protein activates phosphatidylinositol 3-kinase/AKT-dependent survival pathways in Kaposi's sarcoma cells. J Biol Chem **277**(28):25195–202.
173. Ensoli, B., G. Barillari, S. Z. Salahuddin, R. C. Gallo, and F. Wong-Staal. 1990. Tat protein of HIV-1 stimulates growth of cells derived from Kaposi's sarcoma lesions of AIDS patients. Nature **345**(6270):84–6.
174. Barillari, G., and B. Ensoli. 2002. Angiogenic effects of extracellular human immunodeficiency virus type 1 Tat protein and its role in the pathogenesis of AIDS-associated Kaposi's sarcoma. Clin Microbiol Rev **15**(2):310–26.
175. Impola, U., M. A. Cuccuru, M. V. Masala, L. Jeskanen, F. Cottoni, and U. Saarialho-Kere. 2003. Preliminary communication: matrix metalloproteinases in Kaposi's sarcoma. Br J Dermatol **149**(4):905–7.
176. Lafrenie, R. M., L. M. Wahl, J. S. Epstein, I. K. Hewlett, K. M. Yamada, and S. Dhawan. 1996. HIV-1-Tat modulates the function of monocytes and alters their interactions with microvessel endothelial cells. A mechanism of HIV pathogenesis. J Immunol **156**(4):1638–45.
177. Mocroft, A., O. Kirk, N. Clumeck, P. Gargalianos-Kakolyris, H. Trocha, N. Chentsova, F. Antunes, H. J. Stellbrink, A. N. Phillips, and J. D. Lundgren. 2004. The changing pattern of Kaposi sarcoma in patients with HIV, 1994–2003: the EuroSIDA Study. Cancer **100**(12):2644–54.
178. Jones, J. L., D. L. Hanson, M. S. Dworkin, and H. W. Jaffe. 2000. Incidence and trends in Kaposi's sarcoma in the era of effective antiretroviral therapy. J Acquir Immune Defic Syndr **24**(3):270–4.
179. Krown, S. E., M. A. Testa, and J. Huang. 1997. AIDS-related Kaposi's sarcoma: prospective validation of the AIDS Clinical Trials Group staging classification. AIDS Clinical Trials Group Oncology Committee. J Clin Oncol **15**(9):3085–92.
180. Beral, V., T. A. Peterman, R. L. Berkelman, and H. W. Jaffe. 1990. Kaposi's sarcoma among persons with AIDS: a sexually transmitted infection? Lancet **335**(8682):123–8.
181. Hoover, D. R., C. Black, L. P. Jacobson, O. Martinez-Maza, D. Seminara, A. Saah, J. Von Roenn, R. Anderson, and H. K. Armenian. 1993. Epidemiologic analysis of Kaposi's sarcoma as an early and later AIDS outcome in homosexual men. Am J Epidemiol **138**(4):266–78.
182. Campbell, T. B., M. Borok, I. E. White, I. Gudza, B. Ndemera, A. Taziwa, A. Weinberg, and L. Gwanzura. 2003. Relationship of Kaposi sarcoma (KS)-associated herpesvirus viremia and KS disease in Zimbabwe. Clin Infect Dis **36**(9):1144–51.
183. Wabinga, H. R., D. M. Parkin, F. Wabwire-Mangen, and J. W. Mugerwa. 1993. Cancer in Kampala, Uganda, in 1989–91: changes in incidence in the era of AIDS. Int J Cancer **54**(1):26–36.
184. Chokunonga, E., L. M. Levy, M. T. Bassett, B. G. Mauchaza, D. B. Thomas, and D. M. Parkin. 2000. Cancer incidence in the African population of Harare, Zimbabwe: second results from the cancer registry 1993–1995. Int J Cancer **85**(1):54–9.

185. Dezube, B. J. 2000. Acquired immunodeficiency syndrome-related Kaposi's sarcoma: clinical features, staging, and treatment. Semin Oncol **27**(4):424–30.
186. Nasti, G., R. Talamini, A. Antinori, F. Martellotta, G. Jacchetti, F. Chiodo, G. Ballardini, L. Stoppini, G. Di Perri, M. Mena, M. Tavio, E. Vaccher, A. D'Arminio Monforte, and U. Tirelli. 2003. AIDS-related Kaposi's Sarcoma: evaluation of potential new prognostic factors and assessment of the AIDS Clinical Trial Group Staging System in the Haart Era – the Italian Cooperative Group on AIDS and Tumors and the Italian Cohort of Patients Naive From Antiretrovirals. J Clin Oncol **21**(15):2876–82.
187. Dezube, B. J., L. Pantanowitz, and D. M. Aboulafia. 2004. Management of AIDS-related Kaposi sarcoma: advances in target discovery and treatment. AIDS Read **14**(5):236–8, 243–4, 251–3.
188. Danzig, J. B., L. J. Brandt, J. F. Reinus, and R. S. Klein. 1991. Gastrointestinal malignancy in patients with AIDS. Am J Gastroenterol **86**(6):715–8.
189. Aboulafia, D. M. 2000. The epidemiologic, pathologic, and clinical features of AIDS-associated pulmonary Kaposi's sarcoma. Chest **117**(4):1128–45.
190. Gruden, J. F., L. Huang, W. R. Webb, G. Gamsu, P. C. Hopewell, and D. M. Sides. 1995. AIDS-related Kaposi sarcoma of the lung: radiographic findings and staging system with bronchoscopic correlation. Radiology **195**(2):545–52.
191. Boudreaux, A. A., L. L. Smith, C. D. Cosby, M. M. Bason, J. W. Tappero, and B. G. Berger. 1993. Intralesional vinblastine for cutaneous Kaposi's sarcoma associated with acquired immunodeficiency syndrome. A clinical trial to evaluate efficacy and discomfort associated with infection. J Am Acad Dermatol **28**(1):61–5.
192. Pluda, J. M., S. Broder, and R. Yarchoan. 1992. Therapy of AIDS and AIDS-associated neoplasms. Cancer Chemother Biol Response Modif **13**:404–39.
193. Cooper, J. S., A. D. Steinfeld, and I. Lerch. 1991. Intentions and outcomes in the radiotherapeutic management of epidemic Kaposi's sarcoma. Int J Radiat Oncol Biol Phys **20**(3):419–22.
194. de Wit, R., W. G. Smit, K. H. Veenhof, P. J. Bakker, F. Oldenburger, and D. G. Gonzalez. 1990. Palliative radiation therapy for AIDS-associated Kaposi's sarcoma by using a single fraction of 800 cGy. Radiother Oncol **19**(2):131–6.
195. Northfelt, D. W., B. J. Dezube, J. A. Thommes, B. J. Miller, M. A. Fischl, A. Friedman-Kien, L. D. Kaplan, C. Du Mond, R. D. Mamelok, and D. H. Henry. 1998. Pegylated-liposomal doxorubicin versus doxorubicin, bleomycin, and vincristine in the treatment of AIDS-related Kaposi's sarcoma: results of a randomized phase III clinical trial. J Clin Oncol **16**(7):2445–51.
196. Stewart, S., H. Jablonowski, F. D. Goebel, K. Arasteh, M. Spittle, A. Rios, D. Aboulafia, J. Galleshaw, B. J. Dezube. 1998. Randomized comparative trial of pegylated liposomal doxorubicin versus bleomycin and vincristine in the treatment of AIDS-related Kaposi's sarcoma. International Pegylated Liposomal Doxorubicin Study Group. J Clin Oncol **16**(2):683–91.
197. Gill, P. S., J. Wernz, D. T. Scadden, P. Cohen, G. M. Mukwaya, J. H. von Roenn, M. Jacobs, S. Kempin, I. Silverberg, G. Gonzales, M. U. Rarick, A. M. Myers, F. Shepherd, C. Sawka, M. C. Pike, and M. E. Ross. 1996. Randomized phase III trial of liposomal daunorubicin versus doxorubicin, bleomycin, and vincristine in AIDS-related Kaposi's sarcoma. J Clin Oncol **14**(8):2353–64.
198. Welles, L., M. W. Saville, J. Lietzau, J. M. Pluda, K. M. Wyvill, I. Feuerstein, W. D. Figg, R. Lush, J. Odom, W. H. Wilson, M. T. Fajardo, R. W. Humphrey, E. Feigal, D. Tuck, S. M. Steinberg, S. Broder, and R. Yarchoan. 1998. Phase II trial with dose titration of paclitaxel for the therapy of human immunodeficiency virus-associated Kaposi's sarcoma. J Clin Oncol **16**(3):1112–21.
199. Gill, P. S., A. Tulpule, B. M. Espina, S. Cabriales, J. Bresnahan, M. Ilaw, S. Louie, N. F. Gustafson, M. A. Brown, C. Orcutt, B. Winograd, and D. T. Scadden. 1999. Paclitaxel is safe and effective in the treatment of advanced AIDS-related Kaposi's sarcoma. J Clin Oncol **17**(6):1876–83.
200. Tulpule, A., J. Groopman, M. W. Saville, W. Harrington, Jr., A. Friedman-Kien, B. M. Espina, C. Garces, L. Mantelle, K. Mettinger, D. T. Scadden, and P. S. Gill. 2002. Multicenter trial of low-dose paclitaxel in patients with advanced AIDS-related Kaposi sarcoma. Cancer **95**(1):147–54.
201. Stebbing, J., A. Wildfire, S. Portsmouth, T. Powles, C. Thirlwell, P. Hewitt, M. Nelson, S. Patterson, S. Mandalia, F. Gotch, B. G. Gazzard, and M. Bower. 2003. Paclitaxel for anthracycline-resistant AIDS-related Kaposi's sarcoma: clinical and angiogenic correlations. Ann Oncol **14**(11):1660–6.
202. Nasti, G., D. Errante, S. Santarossa, E. Vaccher, and U. Tirelli. 1999. A risk and benefit assessment of treatment for AIDS-related Kaposi's sarcoma. Drug Saf **20**(5):403–25.
203. Bundow, D., and D. M. Aboulafia. 2004. Potential drug interaction with paclitaxel and highly active antiretroviral therapy in two patients with AIDS-associated Kaposi sarcoma. Am J Clin Oncol **27**(1):81–4.
204. Robles, R., D. Lugo, L. Gee, and M. A. Jacobson. 1999. Effect of antiviral drugs used to treat cytomegalovirus end-organ disease on subsequent course of previously diagnosed Kaposi's sarcoma in patients with AIDS. J Acquir Immune Defic Syndr Hum Retrovirol **20**(1):34–8.

205. Martin, D. F., B. D. Kuppermann, R. A. Wolitz, A. G. Palestine, H. Li, and C. A. Robinson. 1999. Oral ganciclovir for patients with cytomegalovirus retinitis treated with a ganciclovir implant. Roche Ganciclovir Study Group. N Engl J Med **340**(14):1063–70.

206. Little, R. F., F. Merced-Galindez, K. Staskus, D. Whitby, Y. Aoki, R. Humphrey, J. M. Pluda, V. Marshall, M. Walters, L. Welles, I. R. Rodriguez-Chavez, S. Pittaluga, G. Tosato, and R. Yarchoan. 2003. A pilot study of cidofovir in patients with kaposi sarcoma. J Infect Dis **187**(1):149–53.

207. Klass, C. M., and M. K. Offermann. 2005. Targeting human herpesvirus-8 for treatment of Kaposi's sarcoma and primary effusion lymphoma. Curr Opin Oncol **17**(5):447–55.

208. Dezube, B. J., J. H. Von Roenn, J. Holden-Wiltse, T. W. Cheung, S. C. Remick, T. P. Cooley, J. Moore, J. P. Sommadossi, S. L. Shriver, C. W. Suckow, and P. S. Gill. 1998. Fumagillin analog in the treatment of Kaposi's sarcoma: a phase I AIDS Clinical Trial Group study. AIDS Clinical Trial Group No. 215 Team. J Clin Oncol **16**(4):1444–9.

209. Little, R. F., K. M. Wyvill, J. M. Pluda, L. Welles, V. Marshall, W. D. Figg, F. M. Newcomb, G. Tosato, E. Feigal, S. M. Steinberg, D. Whitby, J. J. Goedert, and R. Yarchoan. 2000. Activity of thalidomide in AIDS-related Kaposi's sarcoma. J Clin Oncol **18**(13):2593–602.

210. Cianfrocca, M., T. P. Cooley, J. Y. Lee, M. A. Rudek, D. T. Scadden, L. Ratner, J. M. Pluda, W. D. Figg, S. E. Krown, and B. J. Dezube. Matrix metalloproteinase inhibitor COL-3 in the treatment of AIDS-related Kaposi's sarcoma: a phase I AIDS malignancy consortium study. J Clin Oncol **20**(1):153–9.

211. Koon, H. B., G. J. Bubley, L. Pantanowitz, D. Masiello, B. Smith, K. Crosby, J. Proper, W. Weeden, T. E. Miller, P. Chatis, M. J. Egorin, S. R. Tahan, and B. J. Dezube. 2005. Imatinib-induced regression of AIDS-related Kaposi's sarcoma. J Clin Oncol **23**(5):982–9.

212. Noy, A., D. T. Scadden, J. Lee, B. J. Dezube, D. Aboulafia, A. Tulpule, S. Walmsley, and P. Gill. 2005. Angiogenesis inhibitor IM862 is ineffective against AIDS-Kaposi's sarcoma in a phase III trial, but demonstrates sustained, potent effect of highly active antiretroviral therapy: from the AIDS Malignancy Consortium and IM862 Study Team. J Clin Oncol **23**(5):990–8.

213. Sturzl, M., W. K. Roth, N. H. Brockmeyer, C. Zietz, B. Speiser, and P. H. Hofschneider. 1992. Expression of platelet-derived growth factor and its receptor in AIDS-related Kaposi sarcoma in vivo suggests paracrine and autocrine mechanisms of tumor maintenance. Proc Natl Acad Sci USA **89**(15):7046–50.

214. Moses, A. V., M. A. Jarvis, C. Raggo, Y. C. Bell, R. Ruhl, B. G. Luukkonen, D. J. Griffith, C. L. Wait, B. J. Druker, M. C. Heinrich, J. A. Nelson, and K. Fruh. 2002. Kaposi's sarcoma-associated herpesvirus-induced upregulation of the c-kit proto-oncogene, as identified by gene expression profiling, is essential for the transformation of endothelial cells. J Virol **76**(16):8383–99.

215. Scadden, D. T. 2003. AIDS-related malignancies. Annu Rev Med **54**:285–303.

216. Noy, A. 2003. Update in Kaposi sarcoma. Curr Opin Oncol **15**(5):379–81.

217. Sgadari, C., P. Monini, G. Barillari, and B. Ensoli. 2003. Use of HIV protease inhibitors to block Kaposi's sarcoma and tumour growth. Lancet Oncol **4**(9):537–47.

218. Bower, M., P. Fox, K. Fife, J. Gill, M. Nelson, and B. Gazzard. 1999. Highly active anti-retroviral therapy (HAART) prolongs time to treatment failure in Kaposi's sarcoma. AIDS **13**(15):2105–11.

219. Lebbe, C., L. Blum, C. Pellet, G. Blanchard, O. Verola, P. Morel, O. Danne, and F. Calvo. 1998. Clinical and biological impact of antiretroviral therapy with protease inhibitors on HIV-related Kaposi's sarcoma. AIDS **12**(7):F45–9.

220. Tavio, M., G. Nasti, M. Spina, D. Errante, E. Vaccher, and U. Tirelli. 1998. Highly active antiretroviral therapy in HIV-related Kaposi's sarcoma. Ann Oncol **9**(8):923.

221. Levine, A. M., and A. Tulpule. 2001. Clinical aspects and management of AIDS-related Kaposi's sarcoma. Eur J Cancer **37**(10):1288–95.

222. Nasti, G., F. Martellotta, M. Berretta, M. Mena, M. Fasan, G. Di Perri, R. Talamini, G. Pagano, M. Montroni, R. Cinelli, E. Vaccher, A. D'Arminio Monforte, and U. Tirelli. 2003. Impact of highly active antiretroviral therapy on the presenting features and outcome of patients with acquired immunodeficiency syndrome-related Kaposi sarcoma. Cancer **98**(11):2440–6.

223. Tam, H. K., Z. F. Zhang, L. P. Jacobson, J. B. Margolick, J. S. Chmiel, C. Rinaldo, and R. Detels. 2002. Effect of highly active antiretroviral therapy on survival among HIV-infected men with Kaposi sarcoma or non-Hodgkin lymphoma. Int J Cancer **98**(6):916–22.

224. Jung, C., J. R. Bogner, and F. Goebel. 1998. Resolution of severe Kaposi's sarcoma after initiation of anti-retroviral triple therapy. Eur J Med Res **3**(9):439–42.

225. Castleman, B., L. Iverson, and V. P. Menendez. 1956. Localized mediastinal lymphnode hyperplasia resembling thymoma. Cancer **9**(4):822–30.

226. Herrada, J., F. Cabanillas, L. Rice, J. Manning, and W. Pugh. 1998. The clinical behavior of localized and multicentric Castleman disease. Ann Intern Med **128**(8):657–62.

227. Leibetseder, F., and J. Thurner. 1973. Angiofollicular lymph node hyperplasia (onion-skin lymphoma). Med Klin **68**(24):817–20.

228. Gaba, A. R., R. S. Stein, D. L. Sweet, and D. Variakojis. 1978. Multicentric giant lymph node hyperplasia. Am J Clin Pathol **69**(1):86–90.

229. Frizzera, G., B. A. Peterson, E. D. Bayrd, and A. Goldman. 1985. A systemic lymphoproliferative disorder with morphologic features of Castleman's disease: clinical findings and clinicopathologic correlations in 15 patients. J Clin Oncol **3**(9):1202–16.

230. Peterson, B. A., and G. Frizzera G. 1993. Multicentric Castleman's disease. Semin Oncol **20**(6):636–47.

231. Lachant, N. A., N. C. Sun, L. A. Leong, R. S. Oseas, and H. E. Prince. 1985. Multicentric angiofollicular lymph node hyperplasia (Castleman's disease) followed by Kaposi's sarcoma in two homosexual males with the acquired immunodeficiency syndrome (AIDS). Am J Clin Pathol **83**(1):27–33.

232. Oksenhendler, E., E. Boulanger, L. Galicier, M. Q. Du, N. Dupin, T. C. Diss, R. Hamoudi, M. T. Daniel, F. Agbalika, C. Boshoff, J. P. Clauvel, P. G. Isaacson, and V. Meignin. 2002. High incidence of Kaposi sarcoma-associated herpesvirus-related non-Hodgkin lymphoma in patients with HIV infection and multicentric Castleman disease. Blood **99**(7):2331–6.

233. Russo, J. J., R. A. Bohenzky, M. C. Chien, J. Chen, M. Yan, D. Maddalena, J. P. Parry, D. Peruzzi, I. S. Edelman, Y. Chang, and P. S. Moore. 1996. Nucleotide sequence of the Kaposi sarcoma-associated herpesvirus (HHV8). Proc Natl Acad Sci USA **93**(25):14862–7.

234. Foussat, A., R. Fior, T. Girard, F. Boue, J. Wijdenes, P. Galanaud, and D. Emilie. 1999. Involvement of human interleukin-6 in systemic manifestations of human herpesvirus type 8-associated multicentric Castleman's disease. AIDS **13**(1):150–2.

235. Drexler, H. G., C. Meyer, G. Gaidano, and A. Carbone. 1999. Constitutive cytokine production by primary effusion (body cavity-based) lymphoma-derived cell lines. Leukemia **13**(4):634–40.

236. Dupin, N., T. L. Diss, P. Kellam, M. Tulliez, M. Q. Du, D. Sicard, R. A. Weiss, P. G. Isaacson, and C. Boshoff. 2000. HHV-8 is associated with a plasmablastic variant of Castleman disease that is linked to HHV-8-positive plasmablastic lymphoma. Blood **95**(4):1406–12.

237. Oksenhendler, E., M. Duarte, J. Soulier, P. Cacoub, Y. Welker, J. Cadranel, D. Cazals-Hatem, B. Autran, J. P. Clauvel, and M. Raphael. 1996. Multicentric Castleman's disease in HIV infection: a clinical and pathological study of 20 patients. AIDS **10**(1):61–7.

238. Du, M. Q., H. Liu, T. C. Diss, H. Ye, R. A. Hamoudi, N. Dupin, V. Meignin, E. Oksenhendler, C. Boshoff, and P. G. Isaacson. 2001. Kaposi sarcoma-associated herpesvirus infects monotypic (IgM lambda) but polyclonal naive B cells in Castleman disease and associated lymphoproliferative disorders. Blood **97**(7):2130–6.

239. Aaron, L., O. Lidove, C. Yousry, L. Roudiere, B. Dupont, and J. P. Viard. 2002. Human herpesvirus 8-positive Castleman disease in human immunodeficiency virus-infected patients: the impact of highly active antiretroviral therapy. Clin Infect Dis **35**(7):880–2.

240. Frizzera, G., P. M. Banks, G. Massarelli, and J. Rosai. 1983. A systemic lymphoproliferative disorder with morphologic features of Castleman's disease. Pathological findings in 15 patients. Am J Surg Pathol **7**(3):211–31.

241. Kim, J. H., T. G. Jun, S. W. Sung, Y. S. Shim, S. K. Han, Y. W. Kim, C. G. Yoo, J. W. Seo, and J. R. Rho. 1995. Giant lymph node hyperplasia (Castleman's disease) in the chest. Ann Thorac Surg **59**(5):1162–5.

242. Loi, S., D. Goldstein, K. Clezy, S. T. Milliken, J. Hoy, and M. Chipman. 2004. Castleman's disease and HIV infection in Australia. HIV Med **5**(3):157–62.

243. Frizzera, G. 2001. Atypical lymphoproliferative disorders. In: Knowles, D. M., editor. Neoplastic Hematopathology. 2nd ed. Philadelphia: Lippincott Williams & Wilkins; 2001, pp. 569–622.

244. Seirafi, P. A., E. Ferguson, F. H. Edwards. 2003. Thoracoscopic resection of Castleman disease: case report and review. Chest **123**(1):280–2.

245. Chronowski, G. M., C. S. Ha, R. B. Wilder, F. Cabanillas, J. Manning, and J. D. Cox. 2001. Treatment of unicentric and multicentric Castleman disease and the role of radiotherapy. Cancer **92**(3):670–6.

246. Bowne, W. B., J. J. Lewis, D. A. Filippa, R. Niesvizky, A. D. Brooks, M. E. Burt, and M. F. Brennan. 1999. The management of unicentric and multicentric Castleman's disease: a report of 16 cases and a review of the literature. Cancer **85**(3):706–17.

247. Scott, D., L. Cabral, and W. J. Harrington, Jr. 2001. Treatment of HIV-associated multicentric Castleman's disease with oral etoposide. Am J Hematol **66**(2):148–50.

248. Casper, C., W. G. Nichols, M. L. Huang, L. Corey, and A. Wald. 2004. Remission of HHV-8 and HIV-associated multicentric Castleman disease with ganciclovir treatment. Blood **103**(5):1632–4.

249. Senanayake, S., J. Kelly, A. Lloyd, Z. Waliuzzaman, D. Goldstein, and W. Rawlinson. 2003. Multicentric Castleman's disease treated with antivirals and immunosuppressants. J Med Virol **71**(3):399–403.

250. Marcelin, A. G., L. Aaron, C. Mateus, E. Gyan, I. Gorin, J. P. Viard, V. Calvez, and N. Dupin. Rituximab therapy for HIV-associated Castleman disease. Blood **102**(8):2786–8.

251. Corbellino, M., G. Bestetti, C. Scalamogna, S. Calattini, M. Galazzi, L. Meroni, D. Manganaro, M. Fasan, M. Moroni, M. Galli, and C. Parravicini. 2001. Long-term remission of Kaposi sarcoma-associated

herpesvirus-related multicentric Castleman disease with anti-CD20 monoclonal antibody therapy. Blood **98**(12):3473–5.

252. Kumari, P., G. P. Schechter, N. Saini, and D. A. Benator. 2003. Successful treatment of human immunodeficiency virus-related Castleman's disease with interferon-alpha. Clin Infect Dis **31**(2):602–4.

253. Nord, J. A., and D. Karter. 2003. Low dose interferon-alpha therapy for HIV-associated multicentric Castleman's disease. Int J STD AIDS **14**(1):61–2.

254. Yoshizaki, K., T. Matsuda, N. Nishimoto, T. Kuritani, L. Taeho, K. Aozasa, T. Nakahata, H. Kawai, H. Tagoh, T. Komori, et al. 1989. Pathogenic significance of interleukin-6 (IL-6/BSF-2) in Castleman's disease. Blood **74**(4):1360–7.

255. Beck, J. T., S. M. Hsu, J. Wijdenes, R. Bataille, B. Klein, D. Vesole, K. Hayden, S. Jagannath, and B. Barlogie. 1994. Brief report: alleviation of systemic manifestations of Castleman's disease by monoclonal anti-interleukin-6 antibody. N Engl J Med **330**(9):602–5.

256. Nishimoto, N., M. Sasai, Y. Shima, M. Nakagawa, T. Matsumoto, T. Shirai, T. Kishimoto, and K. Yoshizaki. 2000. Improvement in Castleman's disease by humanized anti-interleukin-6 receptor antibody therapy. Blood **95**(1):56–61.

257. Jung, C. P., B. Emmerich, F. D. Goebel, and J. R. Bogner. 2004. Successful treatment of a patient with HIV-associated multicentric Castleman disease (MCD) with thalidomide. Am J Hematol **75**(3):176–7.

258. Boivin, G., A. Gaudreau, and J. P. Routy. 2000. Evaluation of the human herpesvirus 8 DNA load in blood and Kaposi's sarcoma skin lesions from AIDS patients on highly active antiretroviral therapy. AIDS **14**(13):1907–10.

259. de Jong, R. B., P. M. Kluin, S. Rosati, P. L. van Haelst, H. G. Sprenger, and D. J. van Spronsen. 2003. Sustained high levels of serum HHV-8 DNA years before multicentric Castleman's disease despite full suppression of HIV with highly active antiretroviral therapy. AIDS **17**(9):1407–8.

260. Lanzafame, M., G. Carretta, M. Trevenzoli, L. Lazzarini, and S. Vento Ercole Concia. 2000. Successful treatment of Castleman's disease with HAART in two HIV-infected patients. J Infect **40**(1):90–1.

261. Dupin, N., A. Krivine, V. Calvez, I. Gorin, N. Franck, and J. P. Escande. 1997. No effect of protease inhibitor on clinical and virological evolution of Castleman's disease in an HIV-1-infected patient. AIDS **11**(11):1400–1.

262. Knowles, D. M., G. Inghirami, A. Ubriaco, and R. Dalla-Favera. 1989. Molecular genetic analysis of three AIDS-associated neoplasms of uncertain lineage demonstrates their B-cell derivation and the possible pathogenetic role of the Epstein-Barr virus. Blood **73**(3):792–9.

263. Horenstein, M. G., R. G. Nador, A. Chadburn, E. M. Hyjek, G. Inghirami, D. M. Knowles, and E. Cesarman. 1997. Epstein-Barr virus latent gene expression in primary effusion lymphomas containing Kaposi's sarcoma-associated herpesvirus/human herpesvirus-8. Blood **90**(3):1186–91.

264. Wilson, J. B., J. L. Bell, A. J. Levine. 1996. Expression of Epstein-Barr virus nuclear antigen-1 induces B cell neoplasia in transgenic mice. Embo J **15**(12):3117–26.

265. Nador, R. G., E. Cesarman, D. M. Knowles, and J. W. Said. 1995. Herpes-like DNA sequences in a body-cavity-based lymphoma in an HIV-negative patient. N Engl J Med **333**(14):943.

266. Ascoli, V., F. Lo Coco, G. Torelli, D. Vallisa, L. Cavanna, C. Bergonzi, and M. Luppi. 2002. Human herpesvirus 8-associated primary effusion lymphoma in HIV-patients: a clinicopidemiologic variant resembling classic Kaposi's sarcoma. Haematologica **87**(4):339–43.

267. Boulanger, E., L. Gerard, J. Gabarre, J. M. Molina, C. Rapp, J. F. Abino, J. Cadranel, S. Chevret, and E. Oksenhendler. 2005. Prognostic factors and outcome of human herpesvirus 8-associated primary effusion lymphoma in patients with AIDS. J Clin Oncol **23**(19):4372–80.

268. Strauchen, J. A., A. D. Hauser, D. Burstein, R. Jimenez, P. S. Moore, and Y. Chang. 1996. Body cavity-based malignant lymphoma containing Kaposi sarcoma-associated herpesvirus in an HIV-negative man with previous Kaposi sarcoma. Ann Intern Med **125**(10):822–5.

269. Jones, D., M. E. Ballestas, K. M. Kaye, J. M. Gulizia, G. L. Winters, J. Fletcher, D. T. Scadden, and J. C. Aster. 1998. Primary-effusion lymphoma and Kaposi's sarcoma in a cardiac-transplant recipient. N Engl J Med **339**(7):444–9.

270. Cannon, M., and E. Cesarman. 2000. Kaposi's sarcoma-associated herpes virus and acquired immunodeficiency syndrome-related malignancy. Semin Oncol **27**(4):409–19.

271. Parekh, S., and J. Sparano. 2005. Primary Effusion Lymphomas: Biology and Management. In: Volberding, P. A., J. M. Palefsky, editors. Viral and Immunological Malignancies. 1st ed. Hamilton: BC Decker; 2005, pp. 122–7.

272. Simonelli, C., M. Spina, R. Cinelli, R. Talamini, R. Tedeschi, A. Gloghini, E. Vaccher, A. Carbone, and U. Tirelli. 2003. Clinical features and outcome of primary effusion lymphoma in HIV-infected patients: a single-institution study. J Clin Oncol **21**(21):3948–54.

273. Oksenhendler, E., J. P. Clauvel, S. Jouveshomme, F. Davi, and G. Mansour. 1998. Complete remission of a primary effusion lymphoma with antiretroviral therapy. Am J Hematol **57**(3):266.

274. Lowy, D., and P. Howley. 2001. Papillomaviruses. In: Knipe, D., P. Howley, D. Griffin, et al., editors. Fields Virology. 4th ed. Philadelphia: Lippincott Williams & Wilkins; 2001, pp. 2231–64.

275. Bjorge, T., A. Engeland, T. Luostarinen, J. Mork, R. E. Gislefoss, E. Jellum, P. Koskela, M. Lehtinen, E. Pukkala, S. O. Thoresen, and J. Dillner. 2002. Human papillomavirus infection as a risk factor for anal and perianal skin cancer in a prospective study. Br J Cancer **87**(1):61–4.

276. Durante, A. J., A. B. Williams, M. Da Costa, T. M. Darragh, K. Khoshnood, and J. M. Palefsky. 2003. Incidence of anal cytological abnormalities in a cohort of human immunodeficiency virus-infected women. Cancer Epidemiol Biomarkers Prev **12**(7):638–42.

277. Palefsky, J. M., E. A. Holly, J. Gonzales, J. Berline, D. K. Ahn, and J. S. Greenspan. 1991. Detection of human papillomavirus DNA in anal intraepithelial neoplasia and anal cancer. Cancer Res **51**(3):1014–9.

278. Zaki, S. R., R. Judd, L. M. Coffield, P. Greer, F. Rolston, and B. L. Evatt. 1992. Human papillomavirus infection and anal carcinoma. Retrospective analysis by in situ hybridization and the polymerase chain reaction. Am J Pathol **140**(6):1345–55.

279. Petry, K. U., H. Kochel, U. Bode, I. Schedel, S. Niesert, M. Glaubitz, H. Maschek, and H. Kuhnle. 1996. Human papillomavirus is associated with the frequent detection of warty and basaloid high-grade neoplasia of the vulva and cervical neoplasia among immunocompromised women. Gynecol Oncol **60**(1):30–4.

280. Frisch, M., R. J. Biggar, and J. J. Goedert. 2000. Human papillomavirus-associated cancers in patients with human immunodeficiency virus infection and acquired immunodeficiency syndrome. J Natl Cancer Inst **92**(18):1500–10.

281. Ateenyi-Agaba, C. 1995. Conjunctival squamous-cell carcinoma associated with HIV infection in Kampala, Uganda. Lancet **345**(8951):695–6.

282. Waddell, K. M., S. Lewallen, S. B. Lucas, C. Atenyi-Agaba, C. S. Herrington, and G. Liomba. 1996. Carcinoma of the conjunctiva and HIV infection in Uganda and Malawi. Br J Ophthalmol **80**(6): 503–8.

283. Moubayed, P., H. Mwakyoma, and D. T. Schneider. 2004. High frequency of human papillomavirus 6/11, 16, and 18 infections in precancerous lesions and squamous cell carcinoma of the conjunctiva in subtropical Tanzania. Am J Clin Pathol **122**(6):938–43.

284. Barbosa, M. S., and R. Schlegel. 1989. The E6 and E7 genes of HPV-18 are sufficient for inducing two-stage in vitro transformation of human keratinocytes. Oncogene **4**(12):1529–32.

285. Munger, K., W. C. Phelps, V. Bubb, P. M. Howley, and R. Schlegel. 1989. The E6 and E7 genes of the human papillomavirus type 16 together are necessary and sufficient for transformation of primary human keratinocytes. J Virol **63**(10):4417–21.

286. Riley, R. R., S. Duensing, T. Brake, K. Munger, P. F. Lambert, and J. M. Arbeit. 2003. Dissection of human papillomavirus E6 and E7 function in transgenic mouse models of cervical carcinogenesis. Cancer Res **63**(16):4862–71.

287. Tornesello, M. L., F. M. Buonaguro, E. Beth-Giraldo, and G. Giraldo. 1993. Human immunodeficiency virus type 1 tat gene enhances human papillomavirus early gene expression. Intervirology **36**(2):57–64.

288. Vernon, S. D., C. E. Hart, W. C. Reeves, and J. P. Icenogle. 1993. The HIV-1 tat protein enhances E2-dependent human papillomavirus 16 transcription. Virus Res **27**(2):133–45.

289. IARC. 1996. IARC W group on the evaluation of carcinogenic risk to humans. Human Immunodeficinecy viruses and human T-cell lymphotropic viruses; 1996.

290. Serraino, D., L. Dal Maso, C. La Vecchia, and S. Franceschi. 2002. Invasive cervical cancer as an AIDS-defining illness in Europe. AIDS **16**(5):781–6.

291. Serraino, D., P. Carrieri, C. Pradier, E. Bidoli, M. Dorrucci, E. Ghetti, A. Schiesari, R. Zucconi, P. Pezzotti, P. Dellamonica, S. Franceschi, and G. Rezza. 1999. Risk of invasive cervical cancer among women with, or at risk for, HIV infection. Int J Cancer **82**(3):334–7.

292. Sitas, F., R. Pacella-Norman, H. Carrara, M. Patel, P. Ruff, R. Sur, U. Jentsch, M. Hale, P. Rowji, D. Saffer, M. Connor, D. Bull, R. Newton, and V. Beral. 2000. The spectrum of HIV-1 related cancers in South Africa. Int J Cancer **88**(3):489–92.

293. Gichangi, P., H. De Vuyst, B. Estambale, K. Rogo, J. Bwayo, and M. Temmerman. 2002. HIV and cervical cancer in Kenya. Int J Gynaecol Obstet **76**(1):55–63.

294. La Ruche, G., B. You, I. Mensah-Ado, C. Bergeron, C. Montcho, R. Ramon, K. Toure-Coulibaly, C. Welffens-Ekra, F. Dabis, and G. Orth. 1998. Human papillomavirus and human immunodeficiency virus infections: relation with cervical dysplasia–neoplasia in African women. Int J Cancer **76**(4):480–6.

295. Patil, P., B. Elem, and A. Zumla. 1995. Pattern of adult malignancies in Zambia (1980–1989) in light of the human immunodeficiency virus type 1 epidemic. J Trop Med Hyg **98**(4):281–4.

296. Newton, R., A. Grulich, V. Beral, B. Sindikubwabo, P. J. Ngilimana, A. Nganyira, and D. M. Parkin. 1995. Cancer and HIV infection in Rwanda. Lancet **345**(8961):1378–9.

297. International Conference on HIV and Cancer. 2000. Highly active antiretroviral therapy and incidence of cancer in human immunodeficiency virus-infected adults. J Natl Cancer Inst **92**(22):1823–30.

298. Mandelblatt, J. S., P. Kanetsky, L. Eggert, and K. Gold. 1999. Is HIV infection a cofactor for cervical squamous cell neoplasia? Cancer Epidemiol Biomarkers Prev **8**(1):97–106.
299. Massad, L. S., K. A. Riester, K. M. Anastos, R. G. Fruchter, J. M. Palefsky, R. D. Burk, D. Burns, R. M. Greenblatt, L. I. Muderspach, and P. Miotti. 1999. Prevalence and predictors of squamous cell abnormalities in Papanicolaou smears from women infected with HIV-1. Women's Interagency HIV Study Group. J Acquir Immune Defic Syndr **21**(1):33–41.
300. Duerr, A., B. Kieke, D. Warren, K. Shah, R. Burk, J. F. Peipert, P. Schuman, and R. S. Klein. 2001. Human papillomavirus-associated cervical cytologic abnormalities among women with or at risk of infection with human immunodeficiency virus. Am J Obstet Gynecol **184**(4):584–90.
301. Ahdieh, L., A. Munoz, D. Vlahov, C. L. Trimble, L. A. Timpson, and K. Shah. 2000. Cervical neoplasia and repeated positivity of human papillomavirus infection in human immunodeficiency virus-seropositive and -seronegative women. Am J Epidemiol **151**(12):1148–57.
302. US Public Health Service (USPHS) and Infectious Diseases Society of America. Guidelines for Preventing Opportunistic Infections Among HIV-Infected Persons. Accessed May 1, 2006.
303. Kurman, R. J., D. E. Henson, A. L. Herbst, K. L. Noller, M. H. Schiffman. 1994. Interim guidelines for management of abnormal cervical cytology. The 1992 National Cancer Institute Workshop. JAMA **271**(23):1866–9.
304. Tate, D. R., and R. J. Anderson. 2002. Recrudescence of cervical dysplasia among women who are infected with the human immunodeficiency virus: a case-control analysis. Am J Obstet Gynecol **186**(5):880–2.
305. Robinson, W. R., C. A. Hamilton, S. H. Michaels, and P. Kissinger. 2001. Effect of excisional therapy and highly active antiretroviral therapy on cervical intraepithelial neoplasia in women infected with human immunodeficiency virus. Am J Obstet Gynecol **184**(4):538–43.
306. Orlando, G., M. M. Fasolo, M. Schiavini, R. Signori, and A. Cargnel. 1999. Role of highly active antiretroviral therapy in human papillomavirus-induced genital dysplasia in HIV-1-infected patients. AIDS **13**(3):424–5.
307. Lillo, F. B., D. Ferrari, F. Veglia, M. Origoni, M. A. Grasso, S. Lodini, E. Mastrorilli, G. Taccagni, A. Lazzarin, and C. Uberti-Foppa. 2001. Human papillomavirus infection and associated cervical disease in human immunodeficiency virus-infected women: effect of highly active antiretroviral therapy. J Infect Dis **184**(5):547–51.
308. Moore, A. L., C. A. Sabin, A. Madge, A. Mocroft, W. Reid, and M. A. Johnson. 2002. Highly active antiretroviral therapy and cervical intraepithelial neoplasia. AIDS **16**(6):927–9.
309. Heard, I., V. Schmitz, D. Costagliola, G. Orth, and M. D. Kazatchkine. 1998. Early regression of cervical lesions in HIV-seropositive women receiving highly active antiretroviral therapy. AIDS **12**(12):1459–64.
310. Minkoff, H., L. Ahdieh, L. S. Massad, K. Anastos, D. H. Watts, S. Melnick, L. Muderspach, R. Burk, and J. Palefsky. 2001. The effect of highly active antiretroviral therapy on cervical cytologic changes associated with oncogenic HPV among HIV-infected women. AIDS **15**(16):2157–64.
311. Heard, I., J. M. Tassie, M. D. Kazatchkine, and G. Orth. 2002. Highly active antiretroviral therapy enhances regression of cervical intraepithelial neoplasia in HIV-seropositive women. AIDS **16**(13): 1799–1802.
312. Selik, R. M., and C. S. Rabkin. 1998. Cancer death rates associated with human immunodeficiency virus infection in the United States. J Natl Cancer Inst **90**(17):1300–2.
313. Daling, J. R., N. S. Weiss, T. G. Hislop, C. Maden, R. J. Coates, K. J. Sherman, R. L. Ashley, M. Beagrie, J. A. Ryan, and L. Corey. 1987. Sexual practices, sexually transmitted diseases, and the incidence of anal cancer. N Engl J Med **317**(16):973–7.
314. Melbye, M., C. Rabkin, M. Frisch, and R. J. Biggar. 1994. Changing patterns of anal cancer incidence in the United States, 1940–1989. Am J Epidemiol **139**(8):772–80.
315. Johnson, L. G., M. M. Madeleine, L. M. Newcomer, S. M. Schwartz, and J. R. Daling. 2004. Anal cancer incidence and survival: the surveillance, epidemiology, and end results experience, 1973–2000. Cancer **101**(2):281–8.
316. Cress, R. D., and E. A. Holly. 2003. Incidence of anal cancer in California: increased incidence among men in San Francisco, 1973–1999. Prev Med **36**(5):555–60.
317. Grulich, A. E., X. Wan, M. G. Law, M. Coates, and J. M. Kaldor. 1999. Risk of cancer in people with AIDS. AIDS **13**(7):839–43.
318. Clifford, G. M., J. Polesel, M. Rickenbach, L. Dal Maso, O. Keiser, A. Kofler, E. Rapiti, F. Levi, G. Jundt, T. Fisch, A. Bordoni, D. De Weck, S. Franceschi. 2005. Cancer risk in the Swiss HIV Cohort Study: associations with immunodeficiency, smoking, and highly active antiretroviral therapy. J Natl Cancer Inst **97**(6):425–32.
319. Bower, M., T. Powles, T. Newsom-Davis, C. Thirlwell, J. Stebbing, S. Mandalia, M. Nelson, and B. Gazzard. 2004. HIV-Associated Anal Cancer: Has Highly Active Antiretroviral Therapy Reduced the Incidence or Improved the Outcome? J Acquir Immune Defic Syndr **37**(5):1563–5.

320. Palefsky, J. M., E. A. Holly, M. L. Ralston, M. Da Costa, H. Bonner, N. Jay, J. M. Berry, and T. M. Darragh. 2001. Effect of highly active antiretroviral therapy on the natural history of anal squamous intraepithelial lesions and anal human papillomavirus infection. J Acquir Immune Defic Syndr **28**(5):422–8.

321. Piketty, C., T. M. Darragh, I. Heard, M. Da Costa, P. Bruneval, M. D. Kazatchkine, and J. M. Palefsky. 2004. High prevalence of anal squamous intraepithelial lesions in HIV-positive men despite the use of highly active antiretroviral therapy. Sex Transm Dis **31**(2):96–9.

322. Kiviat, N. B., M Redman, S. Hawes, T. Lampinen, P. Nelspon, and C. W. Critchlow. 2002. The effect of HAART on detection of anal HPV and squamous intraepithelial lesions among HIV infected homosexual men. Sixth International Conference on Malignancies in AIDS and Other Immunodeficiencies, Bethesda, MD, USA; 2002.

323. Wilkin, T. J., S. Palmer, K. F. Brudney, M. A. Chiasson, and T. C. Wright. Anal intraepithelial neoplasia in heterosexual and homosexual HIV-positive men with access to antiretroviral therapy. J Infect Dis **190**(9):1685–91.

324. Palefsky, J. M., E. A. Holly, J. T. Efirdc, M. Da Costa, N. Jay, J. M. Berry, and T. M. Darragh. 2005. Anal intraepithelial neoplasia in the highly active antiretroviral therapy era among HIV-positive men who have sex with men. AIDS **19**(13):1407–14.

325. Palefsky, J. M., E. A. Holly, M. L. Ralston, and N. Jay. 1998. Prevalence and risk factors for human papillomavirus infection of the anal canal in human immunodeficiency virus (HIV)-positive and HIV-negative homosexual men. J Infect Dis **177**(2):361–7.

326. Sayers, S. J., A. McMillan, and E. McGoogan. 1998. Anal cytological abnormalities in HIV-infected homosexual men. Int J STD AIDS **9**(1):37–40.

327. Friedman, H. B., A. J. Saah, M. E. Sherman, A. E. Busseniers, W. C. Blackwelder, R. A. Kaslow, A. M. Ghaffari, R. W. Daniel, and K. V. Shah. 1998. Human papillomavirus, anal squamous intraepithelial lesions, and human immunodeficiency virus in a cohort of gay men. J Infect Dis **178**(1): 45–52.

328. Lacey, H. B., G. E. Wilson, P. Tilston, E. G. Wilkins, A. S. Bailey, G. Corbitt, and P. M. Green. 1999. A study of anal intraepithelial neoplasia in HIV positive homosexual men. Sex Transm Infect **75**(3):172–7.

329. Goldstone, S. E., B. Winkler, L. J. Ufford, E. Alt, J. M. Palefsky. 2001. High prevalence of anal squamous intraepithelial lesions and squamous-cell carcinoma in men who have sex with men as seen in a surgical practice. Dis Colon Rectum **44**(5):690–8.

330. Piketty, C., T. M. Darragh, M. Da Costa, P. Bruneval, I. Heard, M. D. Kazatchkine, and J. M. Palefsky. 2003. High prevalence of anal human papillomavirus infection and anal cancer precursors among HIV-infected persons in the absence of anal intercourse. Ann Intern Med **138**(6):453–9.

331. Lee, A., T. Young, D. Hanks, R. Ung, and J. Stansell. 2004. The evaluation of anal dysplasia with anal cytology (PAP) followed by high resolution anoscopy (HRA) and biopsy in HIV-infected men. International Conference on AIDS, Bangkok, Thailand; 2004.

332. Chin-Hong, P. V., M. D. Guimaraes, D. Bonfim, M. Da Casta, T. Darragh, B. Grinsztejn, S. B. May, J. H. Pilotto, and J. M. Palefsky. 2004. Anal squamous intraepithelial lesions and HPV infection in hiv-positive men in Brazil. International Conference on AIDS, Bangkok; 2004.

333. Kreuter, A., N. H. Brockmeyer, B. Hochdorfer, S. J. Weissenborn, M. Stucker, J. Swoboda, P. Altmeyer, H. Pfister, and U. Wieland. 2005. Clinical spectrum and virologic characteristics of anal intraepithelial neoplasia in HIV infection. J Am Acad Dermatol **52**(4):603–8.

334. Palefsky, J. M., E. A. Holly, M. L. Ralston, N. Jay, J. M. Berry, and T. M. Darragh. 1998. High incidence of anal high-grade squamous intra-epithelial lesions among HIV-positive and HIV-negative homosexual and bisexual men. AIDS **12**(5):495–503.

335. Holly, E. A., M. L. Ralston, T. M. Darragh, R. M. Greenblatt, N. Jay, and J. M. Palefsky. 2001. Prevalence and risk factors for anal squamous intraepithelial lesions in women. J Natl Cancer Inst **93**(11):843–9.

336. Palefsky, J. M., E. A. Holly, M. L. Ralston, M. Da Costa, and R. M. Greenblatt. 2001. Prevalence and risk factors for anal human papillomavirus infection in human immunodeficiency virus (HIV)-positive and high-risk HIV-negative women. J Infect Dis **183**(3):383–91.

337. Hillemanns, P., T. V. Ellerbrock, P. Dole, X. W. Sun, M. A. Chiasson, and T. C. Wright. 1996. Anal HPV infection and anal cytologic abnormalities in HIV-seropostiive women. Internation Conference on AIDS, Vancouver, Canada; 1996.

338. Melbye, M., E. Smith, J. Wohlfahrt, A. Osterlind, M. Orholm, O. J. Bergmann, L. Mathiesen, T. M. Darragh, and J. M. Palefsky. 1996. Anal and cervical abnormality in women–prediction by human papillomavirus tests. Int J Cancer **68**(5):559–64.

339. Bosch, F. X., M. M. Manos, N. Munoz, M. Sherman, A. M. Jansen, J. Peto, M. H. Schiffman, V. Moreno, R. Kurman, and K. V. Shah. 1995. Prevalence of human papillomavirus in cervical cancer: a worldwide perspective. International biological study on cervical cancer (IBSCC) Study Group. J Natl Cancer Inst **87**(11):796–802.

340. Beckmann, A. M., J. R. Daling, K. J. Sherman, C. Maden, B. A. Miller, R. J. Coates, N. B. Kiviat, D. Myerson, N. S. Weiss, T. G. Hislop, et al. 1989. Human papillomavirus infection and anal cancer. Int J Cancer **43**(6):1042–9.

341. Darragh, T., G. Birdson, R. Luff, and D. Davey. 2004. Chapter 8: Anal-Rectal Cytology. 2nd ed. New York: Springer-Verlag; 2004.

342. New York State AIDS Malignancy Consortium. 2004. Criteria for the medical care of adults with HIV infection; March 2004.

343. Palefsky, J. M., E. A. Holly, C. J. Hogeboom, J. M. Berry, N. Jay, and T. M. Darragh. 1997. Anal cytology as a screening tool for anal squamous intraepithelial lesions. J Acquir Immune Defic Syndr Hum Retrovirol **14**(5):415–22.

344. Cocchi, V., D. Carretti, S. Fanti, P. Baldazzi, M. T. Casotti, R. Piazzi, L. Prosperi, and A. M. Morselli-Labate. 1997. Intralaboratory quality assurance in cervical/vaginal cytology: evaluation of intercytologist diagnostic reproducibility. Diagn Cytopathol **16**(1):87–92.

345. Stoler, M. H., and M. Schiffman. 2001. Interobserver reproducibility of cervical cytologic and histologic interpretations: realistic estimates from the ASCUS-LSIL Triage Study. JAMA **285**(11):1500–5.

346. Woodhouse, S. L., J. F. Stastny, P. E. Styer, M. Kennedy, A. H. Praestgaard, and D. D. Davey. 1999. Interobserver variability in subclassification of squamous intraepithelial lesions: results of the College of American Pathologists Interlaboratory Comparison Program in Cervicovaginal Cytology. Arch Pathol Lab Med **123**(11):1079–84.

347. Goldie, S. J., K. M. Kuntz, M. C. Weinstein, K. A. Freedberg, M. L. Welton, and J. M. Palefsky. 1999. The clinical effectiveness and cost-effectiveness of screening for anal squamous intraepithelial lesions in homosexual and bisexual HIV-positive men. JAMA **281**(19):1822–9.

348. Arnott, S. J., D. Cunningham, J. Gallagher, et al. 1996. Epidermoid anal cancer: results from the UKC-CCR randomised trial of radiotherapy alone versus radiotherapy, 5-fluorouracil, and mitomycin. UKCCCR Anal Cancer Trial Working Party. UK Co-ordinating Committee on Cancer Research. Lancet **348**(9034):1049–54.

349. Myerson, R. J., F. Kong, E. H. Birnbaum, J. W. Fleshman, I. J. Kodner, J. Picus, G. A. Ratkin, T. E. Read, and B. J. Walz. 2001. Radiation therapy for epidermoid carcinoma of the anal canal, clinical and treatment factors associated with outcome. Radiother Oncol **61**(1):15–22.

350. Pitcher, M. E., T. I. Davidson, C. Fisher, and J. M. Thomas. 1994. Post irradiation sarcoma of soft tissue and bone. Eur J Surg Oncol **20**(1):53–6.

351. Jephcott, C. R., C. Paltiel, and J. Hay. 2004. Quality of life after non-surgical treatment of anal carcinoma: a case control study of long-term survivors. Clin Oncol (R Coll Radiol) **16**(8):530–5.

352. Flam, M., M. John, T. F. Pajak, N. Petrelli, R. Myerson, S. Doggett, J. Quivey, M. Rotman, H. Kerman, L. Coia, and K. Murray. 1996. Role of mitomycin in combination with fluorouracil and radiotherapy, and of salvage chemoradiation in the definitive nonsurgical treatment of epidermoid carcinoma of the anal canal: results of a phase III randomized intergroup study. J Clin Oncol **14**(9):2527–39.

353. Gerard, J. P., L. Ayzac, D. Hun, P. Romestaing, R. Coquard, J. M. Ardiet, and F. Mornex. 1998. Treatment of anal canal carcinoma with high dose radiation therapy and concomitant fluorouracil-cisplatinum. Long-term results in 95 patients. Radiother Oncol **46**(3):249–56.

354. Doci, R., R. Zucali, L. Bombelli, F. Montalto, and G. Lamonica. 1992. Combined chemoradiation therapy for anal cancer. A report of 56 cases. Ann Surg **215**(2):150–6.

355. John, M., M. Flam, and N. Palma. 1996. Ten-year results of chemoradiation for anal cancer: focus on late morbidity. Int J Radiat Oncol Biol Phys **34**(1):65–9.

356. Cleator, S., K. Fife, M. Nelson, B. Gazzard, R. Phillips, and M. Bower. 2000. Treatment of HIV-associated invasive anal cancer with combined chemoradiation. Eur J Cancer **36**(6):754–8.

357. Stadler, R. F., S. G. Gregorcyk, D. M. Euhus, R. J. Place, P. J. Huber, and C. L. Simmang. 2004. Outcome of HIV-infected patients with invasive squamous-cell carcinoma of the anal canal in the era of highly active antiretroviral therapy. Dis Colon Rectum **47**(8):1305–9.

358. Bottomley, D. M., N. Aqel, G. Selvaratnam, and R. H. Phillips. 1996. Epidermoid anal cancer in HIV infected patients. Clin Oncol (R Coll Radiol) **8**(5):319–22.

359. Hocht, S., T. Wiegel, A. J. Kroesen, W. E. Berdel, N. Runkel, and W. Hinkelbein. 1997. Low acute toxicity of radiotherapy and radiochemotherapy in patients with cancer of the anal canal and HIV-infection. Acta Oncol **36**(8):799–802.

360. Peddada, A. V., D. E. Smith, A. R. Rao, D. B. Frost, and A. R. Kagan. 1997. Chemotherapy and low-dose radiotherapy in the treatment of HIV-infected patients with carcinoma of the anal canal. Int J Radiat Oncol Biol Phys **37**(5):1101–5.

361. Hoffman, R., M. L. Welton, B. Klencke, V. Weinberg, and R. Krieg. 1999. The significance of pretreatment CD4 count on the outcome and treatment tolerance of HIV-positive patients with anal cancer. Int J Radiat Oncol Biol Phys **44**(1):127–31.

362. Blazy, A., C. Hennequin, J. M. Gornet, A. Furco, L. Gerard, M. Lemann, and C. Maylin. 2005. Anal carcinomas in HIV-positive patients: high-dose chemoradiotherapy is feasible in the era of highly active antiretroviral therapy. Dis Colon Rectum **48**(6):1176–81.
363. Palefsky, J. M., Ea. Holly, and M. Ralston. 2002. Effect of HAART on incidence of anal intraepithelial neoplasia grade 3 among HIV-positive men who have sex with men. Fourteenth International Conference on AIDS, Barcelona, Spain; 2002.
364. Chang, G. J., J. M. Berry, N. Jay, J. M. Palefsky, and M. L. Welton. 2002. Surgical treatment of high-grade anal squamous intraepithelial lesions: a prospective study. Dis Colon Rectum **45**(4):453–8.
365. Goldstone, S. E., A. Z. Kawalek, and J. W. Huyett. 2005. Infrared coagulator: a useful tool for treating anal squamous intraepithelial lesions. Dis Colon Rectum **48**(5):1042–1054.
366. Kreuter, A., B. Hochdorfer, M. Stucker, P. Altmeyer, U. Weiland, M. A. Conant, and N. H. Brockmeyer. 2004. Treatment of anal intraepithelial neoplasia in patients with acquired HIV with imiquimod 5% cream. J Am Acad Dermatol **50**(6):980–1.
367. Graham, B. D., A. B. Jetmore, J. E. Foote, and L. K. Arnold. 2005. Topical 5-fluorouracil in the management of extensive anal Bowen's disease: a preferred approach. Dis Colon Rectum **48**(3):444–50.
368. Biggar, R. J., E. A. Engls, S, Ly, A. Kahn, M. J. Schymura, J. Sackoff, P. Virgo, and E. M. Pfeiffer. 2005. Survival after cancer diagnosis in persons with AIDS. J Acquir Immune Def Syndr. , Jul 1; **39**(3): 293–9.

3. MOLECULAR BIOLOGY OF KSHV IN RELATION TO AIDS-ASSOCIATED ONCOGENESIS

WHITNEY GREENE, KURT KUHNE, FENGCHUN YE, JIGUO CHEN, FUCHUN ZHOU, XIUFEN LEI, AND SHOU-JIANG GAO

Tumor Virology Program, Children's Cancer Research Institute, Departments of Pediatrics, Microbiology and Immunology, and Molecular Medicine, The University of Texas Health Science Center at San Antonio, San Antonio, TX

INTRODUCTION: KAPOSI'S SARCOMA AND KSHV/HHV-8

In 1872, the famed Hungarian dermatologist Moritz Kaposi characterized an "idiopathic multiple pigmented sarcoma" which is now known as Kaposi's sarcoma (KS).[424] For most of its history, KS has been a rare, slowly progressing neoplasm affecting mainly elderly men of Mediterranean and Eastern European descent. KS was not typically fatal, and patients tended to live ten or more years with the condition, and die of other unrelated ailments.[201] An abrupt increase in the number of cases of KS among previously healthy young homosexual men was first reported in 1981, ushering in a new era of aggressive, rapidly fatal KS.[40,155,180,210,401] Since approximately 30% of AIDS patients presented with KS as their initial symptom of HIV infection, KS evolved into a defining characteristic of one of the most devastating infectious diseases in history.[35,177]

The progressive depletion of cell-mediated immunity and subsequent loss of immune function caused by HIV infection predisposes patients to an array of unusual malignxancies.[177,381] AIDS patients do not have higher incidence of the more common tumors of the breast, colon, or lung than the general population; rather, they exhibit a greatly increased frequency of cancers induced by oncogenic viruses such as Epstein-Barr virus (EBV), human papilloma virus (HPV), and Kaposi's sarcoma-associated virus (KSHV), also known as human herpesvirus 8 (HHV-8).[99,177,381] Impaired immune surveillance promotes a permissive environment for uncontrolled viral replication, the spread of virus to surrounding cells and tissues, and contributes to the multistep process of tumorigenesis.[99,428] In the case of AIDS-KS, interactions between immunosuppression, HIV, and KSHV are required for malignant progression. Although HIV infection is neither necessary nor sufficient for the development of KS, it is associated with a much higher frequency of KS and alteration of its natural course.[99] The HIV epidemic in Africa along with the high prevalence of endemic KSHV infection in the general population has led to an alarming number of cases of KS and KSHV-associated diseases on that continent alone.[118] With the recent advent of HAART (highly active antiretroviral therapy), the overall incidence of AIDS-associated neoplasms has declined;[478] however, factors such as lack of access to treatment, noncompliance with treatment regimens, and the development of drug resistance all contribute to an elevated risk for KS in HIV-infected patients[46,204,492] and ensure that KS will present a continuing major health problem for years to come.[381] Understanding of the molecular biology of KSHV alone and in the context of HIV infection is crucial for the development of therapies to treat and prevent the associated pathological processes.

Discovery

In the 1920s, it was observed that KS occurred more frequently in East and Central Africa. The uneven geographical distribution led to the hypothesis that KS might be caused by an infectious agent.[173,328] In 1990, a landmark epidemiological study from Beral et al. reported that KS was 20,000 times more likely to occur in people with HIV than in the general population.[36] KS was more common in those who had acquired HIV sexually than in those who had acquired it via other routes. The incidence of KS was not related to age or race, but showed a definite geographical distribution, with the highest prevalence in the areas that were the initial foci of the AIDS epidemic.[36]

The accumulation of the epidemiological evidence suggested the involvement of a sexually transmissible agent in the development of KS, which in western countries had spread mainly among homosexual men. Several groups attempted to identify the unknown agent, and in 1994, Yuan Chang and Patrick Moore used representational difference analysis to identify two fragments of a previously unknown herpesvirus in a biopsy sample from an AIDS-KS patient,[89] instigating a new era in KS research.

Diseases Associated with KSHV

Since its discovery, extensive studies have demonstrated an etiologic role for KSHV in the development of KS.[79,113,129,302] The involvement of KSHV in the pathogenesis of KS is now widely accepted based on the following criteria: (1) KSHV genomes are detectable in all clinical forms of KS (classic, endemic, iatrogenic, and AIDS-related);[44,89,91,134,253,301,383] (2) KSHV infection is highly associated with subjects at high risk for developing KS such as HIV-infected gay men;[167,168,229,403]

(3) KSHV infection rate in the general population correlates with KS incidence rate in different geographic regions, e.g., high in some African regions, intermediate in Mediterranean and Eastern European regions, and low in North America;[168] (4) KSHV is detected in the endothelial spindle cells, the neoplastic component of KS lesions;[43,132] and (5) KSHV DNA sequences in peripheral blood and seroconversion to KSHV are detected prior to the development of KS.[167,168,291,304,325,326,337,355,360] In addition to its association with KS, KSHV has also been implicated as the causative agent of two other AIDS-associated malignancies: primary effusion lymphoma (PEL)[80,89] and the plasma cell variant of multicentric Castleman's disease (MCD)[133,415] and may be pathologically involved with other disorders resulting from dysregulation of the immune system.[129,384]

Kaposi's Sarcoma

Four clinical forms of KS have been described. *Classic KS* is an indolent tumor of elderly men, most often found in Mediterranean, Eastern European, and Near East regions.[424] *Endemic KS* is prevalent in Central Africa, where it is one of the most frequently occurring tumors. *Iatrogenic KS* has been identified in transplant recipients undergoing immunosuppressive therapy. In western nations, KS became a hallmark of the AIDS epidemic.[77] The occurrence of KS in young male homosexuals/bisexuals was first reported in 1981.[40,155,180,210,401] *Epidemic AIDS-KS* is the most common neoplastic manifestation of AIDS in the United States and Europe[280] and is one of the diagnostic criteria for AIDS.[1] In contrast to the indolent course of classic KS, AIDS-related KS takes a much more aggressive course.[36] AIDS-related KS tends to disseminate widely to mucous membranes and the visceral organs.[1,424] Despite the different clinical manifestations of KS, the histology of lesions from skin, lymph nodes, respiratory tract, and intestines are very similar.[1] KS is a vascular tumor consisting of interweaving bands of spindle cells, irregular slit-like channels embedded with reticular and collagen fibers, and inflammatory infiltrates of mononuclear cells and plasma cells. The neoplastic components of the KS lesion are the so-called spindle cells due to their characteristic abnormal elongated shapes.[99] The tumor is highly vascular, containing abnormally dense and irregular blood vessels, which leak red blood cells into the surrounding tissue and give the tumor its dark color.

Cutaneous lesions are divided into patch, plaque, and nodular stages. The patch stage demonstrates a proliferation of irregularly branching blood vessels, which may be grouped around normal-appearing blood vessels. Perivascular infiltrating lymphocytes and multiple plasma cells are typically present. The irregular vessels are small, flat and widely spaced and are comprised of apparently normal endothelial cells.[1]

The plaque stage is characterized by the appearance of spindle cells forming bundles in the vascular spaces. Mitoses and nuclear abnormalities are more prominent in both the spindle cells and the endothelial cells. Extravasation of red blood cells and macrophages is often evident.

As the vascular and spindle cells continue to proliferate, the characteristic nodular phase of classic KS develops. The nodular lesions are composed of well-defined, densely packed aggregates of spindle cells and vascular spaces arranged into bundles. The vascular slits are often distended by red blood cells. Mitotic figures, extravasated red blood cells, hemosiderin, and hemosiderophages are often present.[1]

KS lesions are composed of a mixed-cell infiltrate with hyperplastic spindle-shaped cells that resemble cytokine-activated ECs and monocytes/macrophages by their expression of tissue-specific markers (CD34, vascular-endothelial cadherin, endothelial leukocyte adhesion molecule type 1, CD4, CD14, CD68, and PECAM-1).[207,444,472] Immunohistology, in situ hybridization, and in situ PCR studies have shown that the majority of spindle endothelial cells in advanced KS lesions as well as atypical endothelial cells in early lesions are latently infected by KSHV.[43,132,336,353,497] Less than 3% of the KSHV-infected cells in KS lesions have been found to express viral proteins characteristic of viral lytic replication.[101,227,336,420] The low rate of spontaneous lytic replication is postulated to have a role in KS development through an autocrine and paracrine mechanism, as well as ensuring the continued maintenance of KSHV infection.[276]

Though the classic form of KS is usually localized to the lower extremities, AIDS-KS commonly involves many other parts of the body. The skin of the face, the extremities, torso, and mucous membranes of the oral cavity are often affected. In one study, 45% of patients presented with one or more lesions along the gastrointestinal tract.[99]

Primary Effusion Lymphoma

PEL, also called body cavity-based lymphoma, is a rare lymphoma commonly found in HIV-infected patients.[79] This type of lymphoma is characterized as a malignant effusion in the peritoneal, pleural, or pericardial space, usually without a tumor mass.[69,79] The lymphoma cells are usually monoclonal and of B-cell origin, but display only a few markers of B-cell differentiation i.e., CD20, as well as several activation markers such as CD30, CD38, CD71, and epithelial membrane antigen.[68,161] PEL cells typically express a 420 kDa isoform of CD138/syndecan-1, suggesting they are in a preterminal stage of B-cell differentiation close to that of plasma cells.[66,162] KSHV is invariably detected in PEL, and is considered to be a diagnostic criterion for this type of lymphoma.[79,161] KSHV is detected as either monoclonal or oligoclonal episomes in PEL samples.[222] Similar to that seen in KS lesions, the pattern of KSHV gene expression in PEL is mainly latent.[132,227,336,362] The majority of cells express the viral latent proteins LANA-1 (ORF73), vCyclin (ORF72), vFLIP (ORF71), kaposin (ORFK12), and LANA-2 (ORFK10.5) with a small percentage (2–5%) expressing vIL-6 (ORFK2).[132,227,336,362] Proteins of the viral lytic cycle are detected in less than 1% of the tumor cells. Though generally considered to be a lytic protein, vIL-6 expression in PEL may be independent of the lytic replication cascade of gene expression.[92]

Multicentric Castleman's Disease

MCD is a localized lymphoproliferative condition characterized by expanded germinal centers with B-cell proliferation and vascular proliferation. The plasma cell variant of MCD is more commonly seen in AIDS patients and transplant recipients. MCD is frequently but not invariantly associated with KSHV infection.[11,109,133,415] Immunohistochemical analysis of biopsy specimens has shown that 10–50% of B-cells surrounding the follicular centers of MCD are positive for LANA-1.[131,132,227,336] Of the LANA-1 positive B cells, about 5–25% also express the KSHV proteins vIL-6 and vIRF-1 (ORFK9).[227,336] In addition, a small proportion of the mantle zone cells of MCD also express viral proteins associated with lytic replication,[227,336] suggesting that in MCD, KSHV adopts a less restricted pattern of gene expression compared to that seen in either KS or PEL.

Other Possible KSHV-Associated Disorders

Since the isolation of KSHV DNA from KS lesions in 1994, KSHV has been postulated to be involved in several other disease states resulting from immune dysfunction.

Primary pulmonary hypertension is a progressive disorder characterized by elevated mean arterial pressure that may lead to right ventricular failure, and complex, lumen-occluding vascular lesions resulting from dysregulated endothelial cell proliferation. A study published in 2003 reported a possible association between KSHV infection and the nonfamilial form of this disorder, but follow-up studies were not able to confirm this association.[108,250]

Hemophagocytic syndrome, also called macrophage activation syndrome and hemophagocytic lymphohistiocytosis, is a reactive disorder of the mononuclear phagocytic system that is characterized by benign generalized histiocyte proliferation with profound hemophagocytosis resulting in the destruction of the formed elements of the blood. The acquired form of this syndrome is associated with underlying disease such as immunodeficiency, and can be triggered by infection or medication.[142] KSHV infection may be able to trigger episodes of this syndrome, but it is not the only viral cause.[3,384]

Limited evidence indicates a possible role for KSHV infection in other diseases, including pemphigus, salivary gland tumors, bullous phemigoid, nonneoplastic lymphadenopathies, sarcoidosis, Kikuchi's disease, and multiple myeloma.[3]

Interaction of HIV and KSHV

As KSHV is a necessary but not sufficient etiological factor for KS, only a small proportion of infected people ever develop KS or KSHV-induced lymphomas. Cofactors such as HIV infection and iatrogenic immunosuppression dramatically increase the risk of developing a KSHV-related malignancy in infected individuals. Among KSHV-infected individuals, the risk of KS is much higher in those with HIV-1 infection than among those with other types of immunosuppression, suggesting a direct action of HIV-1 on KSHV replication and tumorigenesis. These two viruses may interact at the molecular level in coinfected patients, resulting in increased HIV-1 viral load.[293] Studies have demonstrated that coinfection with KSHV can modulate HIV-1 replication.[73–76] This occurs either through direct interaction between these two viruses or through secondary effects resulting from the release of cellular factors in response to infection. The KSHV ORF50-encoded reactivation and transcriptional activator (RTA) interacts synergistically with HIV-1 Tat protein in the transactivation of HIV-1 LTR, leading to increased cellular susceptibility to HIV infection.[76] LANA-1 interacts with Tat and activates HIV-1 LTR.[211] Expression of RTA increases the efficiency of HIV infection in different cell types.[75] This potentially could result in enhanced HIV spread within the infected organism and faster progression of the disease. On the other hand, HIV-1 infection leads to reactivation of latent KSHV genomes, through Tat.[292] HIV-1-encoded Vpr proteins increase the expression of KSHV genes.[206] Consequently, HIV infection can promote KS progression. In AIDS-KS, HIV and its Tat protein induce inflammatory cytokines, which can further promote the pathogenesis of KS.[137,140] In fact, HIV induced oncostatin-M (OSM) and interferon (IFN)-γ to promote KSHV lytic replication.[294] It was shown that HIV infection of PEL cells triggered KSHV reactivation[292] while KSHV enhanced HIV replication in acutely infected cells and induced reactivation in latently infected cells.[73] HIV-1 Tat also enhanced KSHV infectivity, probably by concentrating virions on cell surface.[16]

In order to elucidate the basis for the increased frequency, enhanced aggressiveness, and disease progression seen in AIDS-KS, understanding of the molecular biology of KSHV is required. The remainder of this chapter focuses on describing the structure and genetics of the virus, infection systems and animal models used to study the virus, the viral lifecycle, impact on host cellular pathways, potential oncogenic mechanisms, and specific gene functions.

BIOLOGY OF KSHV

Virion Structure and Assembly

The *Herpesviridae* family comprises three subfamilies, alpha-, beta-, and gamma-herpesvirus. KSHV belongs to the γ_2-herpesvirus group.[288] KSHV and other members of the herpesvirus family share a characteristic architecture in which the double-stranded DNA genome is surrounded by an icosahedral protein capsid, a thick tegument layer, and a lipid bilayer envelope.[481] Mature KSHV has at least 24 virion-associated proteins. These include five capsid proteins, eight envelope glycoproteins, six tegument proteins, and five proteins whose locations in the virion have not yet been defined.[323,499] The 3D structure of KSHV capsids has been investigated by cryo–electron microscopy.[481] These studies have shown that KSHV has the same $T = 16$ triangulation number and much the same capsid architecture as herpes simplex virus (HSV) and cytomegalovirus despite limited sequence similarity between the corresponding capsid proteins.[481] The principal constituents of the capsids are a major capsid protein (ORF25), two triplex proteins (ORF62 and ORF26), and a small capsid protein of ~19 kDa (ORF65).[323] Viral assembly within infected cells has yet to be studied.

Genome Structure and Organization

KSHV contains a large double-stranded DNA which is a closed circular episome in the nucleus during latency but is linear during lytic replication. Over 90 genes/ORFs are encoded by a 140 kb long unique region (LUR) with 53.5% G+C content, which is flanked by 20–35 kb terminal repeat regions composed of 801 bp terminal repeat units with 84.5% G+C content (Fig. 1). The KSHV genome shares the seven block organization of other herpesviruses with KSHV unique ORFs present between blocks. These unique genes, some of which are homologs to human genes, were designated names with a "K" prefix followed by number.[367] Because of the KSHV genome complexity, the exact number of genes in the genome remains unknown. It is possible that new genes have yet to be discovered. Organization of KSHV genome is closely colinear to that of herpesvirus saimiri (HVS) and rhesus macaque rhadinovirus (RRV) genomes. HVS is the prototypical gammaherpesvirus of the *Rhadinovirus* genus.[144] Sequence analysis indicates HVS shares significant homology with KSHV.[367] RRV, a simian gamma-2 herpesvirus closely related to KSHV, replicates lytically in cultured rhesus monkey fibroblasts and establishes persistence in B cells. The similarity between the genomes of RRV and KSHV is high, and almost all of the ORFs/genes in KSHV have at least one homolog in RRV.[10,388]

Genetic Manipulation of KSHV Genome

Studies of KSHV infection and functions of individual genes have been hampered by the lack of viral mutants. For KSHV regulatory genes whose functions have been characterized, most have been examined in cell culture after gene cloning; however, their biological functions in KSHV infection remain largely unclear.[129,302] The construction of viral mutants is the most straightforward way to examine the functions of individual viral genes, and has been widely used for other herpesviruses. In recent years, the adoption of the bacteria artificial chromosome (BAC) system has accelerated the study of functions of herpesvirus genes because the genome can be easily modified either by transposon or ET cloning in *E. coli* once it is cloned as a BAC. Transfection of the cloned viral DNA into permissive mammalian cells leads to the production of infectious virions without the need for further genetic repair and homologous recombination in the transfected cells.

Recently, Zhou et al. successfully cloned the full-length KSHV genome as a BAC in *E. coli*, and reconstituted it in 293 cells.[498] The sequences of BAC, GFP and

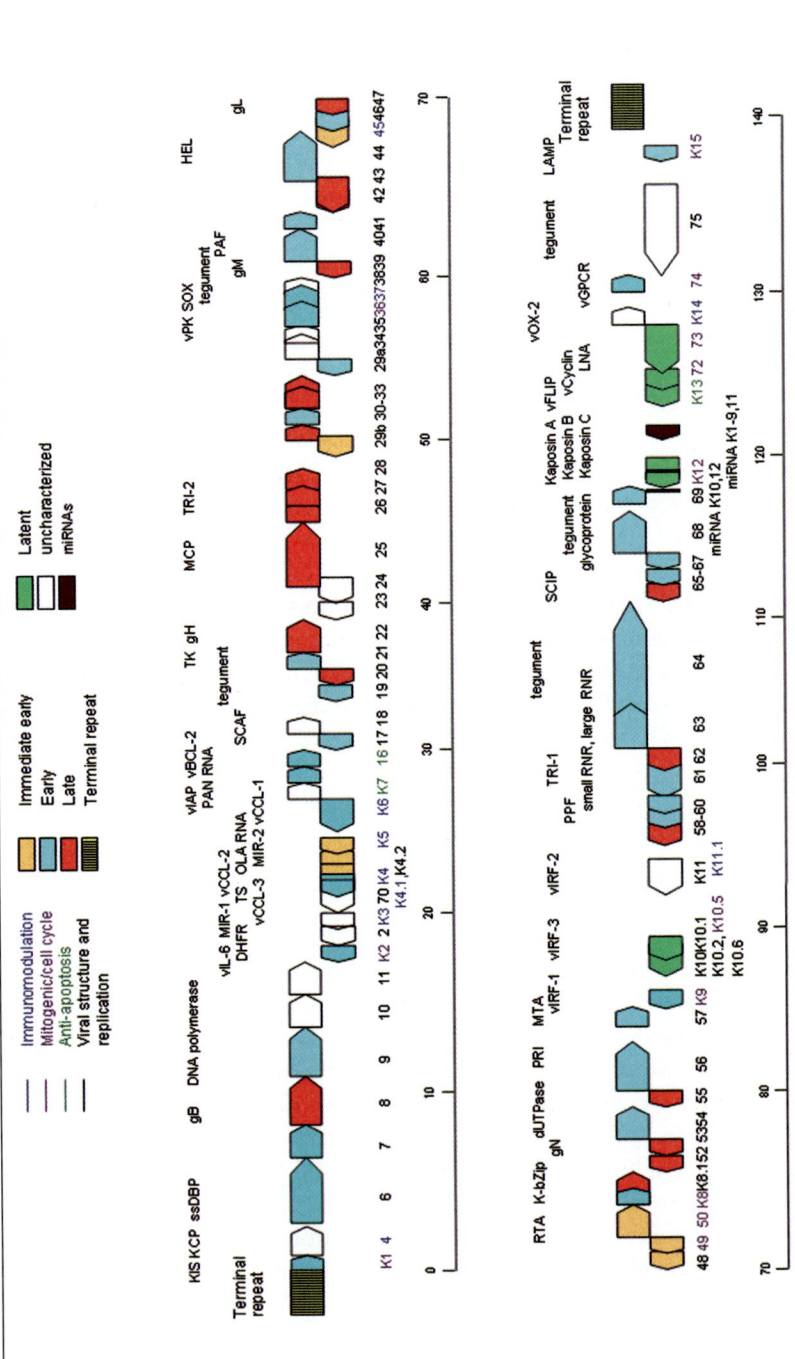

Figure 1. KSHV genome and genes. The names of genes are labeled in color according to gene function (immunomodulation in blue, mitogenic and cell cycle in purple, antiapoptosis in green, and viral structure and replication in black). The blocks representing the ORFs are also labeled in color according to gene class (immediate–early in orange, early in light blue, late in red, latent in green, and unclassified in white).

hygromycin-resistance cassette were inserted into the loci between ORF18 and ORF19 by homologous recombination without disrupting the expression of these two genes. The recombinant virus can be induced into productive lytic replication producing infectious virions that are capable of infecting mammalian cells such as 293 and primary human endothelial cells.[166,498] To elucidate gene function in the context of KSHV infection, the ET cloning system for BAC mutation was developed to generate mutants. The first mutant of BAC36 was obtained by replacing vIRF with a kanamycin-resistance cassette.[498] Later K8.1, RTA, gB, and MTA mutants of BAC36 were made using the same strategy and their functions have been further addressed.[194,242,277,485] Transposition is a powerful tool for modifying DNA molecules and a Tn5-based mutagenesis library of BAC36 has been constructed, in which one mutant with Tn5-disrupted LANA-1 has been used to dissect the function of LANA-1.[487]

Host Range and Cell Tropism

As with all human herpesviruses, KSHV infects a large proportion of the human population and remains in a latent state throughout most of the life of the host. Although a number of cell types are known to harbor the latent virus, the true latent reservoir(s) remain to be defined. Upon reactivation by immunosuppression, such as in HIV infection, the latent virus switches to lytic replication, producing new virus particles that infect and trigger the proliferation of endothelial cells involved in KS. Alternatively, the latent cells might home to the endothelium and proliferate into spindle cells. An early study demonstrated 78 times higher number of KS-like spindle cells detected in the peripheral blood of HIV-1-infected individuals compared to normal controls.[58] These peripheral blood-derived spindle cells (PBsc) were shown to express a variety of endothelial cell markers, such as Ulex europaeus I lectin, EN4, EN2/3, EN7/44, CD13, CD34, CD36, CD54, ELAM-1, and HLA-DR. Consistent with these early observations, recent studies suggested that KSHV could infect human fetal mesenchymal stem cells (MSCs) and CD34$^+$ hematopoietic progenitor cells (HPCs), and the virus may be disseminated following differentiation of infected HPCs into B cell and monocyte lineages.[338,483] Thus, MSCs and CD34$^+$ HPCs may be reservoirs for KSHV infection and provide continuous sources of virally infected cells *in vivo*. Several studies also suggest that the lymphoid system could be the reservoir of latently infected cells from which KSHV reactivates under conditions of immunosuppression.[38,327] Similar to EBV, KSHV displays a tropism for B lymphocytes, since viral DNA has been detected in purified CD19$^+$ B cells but not in CD8$^+$ cells from the peripheral blood of patients with both KS and HIV infection.[38,252,327]

The origin of KS tumor cells has been a matter of controversy and many mesenchymal cell types have been proposed.[34,402] The recent finding of several endothelial markers (including CD31, CD34, CD36, and KDR) on KS tumor cells has supported the theory that endothelial cells are the cells of origin of KS.[472] More support comes from several studies reporting spindle-shape transformation of different types of human endothelial cells following infection with KSHV.[106,166] However, the lack of Pal-E and eNOS expression has brought into question the endothelial origin of this tumor.[473] Recently, several new endothelial markers have been identified and characterized. The expression of two such proteins, VEGFR3 (VEGF-C receptor, flt-4) and podoplanin (a membrane glycoprotein), have been found to be restricted to lymphatic endothelium.[472] KS tumor cells express these two markers of lymphatic endothelium as well as the general endothelial marker CD31, strongly suggesting that KS is derived from lymphatic endothelium.[350,472]

The most common cell type isolated from AIDS-KS has been the spindle-shaped endothelial-like cells that proliferate in response to inflammatory cytokines and have a limited replicating life span.[281,329,372,373] The spindle cells isolated from KS lesions of HIV-1-infected individuals are generally hyperplastic nontransformed cells that proliferate in culture with inflammatory cytokines. These cells initially contain KSHV that is rapidly lost as the cells are passaged in culture. Cell lines composed of transformed cells have also been isolated from advanced KS lesions. KS cell lines have been isolated from KS lesions (KS SLK and KS IMM) from two HIV-1 negative renal transplant recipients.[9,251] These cell lines, unlike AIDS-KS spindle cells, are transformed cells that grow in the absence of inflammatory cytokines, contain cytogenetic abnormalities, and induce durable tumor lesions when inoculated into nude mice.[223,373] Similar to the hyperplastic KS spindle cells, the KS transformed cells have also lost the KSHV genomes. Taken together, these findings indicate that early stage KS is a hyperplasia resulting from proliferation of spindle-shaped endothelial-like cells in response to KSHV-induced angiogenesis and inflammation. However, the spindle-shaped cells, most of which have latent KSHV, may eventually become transformed malignant tumor cells during advanced KS. In support of this concept, it has been experimentally demonstrated that primary human endothelial cells and keratinocytes could be immortalized upon *de novo* KSHV infection.[78,147] Ectopic expression of a number of KSHV genes also leads to immortalization and/or transformation of various cells, further supporting that KSHV is an oncogenic virus.[22,165,311,429]

In cell culture, KSHV is capable of infecting a diverse range of human and animal cell types/lines including primary CD19$^+$ B cells, HPV-transformed human brain endothelial cells BB18 and 181GB1-4, primary neonatal capillary endothelial cells, human embryonic kidney 293 cells, Ln-Cap cells, human lung carcinoma A549 cells, CHELI (Chediak–Higashi syndrome) cells, squamous cell carcinoma SCC15 cells, human fibroblast cells, human bladder carcinoma T24 cells, human prostate carcinoma DU145 cells, human cervical carcinoma HeLa cells, baby hamster kidney BHK-21 cells, owl monkey kidney OMK637 cells, green monkey fibroblasts (Vero) cells, green monkey kidney CV-1 cells, SLK cells (KS-spindle cells), and murine fibroblast 3T3 cells.[31,357]

KSHV Infection Systems

Because of the direct involvement of endothelial cells in KS tumors, many groups have focused on infection of endothelial cells and several reports have documented KSHV infection of human primary endothelial cells.[106,147,166,246,306] In the initial report, KSHV infected only a small number of cells in primary human bone marrow microvascular endothelial cell cultures and primary human umbilical vein endothelial cell (HUVEC) cultures. The KSHV-infected cells acquired a spindle-shape morphology and were maintained for >12 months while the control cultures underwent senescence within 3 weeks of culture.[147] It has been proposed that the small number of infected cells provide a paracrine effect to sustain the cultures. Subsequent study showed that KSHV could infect primary human dermal microvascular endothelial cell (DMVEC) cultures and form colonies or plaques of spindle-shaped cells.[106] Again, the primary-infection efficiency in this system was low, even though the virus eventually infected the entire culture after 2–3 weeks. Paradoxically, to sustain long-term KSHV infection, it was necessary to periodically replenish the cultures with uninfected endothelial cells at a ratio of ten portions of normal cells to one portion of infected cells. To facilitate the manipulation of primary endothelial cells, HPV E6, E7-immortalized DMVEC cultures were used as targets for KSHV infection.[306] In this system, the primary-infection efficiency remained low even though the virus also eventually spread to the entire culture, which could be stably maintained. In contrast, KSHV infection of telomerase-immortalized microvascular endothelial cells was extremely efficient, reaching the entire culture within 2–3 days of infection; however, the infected cells were unable to sustain persistent KSHV infection, and the cultures quickly lost the virus after several passages.[246] The limitations, such as low primary-infection efficiency and/or failure of long-term sustainability for virus growth, of the above-mentioned systems have restricted their use for KSHV characterization, especially virus–cell interactions at the initial stage of infection. Furthermore, even in systems that can sustain persistent KSHV infection, the cultures are predominantly in the viral latent phase.[106] Active viral lytic replication was not observed in any stages of infection in these systems. In contrast to these, efficient infection of HUVEC cultures by recombinant KSHV BAC36 is permissive for lytic replication at the early stage of infection, producing large amounts of infectious virion.[166] Infected cultures form bundles of spindle-shaped cells, which are reminiscent of KS vascular structures, and establish latency at a late stage of infection. The latently infected cells can be induced into lytic replication. Thus, this system can be used for examining KHSV latent and lytic replication, as well as productive primary infection.[166]

Animal Models

Animal models have been developed to investigate the *in vivo* behavior of KSHV-related malignancies. In a number of studies, tumors were induced by injecting KSHV-infected PEL cells into mice.[42,343] In one report, BCBL-1 cells were injected alone or with human peripheral blood mononuclear cells (PBMC) into SCID mice.[343] The lymphomas, which developed at or near the site of injection, appeared to derive exclusively from the injected BCBL-1 cells and not from the injected human PBMC. The tumors induced a marked murine angiogenic response, but known angiogenic cytokines were not detected in the BCBL-1 cells.[343] In a similar more comprehensive study, Boshoff et al. found that injection of either singly (KSHV+) or dually (KSHV+/EBV+) infected PEL cell lines into Nod/SCID mice resulted in a significant pathogenic difference.[42] PEL-like effusions were observed in the mice following intraperitoneal (i.p.) injection of both types of PEL cells, whereas only the dually infected, but not singly infected, PEL produced an effusion following intravenous injection. Singly infected PELs express an array of cell surface homing receptors very different from most lymphomas, with the potential for both positive and negative effects on effusion formation. Dually infected PEL cells expressed adhesion molecules that were very similar to EBV-positive Burkitt's lymphoma (BL) cells. These differences might explain the differential metastasis to solid tumors (as in BLs) between the PEL types.[42] Another study hypothesized that PEL cells secrete VEGF to promote the vascular permeability of peritoneal vessels, leading to effusion rather than to neovascularization of tumors. The ability of various lymphoma lines (including both PELs and BLs) to form effusions following i.p. inoculation correlated directly with their respective magnitudes of VEGF release. Coinjection of antibodies specific for VEGF, but not control antibodies, blocked effusion formation by the PELs.[17] A *de novo* infection model of SCID/hu mice by KSHV demonstrated that the virus remained confined to the grafted CD19+ B cells, and did not spread to mouse tissue based on the detection of the viral DNA and mRNA.[126] In a similar model, six of eight mice developed KS-like lesions with angiogenesis.[151] Similar to human infection, keratinocytes in the epidermis and spindle cells in the dermis supported a largely latent infection, with rare cells expressing lytic genes. The lack of universal infection suggests that the skin model may allow an analysis of viral and host determinants of permissiveness.[126] A study published in 2004 describes the attempt to develop a primate model by infecting rhesus macaque with KSHV, either with or without SIV, the simian form of HIV.[358] Unfortunately, only a very low amount of viral DNA was detectable, and after 27 months postinoculation, KSHV infection did not result in any observable pathology in either SIV-negative or SIV-positive animals. Two macaque homologs of KSHV, retroperitoneal fibromatosis-associated herpesviruses (RFHV), were identified in retroperitoneal fibromatosis, a malignancy closely resembling KS.[366] Unfortunately, attempts to grow these viruses have not been successful so far. The recently identified RRV is closely related to KSHV,[10,388] and when coinfected with SIV, rhesus macaques develop a B-cell hyperplasia similar to MCD occurring in humans with HIV/KSHV.[477]

THE LIFECYCLE OF KSHV

Like all other herpesviruses, KSHV exhibits two distinct phases of infection. Lytic or productive infection results in the replication of viral DNA, the production of infectious virions, and the death of the host cell. During latent infection, the viral genome is maintained as a circular episome within the host cell nucleus and only a fraction of the viral genes are expressed. The viral episome replicates each time the host cell divides, using existing cellular replication machinery. Although infectious virions are not produced during latency, the viral genome retains the ability to be reactivated into lytic replication.[129] The molecular mechanisms involved in latency and lytic reactivation are described in details in the following sections.

Mechanisms of KSHV Latency

Expression of KSHV Latent Genes/Transcripts

Latent infection by KSHV involves the expression of only a few of the 90 KSHV genes. Specifically, all KSHV latently infected cells express vFLIP, vCyclin, and LANA-1.[132,217,377] Interestingly, vFLIP, vCyclin, and LANA-1 are located adjacent to one another in the KSHV genome.[367] They belong to a multicistronic transcriptional unit, known as the latency transcript (LT) cluster.[125] The LT cluster is transcribed from a constitutively active promoter (LTc) which is initiated at nucleotide 127,886, giving rise to a unspliced 5.8-kb mRNA and an alternatively spliced 5.4-kb mRNA, both containing vFLIP, vCyclin, and LANA-1, and a 1.7-kb transcript containing vFLIP and vCyclin (Fig. 1). It is likely that LANA-1 is the principal translation product of the longer mRNAs, whereas both vCyclin and vFLIP are synthesized from the shorter transcript, the latter by way of an internal ribosome entry site upstream of vFLIP.[37,272] These three genes are separated from the K12 gene, which is expressed at low levels during latency, by an ~4.5 kb KSHV sequence that lacks any significant ORFs, representing the largest coding gap within the unique region of the KSHV genome.[367] Surprisingly, 10 of the 12 KSHV microRNAs (miRNAs) identified so far are located within this coding gap, whereas the miR-K10 is found within K12,[63,342,374] and the miR-K12 is located within the 3′-UTR, 265 nt away from the end of the transcript.[186] All 12 KSHV miRNAs are oriented in the same genomic direction.

MicroRNAs are endogenous approximately 22 nucleotide RNAs that play important gene regulatory roles by pairing to the messages of protein coding genes to specify mRNA cleavage or by repressing productive translation.[29] miRNAs have implicated in cell proliferation, cell death, metabolism, and cancer. Recent discovery of virus-encoded miRNAs indicates that viruses also use this fundamental mode of gene regulation.[316] The first reported virus-encoded miRNAs were the viral miRNAs expressed in EBV-infected cells. Among the various families of viruses, the herpesvirus family stands out in establishing long-standing latent infection as a major part in the viral life cycle. It is possible that miRNAs may play a critical role in the establishment and/or maintenance of latent infection initiated by herpesviruses. In fact, the KSHV microRNAs are all expressed in latently infected cells and largely unaffected after induction of lytic replication.[63,342,374] Computationally predicted targets for KSHV miRNAs include viral genes such as ORF23, 27, 31, 52, 49, 61, 68, K7, K13, and K14, and several B-cell-specific genes involved in apoptosis and signaling.[62] Recently, another latent promoter was characterized which is

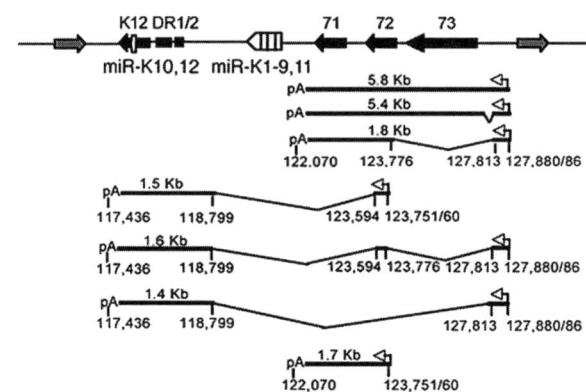

Figure 2. Transcription of the latency-associated region of the KSHV genome. The KSHV latency-associated region encodes four proteins, indicated by black boxes, as well as 12 miRNAs (white boxes) and is flanked by lytic genes (gray boxes). Two latent promoters (white arrows), splice sites, and poly(A) addition sites are indicated. Sequence coordinates are derived from the prototype BC-1 KSHV genome (GenBank accession number NC_003409). Figure is modified from reference 62.

embedded within the ORF71-73 transcription unit specifying transcripts that encode ORF71/72 or K12 (Fig. 2). This promoter is initiated at 123,751 and polyadenylated at 122,070 or 117,436.[62,340] Multiple novel transcripts initiated from LTc and terminated at either 122,070 or 117,436 have also been identified. These transcripts may also be the source of miRNAs.[340] It is believed that the transcripts initiating at the two latent promoters present in the KSHV latency-associated region can undergo two entirely distinct fates, i.e., processing to give a kaposin mRNA and viral microRNAs on the one hand or expression as KSHV vFLIP, vCyclin, or LANA-1 mRNAs on the other, depending on whether the viral polyadenylation site located at position 122,070 is ignored or recognized, respectively.[62]

The KSHV K15 is another gene that has been detected in latently infected cells; however, the expression levels of this transcript increase following induction of lytic replication.[103] The K15 gene is encoded by the last ORF in the KSHV genome, adjacent to the terminal repeats, and is transcribed in a leftward orientation away from the terminal repeat region.[103] The K15 gene contains eight exons which undergo alternative splicing, yielding several isoforms with a predicted common carboxyl-terminal cytoplasmic tail and different numbers of transmembrane domains.[103]

ORFK10.5 encodes latency-associated nuclear antigen 2 (LANA-2/vIRF-3), which is expressed in KSHV-infected hematopoietic tissues, including PEL and MCD but not KS lesions. LANA-2 is abundantly expressed in the nuclei of cultured KSHV-infected PEL cells. Transcription of K10.5 in PEL cell cultures is neither inhibited by DNA polymerase inhibitors nor significantly induced by phorbol ester treatment.[362]

Mechanism of KSHV Episomal Persistence

LANA-1, originally named latent nuclear antigen (LNA), is the dominant protein expressed during latency.[167] LANA-1 is a multifunctional multidomain protein of 1,162 amino acids in length, 222–234 kDa in size, with a characteristic speckled or punctate nuclear localization.[167,168,229] LANA-1 has three domains: (1) a 329 aa N-terminal rich in serine/threonine, proline, and basic residues; (2) a highly polymorphic 534 aa internal repeat domain (IRD) rich in glutamic acid, aspartic acid, glutamine, and leucine and contains a putative leucine zipper motif; and (3) a 227 aa C-terminal domain rich in charged and hydrophobic residues.[169,495] Because the KSHV viral genome does not encode any of its own centromeres, it must have an alternative method for maintaining and replicating its episome from generation to generation. In KSHV, LANA-1 is responsible for episomal maintenance, replication, and segregation into daughter cells.

In order to carry out these functions, LANA-1 tethers the KSHV episome to the chromosome.[25,27,112,398] In KSHV-infected cells, LANA-1 colocalizes with KSHV and binds episomes.[25,238,441] Because of this colocalization, LANA-1 was implicated as an important element in episomal persistence and there is much *in vitro* evidence to support this conclusion. LANA-1 expression is sufficient for maintenance, replication, and segregation of plasmids containing terminal repeat units.[170,184,205] LANA-1 is sufficient and necessary for viral latent DNA replication and efficient episomal segregation during mitosis, which assures that an equal number of KSHV episomes is distributed to each daughter cell.[25,184,205] Transposon-mediated disruption of LANA-1 protein expression in recombinant KSHV BAC36 renders the virus incapable of establishing latency and leads to the loss of the episome, demonstrating that LANA-1 is essential for these functions *in vivo*.[487]

LANA-1 utilizes protein–protein interactions with the folding regions of core histones H2A and H2B to facilitate episomal tethering to the nucleosome during mitosis and interphase.[27] Both the N- and C-terminal domains of LANA-1 interact with the host chromosomes; however, the C-terminal cannot maintain the KSHV episome in N-terminal mutants.[26,27,243,345,398] It is most likely that LANA-1 binds to two unique sites in the long terminal repeat of the KSHV episome with its C-terminal and the chromosomes with the N-terminal while the C-terminal plays a supportive role in binding KSHV episomes to host chromosomes. Nevertheless, the C-terminal is important for LANA-1 oligomerization and evidence suggests that oligomerization is important for efficient tethering of the KSHV episome.[385,447,452] The LANA-1 C-terminus also interacts with a number of chromosome binding and origin recognition complex (ORCs) proteins such as Brd2/RING3, CBP, ORC2, and HBO1 to create an optimal microenvironment for episomal replication.[423,447] When the expression levels of these proteins are knocked down by siRNAs, the efficiency of episomal replication is also reduced, indicating that the C-terminal is likely essential for ensuring episomal replication.[423] Thus, in addition to tethering the episome to the chromosomes, LANA-1 ably hijacks cellular proteins to aid the virus in episomal replication.

Mechanisms of KSHV Reactivation

Expression of KSHV Lytic Genes

A hallmark of herpesvirus life cycle is the expression cascade of genes that can be divided into four categories based on their expression kinetics: latent, immediate-early (IE), early, and late genes. The IE genes are expressed immediately after primary infection or upon reactivation from latency, and do not require *de novo* protein synthesis. IE genes generally encode for regulatory proteins and are critical for initiating viral transcription. RTA is the gene product of the major IE transcript of KSHV, which is necessary and sufficient to drive the switch from latency to lytic gene expression and the production of viral progeny in infected cells. KSHV may also encode several other IE gene products, including ORF45, ORF29b, and ORF4.2.[245] ORF45 is present in purified KSHV virions and appears to be a tegument protein.[499,502] ORF45 was characterized as a phosphorylated protein and may interact with IFN regulatory factor 7 (IRF-7) and inhibit virus-mediated induction of IFN-α/β.[501]

Early gene products are made after the IE genes, but prior to viral DNA synthesis; therefore, their expression is not affected after the inhibition of viral DNA replication by agents such as phosphonoacetic acid (PAA). Several KSHV early genes have also been identified; they include K8 (K-bZIP, also known as RAP), vIRF-1, K1, K3, and K5 (MIR-1 and MIR-2, modulator of immune recognition proteins 1 and 2), ORF57 (a functional homologue of the EBV MTA gene), vIL-6, viral CC chemokine ligands (vCCLs), polyadenylated nuclear (PAN) RNA, vBcl-2, ORF49, K12, K15, viral G protein-coupled receptor (vGPCR), viral dihydrofolate reductase (vDHFR), and thymidylate synthase.[47,48,96,178,195,257,300,340,361,377,431,439,466,467,476]

Late genes whose expression is abolished by PAA usually do not appear until 30 h after induction.[431,500] Examples of these proteins include envelope glycoproteins gB, K8.1, and a small viral capsid antigen encoded by ORF65.[181,377,431]

Molecular Mechanisms Involved in KSHV Reactivation

The switch from latent to lytic infection of KSHV is initiated by a number of stimuli that induce the expression of the key lytic switch protein, RTA. The expression of RTA is necessary and sufficient to trigger the full lytic program resulting in the ordered expression of viral proteins, release of viral progeny, and host cell death.[275,276,430] The expression of RTA precedes the expression of all other cycloheximide-resistant IE genes and cycloheximide-sensitive early genes. In addition, activation of RTA leads to the complete production of nascent, infectious virus particles with kinetics that is consistent with stimulation by chemical inducers.[181]

Following primary infection, KSHV generally establishes latent infection.[31] During latency, only a few genes are transcribed, while the expression of RTA is tightly repressed. However, the cloned promoter region of RTA shows high basal activity,[95,121,370] indicating that epigenetic change and chromatin remodeling of

KSHV genome may be involved in this repression. Epigenetic changes such as DNA methylation act to regulate gene expression in normal mammalian development as well as in cancer through transcriptional silencing of critical growth regulators.[30] With approximately 70% of the CpG sites in the human genome methylated, it is clear that the cellular environment is predisposed toward methylation, and a herpesvirus infecting a host cell must contend with this environment. In fact, before the discovery of KSHV in 1994[85] the genomes of the known gammaherpesviruses (including EBV and various murine, bovine and simian family members) were all shown to be CpG suppressed, suggesting they too have been subjected to heavy methylation.[226] The finding by Chen et al.[95] suggests that hypermethylation in the RTA promoter may regulate its expression and subsequently KSHV reactivation from latency. Other chromatin modifications, such as histone deacetylation and alterations of chromatin-binding proteins, affect local chromatin structure and, in concert with DNA methylation, may also regulate RTA gene transcription.[273] It has been suggested that methylation of the RTA promoter region during latency promotes the association of transcriptional repressors and histone deacetylase (HDAC). Lytic replication of KSHV can be triggered by chemical agents including butyrate and 12-O-tetradecanoylphorbol-13-acetate (TPA). Butyrate, a known activator of lytic replication, is an inhibitor of HDAC, and conversely, TPA is an inducer of histone acetylases (HAT). Therefore, both inducers of KSHV lytic replication are affecting the acetylation state of the RTA promoter that is in turn dependent on methylation. Such findings imply that the control of latency and of switch to lytic replication is a function of chromosomal architecture, and will involve the interplay between viral RTA and host factors that regulate chromatin methylation and acetylation.[474]

RTA can also autoactivate its own expression.[121,370,387] A striking feature of RTA-mediated lytic gene expression is that RTA induced KSHV gene expression in a more powerful and efficient manner than TPA stimulation, indicating that RTA plays a central, leading role in KSHV lytic gene expression.[318]

After activation, RTA is recruited to its responsive elements through direct interaction and transactivation with RBP-Jκ, a notch signal pathway transcription factor, to activate viral lytic gene expression.[83,84,199,262,263,474] Interestingly, RTA also contributes to the establishment of KSHV latency by activating LANA-1 expression through the notch signaling pathway RBP-Jκ. On the other hand, LANA-1 can inhibit viral lytic replication by inhibiting expression as well as antagonizing the function of RTA.[248] The interaction between RTA and LANA-1 provides a feedback mechanism by which these two proteins can regulate each other and is likely to be a key event in the establishment of KSHV latency.[247,249] RTA may also recruit CBP, the SWI/SNF chromatin remodeling complex, and the TRAP/mediator coactivator to its down-stream viral promoters, and that this recruitment is essential for RTA-dependent viral gene expression.[190]

RTA mRNAs and Protein Structure

Multiple transcripts are encoded by the ORF50 region; the main transcript is a 3.6 kb mRNA that encodes the entire RTA. In addition, a downstream gene, K8, was also found to be encoded within the same RNA species. RTA protein is mainly encoded by ORF50 but obtains an additional 60 amino acids for its N-terminal through a splicing event which shifts its start code across ORF49.[96,430] The transcript for RTA can potentially encode for a protein of 691 amino acids, and is predicted to have a molecular mass of 73.7 kDa. However, the expressed protein when analyzed by Western-blotting appears to be about 110 kDa, suggesting that RTA could be modified post-translationally by phosphorylation or by other mechanisms.[275] RTA lacks any significant homology with cellular proteins but is functionally and genetically homologous to the RTA proteins from EBV, HVS, RRV, and MHV68.[116,179,268,482] The RTA protein consists of an N-terminal DNA binding domain, a central dimerization domain, a C-terminal acidic activation domain, and two nuclear localization signals (NLS).[96,275] The DNA-binding domain of RTA is located at the amino terminus from aa 1 to 530. A deletion mutant of the activation domain sequences (aa 531–691), containing only the DNA binding domain, has been shown to be a trans-dominant-negative mutant that maintains DNA-binding activity for RTA responsive promoters but no longer activates lytic gene expression.[275] The activation domain is located at the carboxyl terminus of the protein (aa 486–691), which is highly acidic and contains numerous charged amino acids.[275] This region also contains four repeated units of a highly hydrophobic domain with sequence homology to other transcriptional factors such as VP16 domain A.[275] These characteristics suggest that KSHV RTA is a member of a family of transcriptional factors.[474]

Downstream Targets of RTA

RTA has been shown to regulate and transactivate a number of downstream viral genes that function in lytic replication, including K1, K3, K5, DNA polymerase, vIL-6, vIRFs, vGPCR, K12, K15, etc.[47,48,120,121,178,195,275,370,414,445,476] RTA activates downstream KSHV target genes by at least two mechanisms: direct recognition of RTA response elements (RRE) in the promoters of its target genes and interaction with cellular or viral proteins bound to the promoters.[83,413] For example, RTA directly binds to the promoters of PAN and K12 but does not bind to ORF57 or vCCL-1 promoters. Conversely, RTA transactivates the promoters of ORF57 and vCCL-1 through the binding of a cellular factor, RBP-Jκ protein.[88]

There is no defined consensus sequence for direct RTA binding and there is no significant homology present in the RTA-responsive viral promoters. However, a comparison of the K8 RRE with other viral RRE revealed a pattern of multiple A/T triplets spaced with a periodicity of 10 or 20 bp.[264] The diversity of RTA binding pattern implies that the activation of RTA target gene promoters may result from RTA interaction with other cellular or viral proteins that mediate the DNA-transcriptional complex interaction. One such cellular factor could be

NF-κB.[363] Other cellular factors involved in RTA transactivation may include TATA-binding protein such as the case in the K1 promoter.[48] RTA may regulate vOX-2 (K14) and vGPCR genes through an IFN-stimulated response element (ISRE)-like sequence (K14 ISRE) in the promoter region.[493]

Cellular Factors Involved in RTA-Mediated KSHV Reactivation

Identification of cellular proteins that coordinate with RTA in transactivation and characterization of the mechanisms whereby such host–viral protein complexes mediate the switch from KSHV latency to lytic replication is an important step in understanding the virus life cycle and pathogenesis. Increased viral lytic reactivation has been observed following exposure of latently infected PEL cells to agents such as IL-6, IFN-γ, hypoxia, other viral agents, n-butyrate, ionomycin, 5-azacytidine, and TPA.[86,92,95,117,123,198,292,295,303,359,412,451,504] Cellular pathways involved in the induction of viral lytic reactivation include the NF-κB pathway, and the protein kinase C (PKC) δ, and MEK/ERK, JNK, and p38 MAPK pathways.[107,123,333,392] AP-1, the cellular activator protein complex composed of c-Jun and c-Fos heterodimers, mediates the transcription of MAPK target genes.[463] The RTA promoter contains a consensus AP-1 binding site,[463] and the OriLyts (origins of lytic replication) also contain several putative AP-1 binding sites.[19] The presence of AP-1 binding sites in these regions allows the virus to respond to cellular conditions that may not be favorable for latency. NF-κB has been shown to be constitutively active in PEL cells[231] and promotes their survival. The involvement of NF-κB in lytic reactivation is controversial.[56,392] One study reported that inhibition of NF-κB led to increased viral lytic protein synthesis in KSHV-infected epithelial cells and PEL cells;[56] however, another study demonstrated a requirement for NF-κB activation for the production of infectious virions.[392] In the latter study, virion production was not diminished by suppression of NF-κB, but infectivity of the virions was decreased, suggesting NF-κB may be involved in multiple aspects of lytic reactivation and viral production. In addition, cellular pathways contributing to epigenetic effects such as DNA methylation and histone acetylation may also participate in the reactivation of KSHV.[95,273]

The direct interaction of CREB-binding protein (CBP) and p300 with KSHV RTA in the activation of lytic replication has recently been reported by Hwang et al.[191,209] HDAC was shown to repress RTA activity by binding directly to the proline-rich sequences in the RTA central domain aa 301–449. Specific inhibition of HDAC by trichostatin A (TSA) reversed the inhibition of RTA and stimulated RTA-directed gene expression. KSHV RTA strongly induces CD21 and CD23a expression through RBP-Jκ binding sites and regulates RBP-Jκ-mediated cellular gene expression, which ultimately provides a favorable milieu for viral reproduction in the infected host.[84] RTA was also shown to interact with other cellular factors such as RAP, C/EBPα, CBP, STATs, RBP-Jκ, SWI/SNF, TRAP230, PKC, MAP kinase, AP-1, NF-κB[190,191,193,209, 258,265,423,462,464,480,484,488] and transactivates target genes through Oct-1 and Sp-1.[370,445]

Recently, Wang et al. reported that IRF-7 negatively regulates KSHV lytic reactivation by competing with RTA for binding to the RTA response element in the ORF57 promoter to down regulate RTA–induced gene expression.[458] Yu et al. reported that IRF-7 is targeted for degradation through an ubiquitination-dependent fashion, and RTA itself acts as a ubiquitin E3 ligase.[491] Previously, Zhu et al. reported that ORF45 blocks the phosphorylation and nuclear translocation of IRF-7 and efficiently inhibits the activation of IFN-α/β genes during viral infection.[501]

KSHV Primary Infection

Characterization of KSHV primary infection relies heavily on the development of systems with high infection efficiency. KSHV isolated from PEL and sometimes KS lesions is able to infect various cell types but with limited primary infection efficiency or unsustainable infection culture (see page 79). Of all the systems examined so far, KSHV eventually establishes latency after primary infection. In some studies, KSHV was found to immediately enter latency. However, other studies have indicated that KSHV has an early full productive replication phase during infection in at least some cell types or infection conditions.[124,149,166] In fact, in the first description of KSHV infection and transmission system in 293 cells, cytopathic effect (CPE) was observed after primary infection, an indication of lytic replication.[152] In both MVDEC and HUVEC, KSHV linear genomes and lytic transcripts were detected several days after primary infection, again indicating productive primary infection.[124] Strikingly, efficient infection of HUVEC by recombinant KSHV BAC36 is lytic replication-permissive at the early stage of infection, producing large amount of infectious virions.[166] Examination of the expression of lytic proteins and transcripts further confirmed KSHV productive lytic replication in this system.[166,489] BAC36 infection of HUVEC displayed two phases: an early productive phase, in which the virus actively replicates producing large number of virions, and concomitantly resulting in massive cell death; and a latent phase, in which the virus switches into latent infection in the surviving cells.[166] Similar results were also observed in a separate study.[149] The different results from these studies point to the variations among cell types in supporting KSHV productive primary infection. The determination of the cell types and/or conditions that can support KSHV productive primary infection and the delineation of the underlying molecular basis could help understand the mechanism controlling KSHV replication and latency.

Attachment, Entry, and Cellular Receptors

Enveloped viruses infect host cells in two steps. The first step is attachment, or binding, of the virus particle to host cell receptors. This step is mediated by the interaction of viral proteins with cell surface molecules such as glycosaminoglycans (i.e., heparan sulfate).[416] Attachment allows other viral proteins to contact cellular coreceptors, which will then stimulate the second step, entry, by either a fusion event between viral envelope and cell membrane, or receptor-mediated endocytosis.[416]

Similar to other herpesviruses, KSHV expresses several transmembrane glyco-proteins that are virion associated, and involved in the attachment and entry of tar-get cells. KSHV glycoproteins gB (ORF8), gH (ORF22), gL (ORF47), gM (ORF39), and gN (ORF53) are all conserved among the herpesviruses. In addition, KSHV encodes several unique glycoproteins K1, K8.1A, K8.1B, and vOX-2 that share no significant homology with glycoproteins of other herpesviruses.[82,105,240,242,261,277]

The ability of KSHV to infect a variety of cell types *in vivo* and *in vitro*[7,31,164,303,357] indicates that it must be able to recognize either ubiquitously expressed cell surface receptors, or more than one type of receptor. To date, KHV has been shown to attach to the cell surface molecules heparan sulfate, integrin $\alpha 3\beta 1$, and DC-SIGN.[6–8,39,354,454] K8.1A and gB have both been shown to interact with the ubiq-uitously expressed heparan sulfate.[6,8,39,454] Heparan sulfate is a linear carbohydrate that is localized at the extracellular cell surface, typically covalently bound to a core proteoglycan imbedded in the cell membrane. Heparan sulfate proteoglycans participate in many biological processes including cell–matrix interactions, activa-tion of chemokines, enzymes, and growth factors,[440] and is well established to be important for the cell attachment of many other herpesviruses.[416]

Although conserved among herpesviruses, only KSHV gB contains an RGD motif, which is the minimal peptide region known to interact with integrins in the cell membrane.[5,7,455] KSHV gB specifically binds integrin $\alpha 3\beta 1$ through its unique RGD motif.[7,494] Integrins are a large family of heterodimeric receptors that contain two transmembrane glycoprotein subunits, α and β. There are 24α and 9β subunits identified so far, with more than 24 known combinations of these subunits.[172,346] Each cell expresses several combinations of $\alpha\beta$ integrins, and each combination has its own binding specificity and signaling properties.[172,346,380,382] Integrin $\alpha 3\beta 1$ is a receptor for laminin 5 and fibronectin, and is expressed at high levels in endothelial cells.[220,479,494]

DC-SIGN (dendritic cell-specific ICAM-3-grabbing nonintegrin) is a type II C-type lectin that is found on myeloid dendritic cells in the dermis, mucosa, lymph nodes, lung, and thymus, and IL-4-stimulated monocyte-derived dendritic cells.[356,409,443] It is also expressed on lung alveolar macrophages,[409] placenta,[408] inflammatory lesions[411] and IL-13-stimulated, monocyte derived macrophages,[94,409–411] DC-SIGN acts as a pathogen recognition receptor in these cells of the immune system, activating macrophages and dendritic cells to ingest and process pathogens for antigen presentation to T cells.[94,171] KSHV may have evolved the ability to exploit DC-SIGN in order to infect dendritic cells and macrophages, and interfere with their antigen presentation functions.[354]

Following attachment to the cell, an enveloped virus such as KSHV can gain entry to the cell by either direct fusion of its envelope with the plasma membrane or by endocytosis, followed by fusion with the endosomal membrane. Evidence for both routes of entry has been published, and KSHV may use more than one mech-anism, depending upon the cell type and which receptors are expressed.[5,124,212,341]

Endocytosis provides rapid and convenient transport of virion across the plasma membrane, through the cytoplasm, and delivery of the viral cargo to the perinuclear region. Endocytosis can occur through four major mechanisms: clathrin-coated vesicles, the caveolar pathway, macropinocytosis, and a poorly characterized nonclathrin, noncaveolar dependent form of endocytosis.[232,289,400] KSHV is able to infect human fibroblasts via endocytosis as demonstrated by the presence of virus particles inside endocytic vesicles within 5 min following attachment.[5] The presence of transferrin within the vesicles implicates clathrin–mediated endocytosis as the means used by the virus to enter human fibroblasts. However, it has been demonstrated for other herpesviruses, i.e., HSV-1, that entry via endocytosis results in nonproductive infection with minimal infection.[237]

Fusion of the viral envelope at the plasma membrane has been well established for other members of the herpesvirus family.[416] Some studies indicate that KSHV fuses with the plasma membrane to enter 293 cells and MVDEC.[124,212,341] Expression of KSHV envelope glycoproteins gB, gH, and gL in mammalian cells induced cell–cell fusion.[341] Expression of the cellular membrane protein xCT facilitates the fusion of KSHV-negative cells with KSHV-positive BCBL-1 cells that have been stimulated to express viral lytic proteins, specifically glycoproteins at the cell plasma membrane.[225] Transfection of xCT into cell lines that normally are resistant to KSHV infection rendered them permissive to fusion and infection by KSHV. The xCT protein is a transmembrane protein that is upregulated in response to stress induced by the production of reactive oxygen species.[351,379] Exposure of endothelial cells to ROS may enhance the infectivity of KSHV.[459] In addition, the HIV protein Tat has been shown to stimulate the expression of xCT,[51] and Tat can also enhance the infectivity of KSHV.[16]

Limited evidence demonstrates that KSHV adheres to the general dogma of herpesvirus family members once inside the host cell. In MVDEC, following envelope fusion with either the plasma membrane or the endosomal membrane, the viral tegument proteins and nucleocapsid are released into the cytoplasm.[124] The nucleocapsid is then degraded, and the tightly packaged linear viral genome decondenses and is delivered to the nucleus. Within 8 h post infection, the viral genome circularizes, which is the typical conformation of the genome during latency. However by 72 h post infection, both circular and linear genomes can be detected, indicating a mixed population of latent and lytic infection.[124] In fibroblasts, nuclear trafficking appears to be mediated by microtubules and the associated dynein motors.[321] These observations remain to be confirmed in endothelial cells.

Regulation of Cellular Signaling Pathways

The events that lead to successful infection by KSHV, i.e., attachment, entry, nuclear trafficking, and expression of viral genes, cannot occur without careful manipulation of preexisting host cell signaling pathways and machinery, as well as suppression or evasion of host defenses.

The interaction of glycoprotein gB (discussed above) with cellular integrin $\alpha3\beta1$ activates focal adhesion kinase (FAK) and the MEK-ERK1/2 pathway.[7,393,394] Primary infection of HUVEC cells causes a conversion from the normal flat "cobblestone" cell morphology to a "spindle cell" typical of the cells seen in KS lesions.[166] Spindle-cell conversion occurs as early as 6 h post infection and requires viral modulation of host cytoskeletal apparatus. Glycoprotein gB mediates this extensive cytoskeletal rearrangement by modulating the FAK-Src-PI3-kinase-RhoGTPase pathway.[320,393] It has also been reported that glycoprotein gB can activate VEGFR-3 on the microvascular endothelium and trigger a migratory and proliferative response in these cells.[494] Primary KSHV infection also activates and is dependent on the JNK and p38 MAPK pathways in addition to the MEK-ERK1/2 pathway.[333,484] The activation of the multiple MAPK pathways is instrumental in the activation of the transcription factor AP-1. AP-1 regulates the expression of a variety of cellular genes, including IL-6, and in fact is required for the transcription of KSHV genes. KSHV induction of AP-1 could also contribute to a variety of KSHV-induced malignant phenotypes such as cellular proliferation, angiogenesis, inflammatory cytokine production, and dissemination of tumor cells.[484] During primary infection, KSHV-encoded host-modulating genes are expressed which could impact the expression and function of host genes.[241,489] Analysis of cellular transcripts during primary infection reveals that KSHV is able to significantly upregulate expression of cellular genes that are implicated in cell growth and survival, inflammation and angiogenesis, and immune responses.[322]

Viral Gene Expression During Primary Infection

The expression profiles of KSHV transcripts during KSHV primary infection depend on the infection systems analyzed. In the productive HUVEC primary infection system, the expression of latency-associated genes is generally expressed first preceding the cascade of lytic genes and onset of lytic replication. The transcription of lytic genes peaks around 54 h post infection followed the production of infectious virions.[166,489] While after 54 h post infection, the expression of lytic genes declines, latency-associated transcripts continue to increase, indicating that following the permissive phase of lytic replication, surviving cells express the latency gene cluster, specifically ORF71/ORF72/ORF73, and have the tendency to switch to latency.[489] Similarly, KSHV-encoded genes with host modulating functions, including mitogenic and cell cycle-regulatory, immune-modulating, and anti-apoptotic genes, are expressed before those encoding viral structure and replication genes, and sustained at high levels throughout the infection, suggesting KSHV manipulation of host environment to facilitate infection.[489] In the default latency infection systems of fibroblasts and MVDEC, the latent genes are also expressed throughout the infection process; however, lytic genes such as RTA are only transiently expressed, which is consistent with the lack of productive viral replication in these systems. Nevertheless, the early expression of lytic genes involved in immunomodulation and resistance to apoptosis also suggests KSHV manipulation of host defenses to facilitate infection.[241]

MECHANISMS OF KSHV-INDUCED PATHOGENESIS

Introduction: KSHV Infection Promotes Oncogenesis

Although it remains controversial whether KS is a malignant neoplasm, it is well established that the late stage KS lesions are true clonal cancers, probably evolved from a constant reactive, inflammatory/angioproliferative process in immune compromised patients.[77,138,139] Given the association of KSHV with two different human malignancies (KS and PEL), KSHV is considered to be a human oncogenic virus. Distinctive from other oncogenic viruses, KSHV is a complex DNA virus that not only has the ability to promote cellular growth and survival for tumor development but also can provoke deregulated angiogenesis, inflammation, and modulate the patient's immune system in favor of tumor growth. Nevertheless, not all individuals harboring KSHV develop KS. The presence of KSHV DNA in healthy individuals indicates that KSHV alone may not be sufficient to cause clinical KS. Since KS is most commonly found in immunosuppressed individuals, it has been suggested that immune deficiency is an important factor in the pathogenesis of AIDS-KS and that HIV infection may be an important cofactor in disease progression.[54,138,139,176]

The majority of spindle-shaped cells in KS lesions are latently infected by KSHV indicating that latent infection plays an essential role in KSHV-induced malignancy and pathogenesis.[122,132,141] Nevertheless, a small subset of KS cells also undergoes spontaneous viral lytic replication indicating that KSHV lytic replication might also be important for KS development. There is strong correlation between viral load and progression of KS tumor.[55,115,135,146,347] Several drugs that target herpesvirus replication effectively inhibit KS tumor growth.[45,221,296,365] A number of viral lytic genes have been linked to KSHV-induced malignancy and pathogenesis.[77] The production of infectious virions should lead to new infection, which could also produce virus-encoded cytokines and induce cellular cytokines.

Promotion of Cellular Growth and Survival

Although the pathogenesis of KSHV-induced malignancies is still not completely understood, it appears that the virus targets multiple pathways to promote cell proliferation and survival to facilitate tumor development. Several studies have shown that genetic instability is present in KS tumors and PEL.[72,119,160,163,348] KSHV infection is sufficient to induce chromosome instability,[334] for which LANA-1 and vCyclin might be partially responsible.[399,450] Like other oncogenic DNA viruses, KSHV targets both p53 and pRb tumor suppressor pathways. At least five KSHV genes, LANA-1, RTA, K-bZIP, LANA-2, and vIRF-1, interact with and suppress the functions of p53 and pRb.[154,192,317,352,362,391,397] Loss or dysfunction of tumor suppressor genes will inhibit the host cell's ability to repair damaged DNA, eliminate p53-dependent cell death, and dictate cell cycle for uncontrolled cell proliferation, hence contributing to KSHV-induced oncogenesis. Furthermore,

KSHV can regulate cell cycle progression through vCyclin.[90,260] Expressed in latently infected cells and in both KS and PEL cells, vCyclin can interact with and phosphorylate cyclin-dependent kinase 6 (Cdk6) to promote cell cycle progression from G_1 to S-phase and accelerate cellular proliferation.[90,100,175,224,260,376,432]

KSHV uses multiple mechanisms to promote cell survival. The KSHV genome encodes several virus homologues of human antiapoptosis proteins. For instance, vBcl-2 is a KSHV homolog of human antiapoptosis protein Bcl-2.[98,378] KSHV also encodes several cellular IRF homologues vIRFs that inhibit apoptosis.[60,148,165,233,259,274,317,390,391] In addition, the KSHV K7 gene encodes a viral inhibitor of apoptosis (vIAP) survivin-ΔEx3 to inhibit apoptosis.[456] Another unique mechanism that KSHV utilizes to promote tumor growth and cell survival is through activation of the NF-κB pathway. The NF-κB pathway is constitutively active in both KS and PEL,[21,284] and treatment with inhibitors of NF-κB pathway lead to a complete regression of PEL tumors in a mouse model.[230] Two KSHV genes, vGPCR and vFLIP, have been shown to enhance cell growth and survival by activating the NF-κB pathway.[93,145,270,284,386] vGPCR seems to play an important role in promoting endothelial cell proliferation and transformation.[18,22,114,182,298,310,339,386,404,405] Endothelial cells ectopically expressing vGPCR have constitutively active VEGF receptors and can proliferate to form both foci in culture and tumors in nude mice independently of VEGF stimulation.[23,182] KSHV-encoded vIL-6 also promotes cell survival.[15,202,236,297,308,324,332,453] Finally, it is believed that a variety of cellular growth factors and inflammatory cytokines secreted by KSHV-infected endothelial cells and tumor-interacting stromal cells play a pivotal role in the development of KS. It is well documented that KSHV infection induces the secretion of various growth factors such as IL-6, IL-8, Gro-α, VEGF, and bFGF,[71,78,282,322,457,484] which could promote cell proliferation through autocrine and/or paracrine signaling. It is important to emphasize that the interaction between HIV and KSHV in AIDS patients plays a significant role in tumor growth. The aggressiveness of AIDS-related KS implicates HIV-1 infection as an important cofactor in rapid KS progression; indeed, the time of KSHV seroconversion until the onset of KS may be years to decades in classic KS while it is only several months in most of the AIDS-related KS cases.[167,168,291,325,337,355,360] HIV-1 not only activates lytic replication of KSHV[206] but also induces secretion of a number of cytokines and growth factor from infected macrophages,[189,215,256] further promoting tumor cell proliferation and survival, as well as tumor angiogenesis.

Regulation of Angiogenesis

KS tumors are highly angiogenic with abnormally dense and irregular blood vessels. The importance of angiogenesis in KS tumor development is further highlighted by the fact that early stage KS might not yet be a true tumor but a neoplasm of proliferative spindle-shaped cells driven by angiogenesis and inflammation. Angiogenesis is the formation of new blood vessels from existing capillary beds, which, with the exception of wound healing and female reproductive cycle, is a rare event in adults.[70,150] Pathological angiogenesis, however, correlates with tumor growth and metastasis.[150] Angiogenesis is a complicated process that involves a number of different angiogenic factors. Some angiogenic factors initiate blood vessel remodeling by disrupting the existing blood vessels, while others are responsible for promoting migration, adhesion, and proliferation of endothelial cells for new blood vessel growth and maturation.[150]

The mechanisms of KSHV-induced angiogenesis remain to be further elucidated. KSHV-induced angiogenic factors and inflammatory cytokines likely play essential roles. In a SCID mouse model with human skin grafts, it was demonstrated that VEGF is essential for the inoculated early-stage KS cells to grow into KS-like tumors.[373] Many of the cytokines, including VEGF, bFGF, IL-6, IL-8, GRO-α, TNF-β, and ephrin B2, induced by KSHV infection are angiogenic.[282,283,322,457,484] Higher levels of serum VEGF were also seen in HIV-1 infected persons with KS compared with HIV-1 infected persons without KS,[469] and mRNA levels of VEGF and angiopoietins, including both Ang-1 and Ang-2, were also detected at higher levels in KS lesions compared to the adjacent normal tissues.[57] KSHV infection induces expression of cyclooxygenase-2 (COX-2)[322] and heme oxygenase-1,[286] both of which are important players in angiogenesis. In addition to the host factors, KSHV vIL-6 expressed during early infection has been shown to stimulate hematopoiesis, plasmacytosis, and angiogenesis.[15,269,290] A number of KSHV proteins vIL-6, vGPCR, vCCL-1, and vCCL-II have been shown to promote angiogenesis by inducing VEGF expression through different signaling pathways.[15,41,407,425] vGPCR alone can activate PKC, protein kinase B (PKB), Akt, NF-κB, and MAP kinases to regulate the expression of growth factors and cytokines.[18,22,114,298,310,339,386,404] Transgenic mice expressing vGPCR display multifocal and angioproliferative KS-like lesions.[189,218,486]

Immune Modulation

Like other viruses, KSHV must evade the host innate immunity that involves type 1 IFNs, phagocytes, natural killer (NK) cells, and complement, as well as the adaptive immune responses that include humoral and cell-mediated immunity. Viral evasion of host immunity not only safeguards virus persistence but also contributes to viral oncogenesis and pathogenesis. KSHV utilizes two major strategies to achieve successful infections. The first is the so-called passive strategy. After a successful entry into host cell, KSHV establishes latency, during which only a minimum number of virus genes are expressed, thus reducing the number of antigens that are exposed to the immune systems. Secondly, during lytic replication or *de novo* infection, when most viral proteins are expressed and are susceptible to immune surveillance, the virus utilizes an active strategy that involves a number of its own unique genes to modulate the host immune response. Thus, KSHV has evolved multiple mechanisms to not only evade the host immune response but also manipulate existing cytokine regulatory networks to facilitate viral replication and maintenance. In addition to exploiting the cytokine networks, KSHV also expresses genes that modulate the T-cell response, B-cell activation, as well as regulate the complement cascade.

KSHV has a large and complicated genome. It encodes not only proteins necessary for the reproduction and persistence of the virus but also pathogenic factors to promote tumor cell proliferation and survival, regulate angiogenesis, and modulate host immune response in favor of tumor growth (Fig. 1). The function and roles of each of these factors in relation to KSHV-induced malignancy and pathogenesis are further discussed the following sections.

Functions of KSHV Genes

LANA-1/LNA

In addition to its role in episomal persistence (maintenance, replication, and segregation), LANA-1 is a promiscuous protein that has many cellular interaction partners and modulates p53, STAT3, Rb, and GSK-3β pathways with integral roles in apoptosis, immune response, development, cell growth, proliferation, and transformation. LANA-1 acts directly or indirectly as a general transcriptional repressor or occasionally as a transcriptional activator. LANA-1 modulation of cellular functions might ultimately contribute to viral latency in KSHV-infected cells.

Much work has been done to determine the role of LANA-1 in KSHV-mediated transformation. In conjunction with the expression of Hras, LANA-1 transforms primary rat fibroblasts.[352] Overexpression of LANA-1 increases cellular proliferation and lifespan of primary HUVEC.[399,468] Nevertheless, overexpression of LANA-1 cannot induce anchorage independent growth, a hallmark of cellular transformation, in NIH3T3 cells, indicating that other viral genes are likely necessary for KSHV-induced malignant transformation.[468]

The mechanism of LANA-1-mediated cellular transformation has been extensively investigated. LANA-1 targets the Rb/E2F pathway and alleviates Rb-mediated transcriptional repression of the E2F pathway.[352] LANA-1 upregulates the expression of telomerase by interacting with the Sp1 transcription factor.[446] In latently infected PEL cells, LANA-1 blocks p53-mediated apoptosis by binding to and inhibiting its transcriptional activity.[154] In KS tumors, apoptosis is not correlated with high expression levels of p53, rather, it is negatively correlated with the expression of LANA-1.[228] LANA-1 inactivation of p53 activity has also been associated with genomic instability.[399] HeLa and Rat-1 cell lines stably expressing LANA-1 have increased multinucleation, micronuclei, mitotic bridges, and abnormal formation of centrosomes.[399]

β-Catenin is overexpressed in PEL cells and KS tumors.[156–158] Normally, β-catenin is degraded in the cytoplasm by its negative regulator GSK3β. LANA-1 causes the excess accumulation of β-catenin by binding and relocating GSK3β to the nucleus where it can no longer regulate β-catenin. Excess β-catenin accumulates and ultimately translocates to the nucleus where it interacts with Tcf/Lef transcription factors to regulate the expression of target genes. β-Catenin targets many genes that have important roles in cell cycle progression, cell growth, and cell proliferation.[28,50] Notable β-catenin target genes are myc, jun, and cellular cyclin D1.[158]

Overexpression of LANA-1 induces the expression of IL-6 by activating the AP-1 pathway.[12] One consequence of the autocrine and paracrine effects of IL-6 signaling is activation of the STAT3 signaling pathway. LANA-1 itself also binds STAT3 in KSHV-infected PEL cells and augments STAT3-induced transcription in reporter assays.[315]

LANA-1 positively and negatively regulates transcription through a variety of mechanisms. In *in vitro* experiments, the N- and C-terminals of LANA-1 repress transcription when fused to a heterologous DNA binding domain.[385] Recruitment and interaction with members of the mSin3 transcriptional repressor complex is one mechanism by which LANA-1 represses transcription.[244] LANA-1 represses the ATF4/CREB transcriptional activator without interfering with its DNA binding ability.[266] LANA-1 inhibits the CBP histone acetyltransferase activity and represses CBP transcriptional activation by preventing its dimerization with c-Fos.[244] LANA-1 interacts with a potent transcriptional repressor KLIP1 and relieves its transcriptional repression effect.[335,496] LANA-1 activates the AP-1 pathway by binding to Jun and facilitating the formation of Jun–Fos heterodimers.[13] LANA-1 also upregulates Id-1, which is a member of a family of basic helix–loop–helix containing proteins that have key roles in inhibiting cell differentiation, cell cycle regulation, and tumorigenesis.[438] Id proteins are expressed at high levels in KS tumor cells, but their roles in KSHV-mediated tumorigenesis are not well understood because LANA-1 induced increases in Id protein expression has no discernable effect on the proliferative capacity of cells.[438]

LANA-1-mediated transcriptional regulation could contribute to viral latency. LANA-1 form complexes with RBP-Jκ to repress RTA expression.[247] In addition, LANA-1 interacts with RTA and antagonizes its activation of the RTA promoter.[247,248] LANA-1 also inhibits transcription of nonlatency-associated genes through its interaction with SUV39H1 and recruitment of HP-1.[371] SUV39H1 is a histone-H3 methyltransferase. Upon methylation of histone-H3 by SUV39H1, HP-1 is recruited and forms condensed, transcriptionally inactive heterochromatin. Recruitment of these two factors by LANA-1 prompts condensation of the KSHV episome and consequently, transcriptional repression of all viral genes except those within the latency locus.[371] Finally, LANA-1 autoregulates its promoter and induces transcription of itself.[219]

vCyclin

The decision to enter into the cell cycle is made in G0/G1 when cells are receptive to growth and differentiation signals.[208] Cdks act in combination with cellular cyclins A, D, and E to accelerate cells through the cell cycle.[260] The KSHV vCyclin gene encodes a 257 amino acid protein with a molecular weight of 29 kDa.

vCyclin has 32% identity and 54% similarity to mammalian cyclin D2 with the homology extending over the full length of the protein.[260]

vCyclin strongly binds to Cdk6 and weakly binds to Cdk2, Cdk3, Cdk4, and Cdk5; however, only the interaction with Cdk6 has functional significance.[136,260,279,330] Cdk6 is the catalytic subunit of the protein complex.[216] vCyclin binds tightly to Cdk6 to form a potent phosphorylation complex that phosphorylates many more targets, including p27KIP1, Id-2, Cdc25a, histone H1, Cdc6, Orc1, and Bcl-2[260,279,449] than the Cdk6–cellular cyclin D2 protein complex.[260] In addition, the vCyclin–Cdk6 protein complex is resistant to many cell cycle checkpoints, which contributes to increased rates of cell proliferation.[330] These two characteristics have important implications for KSHV-infected cells. Increased cellular proliferation, apoptosis,[330,450] and chromosomal abnormalities[450] are three main consequences of overexpressing vCyclin.[136,175,279,376,432] The vCyclin–Cdk6 complex phosphorylates and inactivates Bcl-2, thus contributing to vCyclin-mediated apoptosis.[330,331]

Rb serves as an important cell cycle checkpoint protein by binding to the transactivation domain of E2F transcription factors to inhibit their functions and prevent cell cycle progression.[130,200] Rb interacts with a number of chromosome remodeling proteins like HDAC and the SWI/SNF complex to regulate gene expression.[239] Cdk6–vCyclin phosphorylates and inactivates Rb. Consequently, KSHV-infected cells avoid Rb-induced cell cycle arrest.[90,175,260]

p27KIP1 is another important regulator of cell cycle. p27KIP1 mutation or inactivity is often associated with poor prognosis in tumor progression. vCyclin can overcome p27KIP1-mediated cell cycle arrest.[136,175,279,376,432] When vCyclin is overexpressed in cells, the vCyclin–Cdk6 complex interacts with and phosphorylates p27KIP1 on Thr187 inducing its degradation via the proteasome dependent pathway;[136,376] however, the situation may be more complicated *in vivo*. Unexpectedly, p27KIP1 is more highly expressed in PEL cell lines when compared to other lymphomas.[67,216,376] In PEL cells, the Cdk6–vCyclin complex prevents p27KIP1 from carrying out its function in two ways. First, it forms a stable complex with p27KIP1, thereby sequestering it and preventing it from carrying out its normal function.[216] Second, the Cdk6–vCyclin complex phosphorylates p27KIP1 on Ser10, which causes p27KIP1 to undergo relocation to the cytoplasm. Upon lytic reactivation, though, the Cdk6–vCyclin complex does induce degradation of p27KIP1 via phosphorylation of Thr187.[376] In addition to resistance to p27KIP1 inhibition, Cdk6–vCyclin resistance to cell cycle blocks by checkpoint proteins p16(Ink4a) and p21Cip1 has also been shown.[279]

K13/vFLIP

K13 encodes viral FLICE (FADD[Fas-associated death domain]-like IL-1beta-converting enzyme)-inhibitory protein (vFLIP), which is a viral homologue of cellular FLIP proteins.[128] vFLIP utilizes its death effector domain and TRAF-binding domains to perform two principal functions, inhibition of apoptosis and constitutive activation of the NF-κB pathway. vFLIP is coded by either tricistronic transcripts or a bicistronic transcript shared by two other latent genes, LANA-1 and vCyclin.[125,185,434] The translation of the vFLIP protein is most initiated from IRES present in the vCyclin coding region, although a minor amount comes from reinitiation of translation.[185,272]

In the cytosol, vFLIP interacts with procaspase 8 to prevent the cleavage of procaspase 8 into its active subunits.[32] vFLIP also inhibits caspase 3 and caspase 9 like activities in Fas-triggered A20 cells.[32] In addition to its role in inhibiting Fas-induced apoptosis, vFLIP constitutively activates both canonical and noncanonical NF-κB pathways. vFLIP utilizes its TRAF-interacting domain to interact with the IκB kinase complex via TRAF2.[188] IκB sequesters inactive NF-κB in the cytoplasm.[20,24] Activation by cytokines induces recruitment of the IκB kinase complex, composed of IKKα, IKKβ, and IKKγ, which phosphorylates and inactivates IκB.[20,24] This IκB inactivation allows NF-κB to translocate to the nucleus and stimulate the expression of NF-κB response genes. In PEL cells, vFLIP interacts with the IκB kinase complex and RIP, a protein kinase that recruits and activates MAPKKK proteins that in turn activate the IκB kinase complex.[145,270] Hsp90 is an important member of this protein complex. Treatment of PEL cells with Hsp90 inhibitor ablates NF-κB activation.[145]

The implications of vFLIP-induced constitutive activation of the NF-κB pathway are significant. NF-κB activates many mitogenic pathways. Constitutive activation of the NF-κB pathways has been observed in KSHV PEL cells.[187,270,284] RNAi-mediated knock-down of vFLIP reduced NF-κB activity by approximately 80% and sensitized PEL cells to external apoptotic signals, subsequently increasing apoptosis levels.[187] The transforming ability of vFLIP derives from activation of the NF-κB pathway. Single gene expression studies of vFLIP in a Rat-1 cell line demonstrate that stable vFLIP expression increases cell growth and proliferation, but also facilitates loss of contact inhibition, anchorage independent growth in soft agar assays and induction of tumors in nude mice.[429] A20 cells, a murine B-lymphoma cell line, transduced with vFLIP underwent cell proliferation despite constant stimulation with death signals.[128] Ninety percent of immunocompetent mice injected with vFLIP-transduced A20 cells developed aggressive tumors. As mentioned above, IL-6 is a necessary growth factor for KSHV PEL cells. vFLIP upregulates the expression of cellular IL-6 by activating the JNK/AP-1 pathway.[14] *In situ* hybridization analysis shows that vFLIP transcripts express at low levels in early KS lesions, but at dramatically higher rates in late stage KS tumors.[427] vFLIP expression is inversely correlated with apoptosis.[426]

K12/Kaposin

Along with the proteins of the latency locus, K12 is expressed in most KSHV-infected spindle tumor cells.[235] Moreover, K12 expression is found in all stages of KSHV tumorigenesis. In advanced spindle cell tumors, K12 expression is detected in approximately 85% of the spindle cells.[257,311] K12 expression during latency varies from tumor to tumor, and cell line to cell line;[257] however, its expression is strongly upregulated in the viral lytic cycle.[87,413] Initial characterization of the K12 transcript identified a 700 bp gene product that coded for a protein with a predicted molecular weight of 6 kDa, now known as Kaposin.[368] Later, detailed transcript analysis determined that the most common mRNA transcript encoded by the K12 locus is actually 1.5–2.3 kb in size.[235,368] Consequently, the translation program for K12 was determined to be more complex than originally thought, utilizing canonical and alternative translation initiation codons to encode three proteins instead of one.[368] These proteins were assigned the names, Kaposin A, Kaposin B, and Kaposin C. While functional characterization of Kaposin A and Kaposin B has been carried out, to date no function has been ascribed to Kaposin C.

Most of the initial characterization on K12 was done with the predominant protein product, Kaposin A.[235] Kaposin A has a predicted molecular weight of 6 kDa; however, it is often detected on Western blots as 16–18 kDa bands and sometimes higher,[235,312] hinting that the protein undergoes extensive posttranslational modification.[312] Kaposin A is a highly hydrophobic type II transmembrane protein and is expressed on the cell surface.[235,311] Initial characterization revealed the transformation nature of Kaposin A. In NIH3T3 and Rat-3 cell lines, overexpression of Kaposin A caused increased foci formation, loss of stress fibers, and loss of contact inhibition.[235,311] When focally transformed Rat-3 cell lines were injected into nude mice; highly vascular and undifferentiated fibrosarcomas formed within 1–2 weeks.[311] Though the level of transforming activity found in the Rat-3 assay was less than what is observed in typical cellular and retroviral oncogenes, it was similar to levels found in other herpesviruses.[311] Kaposin A might mediate mitogenic pathways to achieve cellular transformation. Kaposin A activates the ERK-1/2 MAPK pathway and downstream AP-1[235] through its interaction with cytohesin-1, a guanine nucleotide exchange factor that regulates the function of β-integrins. Cells transformed by Kaposin A show elevated activities of cellular serine-threonine kinases such as PKC, CAM kinase I, and MLCK.[312] Kaposin A possesses a LXXLL nuclear receptor-binding motif. Mutations in this LXXLL motif ablated Kaposin A's transforming ability.[312]

Most cells that express Kaposin A also express Kaposin B. Kaposin B is a 38 kDa protein[442] coded by a sequence upstream of the Kaposin A translation initiation site. The Kaposin B transcript utilizes the alternative start codon, CUG, and encodes a series of 23 nucleotide GC rich repeats that vary in different KSHV isolates.[257,368] Kaposin B upregulates cytokine expression by blocking the degradation of their mRNA transcripts.[287] Kaposin B binds to and activates MK2 resulting in the inhibition of AU-rich elements (AREs)-mediated mRNA degradation.[287]

K15

K15 has a complex expression profile.[103] Isolates from BCBL-1 cells show differentially spliced protein products ranging from 281 to 489 aa in length, and 23–55 kDa in size that localize to the endoplasmic reticulum, mitochondria, and plasma membrane surface.[103,395] K15 can also undergo proteolytic cleavage.[395] K15 is weakly expressed during latency, but is strongly upregulated in viral lytic cycle.[476] K15 promoter is activated by RTA.[103,174,476] Although there is no sequence homology, K15 has the same structural characteristics as EBV LMP2A including a highly hydrophobic N-terminus, a variable number of transmembrane domains, and a C-terminus containing motifs with high sequence identity to putative SH2 and SH3 src kinase binding domains.[52,103,476] Unlike LMP2A, the cytoplasmic domain of K15 does not possess signal transduction capability even though the domain is readily phoshphorylated[103]; however, like LMP2A, it inhibits B-cell receptor (BCR) activation.[103] K15 has been shown to localize in the lipid rafts, common carriers of signal transducers like BCR.[52] The K15 C-terminus binds TRAF-1, TRAF-2, and TRAF-3.[52,174] K15 interaction with TRAF-2 strongly activates the NF-κB pathway and the AP-1 transcription factor through the JNK1 and ERK MAP kinase.[52] All K15 spliced variants have a common C-terminus and high sequence variation in the N-terminal hydrophobic regions.[52,103,476] Even though the C-terminus cannot function as a signal transducer, phosphorylation in its SH2 domain has been detected.[52]

Viral Interferon Regulatory Factors

KSHV encodes three homologues of human IFN regulatory factors (vIRFs) including vIRF, vIRF-2, and vIRF-3 or LANA-2.[61,165,300,362] Cellular IRFs are known to participate in IFN signal transduction. The diverse cellular effects of IFN, including tumor suppression through induction of p21 and histone H4 gene, induction of apoptosis, upregulation of MHC class I and II antigen expression, and regulation of the natural killer cell development, are mediated directly by IRFs or indirectly through interaction with various cellular factors including STAT proteins, p300/CBP, and NF-κB.[203,278,307] The most common IRF-targeting promoter sites are ISRE, GAS, PRD-I, PRD-II, PRD-III, PRD-IV, and NF-κB-binding sites. Some IRFs positively regulate the transcription of IFN-responsive genes[53,196,344,369] while others negatively regulate IFN signaling.[470,471] For example, IRF-1 and IRF-2 compete for the identical ISRE, accounting for their antagonistic activities.[196] IRF-1 positively reinforces IFN-β signal transduction, whereas IRF-2 is a negative regulator. Furthermore, IRF-1 is a tumor suppressor while IRF-2 is an oncogene.[196,436,437] IRF-1 overexpression promotes cell cycle arrest through p53-independent pathways by direct induction of p21.[435,436] IRF-2 overexpression inhibits these IFN-mediated effects leading to cellular transformation and tumor formation in nude mice.[197] vIRF is the first KSHV oncogene identified.[165] Overexpression of vIRF in rodent fibroblast cells causes cellular transformation and downregulation of p21 and MHC class I antigen through the inhibition of cellular IFN signal transduction pathways.[148,165,259,503] In this regard, vIRF behaves similarly to IRF-2. vIRF has also been shown to be a transcriptional regulator and regulates the expression of other KSHV genes.[259,364] vIRF probably achieves these functions through interactions with cellular IRF-1,

IRF-3, ICSBP, p300, and CBP.[60,165,258,267,319,389] It appears that vIRF represses the IFN antiviral response by blocking IRF-3 recruitment of the CBP/p300 coactivators.[267] Furthermore, vIRF interacts with p53, and inhibits p53-dependent apoptosis.[317,391] vIRF also autoactivates its own expression through two *cis*-elements, named *Vac1* and *Vac2*, in the vIRF promoter that do not contain any ISRE-like sequences and are unresponsive to induction with IFN-α/β.[465] Thus, vIRF might be able to target other viral and cellular genes in an IFN-independent mechanism. vIRF is an early lytic gene expressed in KS lesions.[127,467] Although the expression of vIRF is significantly increased by TPA, low levels of vIRF expression are also detected in uninduced KSHV-infected cell lines which may be either due to the presence of a minor latent transcript or an undefined expression mechanism of regulation that is not tied to viral lytic replication.[96,467] The control of vIRF expression during latency is mostly mediated by a transcriptional silencer *Tis* in the vIRF promoter.[466] Similar mechanisms of regulation may also be present for other KSHV lytic genes in addition to LANA-1-mediated gene repression.

vIRF-2 encodes a protein of 18 kDa. Similar to vIRF, vIRF-2 does not bind to ISRE, rather it binds to the NF-κB-binding site.[61] vIRF-2 interacts with IRF-1, IRF-2, ICSBP, RelA, and p300 but not IRF-3 *in vitro*. Consequently, vIRF-2 inhibits the IRF-1- or IRF-3-mediated transcriptional activation of IFN-α promoter as well as RelA-stimulation of HIV LTR.[61] vIRF-2 inhibits the antiviral effect of IFN by interacting with (double-stranded RNA-activated protein kinase) PKR and inhibiting its autophosphorylation, and the phosphorylation of PKR substrates histone 2A and eukaryotic translation initiation factor 2α.[59] Interestingly, vIRF-2 is expressed at low levels in BCBL-1 cells and is not induced by TPA, suggesting that it may be a latent gene.[59] The expression profile of vIRF-2 in KS lesions is unknown.

LANA-2 is abundantly expressed in the nuclei of PEL cells and has been described as a latent gene. However, because its expression levels can be greatly increased following TPA treatment, and is not expressed in KS tumors, whether it is truly latent remains controversial.[274,362] LANA-2 interacts with p53 to mediate gene transcription, and recruits IRF3 and IRF7 to inhibit IFN signaling.[274,362]

vGPCR

vGPCR is a 7-transmembrane, IL-8 receptor homologue that constitutively signals several pathways downstream of multiple G protein subunits in a phospholipase C- and phosphatidylinositol 3-kinase (PI-3K)-dependent manner.[65] Signaling through PKC, PKB, Akt, NF-κB, and MAPK pathways increases expression of genes involved in cellular proliferation, cell survival, and transformation, rendering vGPCR as one of the potential key factors involved in KSHV-induced malignancies.[18,22,114,183,298,299,310,339,386,404,406,407] vGPCR increases expression of a number of autocrine and paracrine cytokines and growth factors such as IL-1β, IL-6, IL-8, GM-CSF, TNF-α, bFGF, Gro-α, VEGF, and MCP-1, which might be essential for early KS cell proliferation and KS tumor angiogenesis and inflammation.[22,299,339,386,396,407] Thus, vGPCR may contribute to KS development via an autocrine and paracrine mechanism albeit it is a viral lytic gene. Expression of vGPCR in endothelial cells leads to constitutive activation of VEGF receptor and cell immortalization,[23,183,406] and transgenic mice expressing vGPCR developed highly vascular "KS-like" lesions.[218,486] Interestingly, vGPCR may also regulate the expression of both latent and lytic genes.[64,421]

vIL-6

vIL-6 is a viral early gene. As a homologue of cellular IL-6, vIL-6 can stimulate multiple cellular pathways to promote cellular proliferation, cell survival, and extrahepatic acute-phase response through engagement of the gp130 coreceptor independently of IL-6 receptor gp80.[104,202,236,297,300,308,324,332,453] vIL-6 also plays an important role in KSHV immune modulation, protecting PEL cells from IFN-α-induced antiviral response.[92] In addition, vIL-6 induces the secretion of cellular IL-6 and VEGF to support cell growth of IL-6-dependent cells, and contribute to hematopoiesis, tumorigenesis, and angiogenesis.[15,153,159,269,305]

K1 (KIS)

K1, the first ORF in the KSHV genome, encodes a type 1 membrane glycoprotein that is significantly induced during viral lytic replication.[254,448] K1 can activate B cells and induce inflammatory cytokines to regulate immune responses.[254] K1 promotes cell growth and causes immortalization of primary endothelial cells.[255,460] Expression of K1 in rodent fibroblasts produces morphological changes and foci formation, characteristics of cellular transformation.[255] K1 transgenic mice develop sarcomas and plasmablastic lymphomas and display constitutive activation of NF-κB, Oct-2, and Lyn.[349] In addition, it has been shown that K1 can induce matrix metalloproteinase-9 and VEGF in endothelial cells,[457,461] two factors implicated in angiogenesis and cell invasion.

K7/vIAP

vIAP is a homolog of the cellular protein survivin-ΔEx3 and is not found in any other herpesviruses. Both the viral and cellular survivin-ΔEx3 contain a conserved BH2 domain and a partial IAP repeat domain.[456] vIAP is a 19–21 kDa glycoprotein that localizes to the host cell mitochondrial membrane and inhibits apoptosis induced by the Fas and TRAIL pathways, Bax, TNF-α as well as cycloheximide, staurosporine, and ceramide.[143,456,490] vIAP protects KSHV-infected cells from apoptotic death in two distinct ways. First, it enhances cytosolic Ca^{2+} influx and prevents mitochondrial damage.[143] Second, it acts as protein bridge between the host mitochondrial protein Bcl-2 and caspase 3 to effectively suppress caspase 3 function.[456]

MIR-1/K3 and MIR-2/K5

Two similar proteins, MIR-1 and MIR-2, have been shown to actively prevent the expression of cell surface MHC Class I antigen that serve to alert the cytolytic T-lymphocytes (CTLs), the main protagonists of virus elimination.[110,111,214] Most virus-specific CTLs are CD8[+] T cells that recognize cytosolic, usually endogenously synthesized, viral antigens in association with Class I MHC on any nucleated cell. MIR-1 and MIR-2, expressed during lytic replication and localized in the endoplasmic reticulum (ER), negatively modulate the CTLs-dependent immune response by triggering endocytosis of the surface MHC Class I molecules for degradation.[110,214] This function of MIR-2 is specific for human leukocyte antigen (HLA)-A and HLA-B, which are responsible for presenting antigens to CTLs, and to a lesser extent, HLA-E, but not HLA-C.[110,214] In contrast to MIR-2, MIR-1 downregulates HLA-A, -B, -C, and -E.[214] Consequently, KSHV-infected cells expressing reduced levels of MHC Class I molecules, are deficient in antigen processing and presentation, and are therefore less susceptible to CTL killing.[49] In addition, MIR-2 also reduces the expression of molecules involved in T-cell activation, reducing the visibility of KSHV-infected B cell to T-helper cells, thereby decreasing T-cell responses, cytokine release and the production of co-stimulatory signals for CTL generation.[111] Both MIR-1 and MIR-2 may also help KSHV evade natural killer cell immune responses, which cause immediate killing of virus-infected cells through exocytosis of perforin- and granzyme-containing granules and secreting cytokines, mainly IFN-γ, to activate CTLs and macrophages.[213,375]

vCCL-1, vCCL-2, and vCCL-3

vCCLs are three KSHV lytic proteins formerly known as vMIPs. vCCL-1 (K6), vCCL-2 (K4), and vCCL-3 (K4.1) share 25–40% homology with human MIP-1α (macrophage inflammatory protein1α) a βCC chemokine.[300,324] Chemokines are molecules that interact with G-protein coupled chemokine receptors and recruit leukocytes to sites of inflammation, and also are involved in angiogenesis, haematopoiesis, and lymphocyte development.[4,102,271,313,314] All three vCCLs are able to induce angiogenesis in the chorioalantonic membranes of chicken eggs[41] suggesting that they may act synergistically with host factors such as VEGF, bFGF, and IL-6 to promote the abnormal angiogenesis in KS lesions. vCCL-1 is able to bind to a broad range of chemokine receptors including CCR1, CCR2, CCR5, CXCR4, and CX3CR1.[234] vCCL-1 may act as a competitive antagonist with the host chemokines, since it does not induce a Ca[2+] influx after binding these receptors. *In vitro*, vCCL-1 blocked the chemoattractive properties of host CC and CXC chemokines,[41,97,234,300] while in a rat model, vCCL-1 suppressed the host inflammatory response and decreased the infiltration of inflammatory leukocytes.[97] vCCL-1 and vCCL-2 bind to CCR-8, and vCCL-1 is able to induce the secretion of VEGF-A in PEL cells.[269] Both vCCL-1 and vCCL-2 are able to block dexamethasone-induced apoptosis in PEL cells.[269] vCCL-3 binds to CCR-8 and CCR-4, and has been shown to be selectively chemoattractant for Th2 cells.[425] KSHV vCCLs can polarize the adaptive immune response toward a predominantly Th2-type (i.e., humoral) response at sites of KSHV infection, potentially reducing the efficacy of the antiviral response.

KCP (ORF4)

Complement bridges both innate and adaptive immune responses and humoral and cell-mediated immunity. The ORF4-encoded complement-control protein (KCP) inhibits the activation of the complement cascade.[309,417–419] KCP is also associated with the envelope of purified KSHV virions where it potentially protects them from complement-mediated immune response.[418]

K14 (vOX2)

vOX2 is a homolog of the cellular protein OX2, a glycosylated cell surface protein and member of the immunoglobulin superfamily that restricts cytokine production in a paracrine fashion. vOX2 has structure similar to cellular OX2; however, in contrast to cellular OX2, it potently activates inflammatory cytokine production.[105] vOX2 induces the production of IL-1β, TNF-α, and IL-6 from various cell types, including monocytes/macrophages, and dendritic cells, and cooperates with IFN-γ in a paracrine manner to induce cytokine production in a B-cell line.[105]

vBcl-2 (ORF16)

vBcl-2 is a viral lytic protein expressed in both spindle cells and monocytes in KS lesions.[378,475] vBcl-2 only displays 15–20% amino acid identity to cellular Bcl-2, but it contains the critical BH1 and BH2 domains required to heterodimerize with cellular Bcl-2, and potently suppress caspase 3 death effector functions.[98,378] vBcl-2 is an even more potent suppressor of apoptosis than its cellular homolog, Bcl-2. vBcl-2 has been shown to prevent apoptosis induced by vCyclin, and unlike cellular Bcl-2, vBcl-2 cannot be converted to a pro-apoptotic form by caspase-mediated cleavage.[33,98]

SUMMARY

KSHV has been established as the causative agent of KS, PEL, and MCD, malignancies occurring more frequently in AIDS patients. The aggressive nature of KSHV in the context of HIV infection suggests that interactions between the two viruses enhance pathogenesis. KSHV latent infection and lytic reactivation are characterized by distinct gene expression profiles, and both latency and lytic reactivation seem to be required for malignant progression. As a sophisticated oncogenic virus, KSHV has evolved to possess a formidable repertoire of potent mechanisms that enable it to target and manipulate host cell pathways, leading to increased cell proliferation, increased cell survival, dysregulated angiogenesis, evasion of immunity, and malignant progression in the immunocompromised host.

Worldwide, approximately 40.3 million people are currently living with HIV infection.[2] Of these, a significant number are coinfected with KSHV.[118,129,381] The complex interplay between the two viruses dramatically elevates the risk for development of KSHV-induced malignancies, KS, PEL, and MCD. Although HAART significantly reduces HIV viral load, the entire T-cell repertoire and immune function may not be completely restored.[99] In fact, clinically significant immune deficiency is not necessary for the induction of KSHV-related malignancy.[81] Because of variables such as lack of access to therapy, noncompliance with prescribed treatment, failure to respond to treatment and the development of drug-resistant strains of HIV, KSHV-induced malignancies will continue to present as major health concerns.

REFERENCES

1. Leads from the MMWR. 1986. Classification system for human T-lymphotropic virus type III/lymphadenopathy-associated virus infections. JAMA **256**:20–1, 24–5.
2. 2006. UNAIDS Reference Group on estimates, modelling and projections – statement on the use of the BED assay for the estimation of HIV-1 incidence for surveillance or epidemic monitoring. Wkly Epidemiol Rec **81**:40.
3. Ablashi, D.V., L. G. Chatlynne, J. E. Whitman, Jr., and E. Cesarman. 1997. Spectrum of Kaposi's sarcoma-associated herpesvirus, or human herpesvirus 8, diseases. Clin Microbiol Rev **15**:439–64.
4. Adams, D. H., and A. R. Lloyd. 1997. Chemokines: leukocyte recruitment and activation cytokines. Lancet **349**:490–5.
5. Akula, S. M., P. P. Naranatt, N. S. Walia, F. Z. Wang, B. Fegley, and B. Chandran. 2003. Kaposi's sarcoma-associated herpesvirus (human herpesvirus 8) infection of human fibroblast cells occurs through endocytosis. J Virol **77**:7978–90.
6. Akula, S. M., N. P. Pramod, F. Z. Wang, and B. Chandran. 2001. Human herpesvirus 8 envelope-associated glycoprotein B interacts with heparan sulfate-like moieties. Virology **284**:235–49.
7. Akula, S. M., N. P. Pramod, F. Z. Wang, and B. Chandran. 2002. Integrin alpha3beta1 (CD 49c/29) is a cellular receptor for Kaposi's sarcoma-associated herpesvirus (KSHV/HHV-8) entry into the target cells. Cell **108**:407–19.
8. Akula, S. M., F. Z. Wang, J. Vieira, and B. Chandran. 2001. Human herpesvirus 8 interaction with target cells involves heparan sulfate. Virology **282**:245–55.
9. Albini, A., I. Paglieri, G. Orengo, S. Carlone, M. G. Aluigi, R. DeMarchi, C. Matteucci, A. Mantovani, F. Carozzi, S. Donini, and R. Benelli. 1997. The beta-core fragment of human chorionic gonadotrophin inhibits growth of Kaposi's sarcoma-derived cells and a new immortalized Kaposi's sarcoma cell line. AIDS **11**:713–21.
10. Alexander, L., L. Denekamp, A. Knapp, M. R. Auerbach, B. Damania, and R. C. Desrosiers. 2000. The primary sequence of rhesus monkey rhadinovirus isolate 26–95: sequence similarities to Kaposi's sarcoma-associated herpesvirus and rhesus monkey rhadinovirus isolate 17577. J Virol **74**:3388–98.
11. Amin, H. M., L. J. Medeiros, J. T. Manning, and D. Jones. 2003. Dissolution of the lymphoid follicle is a feature of the HHV8+ variant of plasma cell Castleman's disease. Am J Surg Pathol **27**:91–100.
12. An, J., A. K. Lichtenstein, G. Brent, and M. B. Rettig. 2002. The Kaposi's sarcoma-associated herpesvirus (KSHV) induces cellular interleukin 6 expression: role of the KSHV latency-associated nuclear antigen and the AP1 response element. Blood **99**:649–54.
13. An, J., Y. Sun, and M. B. Rettig. 2004. Transcriptional coactivation of c-Jun by the KSHV-encoded LANA. Blood **103**:222–8.
14. An, J., Y. Sun, R. Sun, and M. B. Rettig. 2003. Kaposi's sarcoma-associated herpesvirus encoded vFLIP induces cellular IL-6 expression: the role of the NF-kappaB and JNK/AP1 pathways. Oncogene **22**:3371–85.
15. Aoki, Y., E. S. Jaffe, Y. Chang, K. Jones, J. Teruya-Feldstein, P. S. Moore, and G. Tosato. 1999. Angiogenesis and hematopoiesis induced by Kaposi's sarcoma-associated herpesvirus-encoded interleukin-6. Blood **93**:4034–43.
16. Aoki, Y., and G. Tosato. 2004. HIV-1 Tat enhances Kaposi's sarcoma-associated herpesvirus (KSHV) infectivity. Blood **104**:810–4.
17. Aoki, Y., and G. Tosato. 1999. Role of vascular endothelial growth factor/vascular permeability factor in the pathogenesis of Kaposi's sarcoma-associated herpesvirus-infected primary effusion lymphomas. Blood **94**:4247–54.
18. Arvanitakis, L., E. Geras-Raaka, A. Varma, M. C. Gershengorn, and E. Cesarman. 1997. Human herpesvirus KSHV encodes a constitutively active G-protein-coupled receptor linked to cell proliferation. Nature **385**:347–50.
19. AuCoin, D. P., K. S. Colletti, S. A. Cei, I. Papouskova, M. Tarrant, and G. S. Pari. 2004. Amplification of the Kaposi's sarcoma-associated herpesvirus/human herpesvirus 8 lytic origin of DNA replication is

dependent upon a cis-acting AT-rich region and an ORF50 response element and the trans-acting factors ORF50 (K-Rta) and K8 (K-bZIP). Virology 318:542–55.

20. Baeuerle, P. A., and D. Baltimore. 1996. NF-kappa B: ten years after. Cell 87:13–20.

21. Bailer, R. T., C. L. Ng-Bautista, G. M. Ness, and S. R. Mallery. 1997. Expression of interleukin-6 receptors and NF-kappa B in AIDS-related Kaposi's sarcoma cell strains. Lymphology 30:63–76.

22. Bais, C., B. Santomasso, O. Coso, L. Arvanitakis, E. G. Raaka, J. S. Gutkind, A. S. Asch, E. Cesarman, M. C. Gershengorn, E. A. and Mesri. 1998. G-protein-coupled receptor of Kaposi's sarcoma-associated herpesvirus is a viral oncogene and angiogenesis activator. Nature 391:86–9.

23. Bais, C., A. Van Geelen, P. Eroles, A. Mutlu, C. Chiozzini, S. Dias, R. L. Silverstein, S. Rafii, and E. A. Mesri. 2003. Kaposi's sarcoma associated herpesvirus G protein-coupled receptor immortalizes human endothelial cells by activation of the VEGF receptor-2/KDR. Cancer Cell 3:131–43.

24. Baldwin, A. S. Jr. 1996. The NF-kappa B and I kappa B proteins: new discoveries and insights. Annu Rev Immunol 14:649–83.

25. Ballestas, M. E., P. A. Chatis, and K. M. Kaye. 1999. Efficient persistence of extrachromosomal KSHV DNA mediated by latency-associated nuclear antigen. Science 284:641–4.

26. Barbera, A. J., M. E. Ballestas, and K. M. Kaye. 2004. The Kaposi's sarcoma-associated herpesvirus latency-associated nuclear antigen 1 N terminus is essential for chromosome association, DNA replication, and episome persistence. J Virol 78:294–301.

27. Barbera, A. J., J. V. Chodaparambil, B. Kelley-Clarke, V. Joukov, J. C. Walter, K. Luger, and K. M. Kaye. 2006. The nucleosomal surface as a docking station for Kaposi's sarcoma herpesvirus LANA. Science 311:856–61.

28. Barker, N., P. J. Morin, and H. Clevers. 2000. The Yin-Yang of TCF/beta-catenin signaling. Adv Cancer Res 77:1–24.

29. Bartel, D. P. 2004. MicroRNAs: genomics, biogenesis, mechanism, and function. Cell 116:281–97.

30. Baylin, S. B. 2005. DNA methylation and gene silencing in cancer. Nat Clin Pract Oncol 2(Suppl 1):S4–11.

31. Bechtel, J. T., Y. Liang, J. Hvidding, and D. Ganem. 2003. Host range of Kaposi's sarcoma-associated herpesvirus in cultured cells. J Virol 77:6474–81.

32. Belanger, C., A. Gravel, A. Tomoiu, M. E. Janelle, J. Gosselin, M. J. Tremblay, and L. Flamand. 2001. Human herpesvirus 8 viral FLICE-inhibitory protein inhibits Fas-mediated apoptosis through binding and prevention of procaspase-8 maturation. J Hum Virol 4:62–73.

33. Bellows, D. S., B. N. Chau, P. Lee, Y. Lazebnik, W. H. Burns, and J. M. Hardwick. 2000. Antiapoptotic herpesvirus Bcl-2 homologs escape caspase-mediated conversion to proapoptotic proteins. J Virol 74:5024–31.

34. Benelli, R., L. Repetto, S. Carlone, C. Parravicini, and A. Albini. 1994. Establishment and characterization of two new Kaposi's sarcoma cell cultures from an AIDS and a non-AIDS patient. Res Virol 145:251–9.

35. Beral, V. 1991. Epidemiology of Kaposi's sarcoma. Cancer Surv 10:5–22.

36. Beral, V., T. A. Peterman, R. L. Berkelman, and H. W. Jaffe. 1990. Kaposi's sarcoma among persons with AIDS: a sexually transmitted infection? Lancet 335:123–8.

37. Bieleski, L., and S. J. Talbot. 2001. Kaposi's sarcoma-associated herpesvirus vCyclin open reading frame contains an internal ribosome entry site. J Virol 75:1864–9.

38. Bigoni, B., R. Dolcetti, L. de Lellis, A. Carbone, M. Boiocchi, E. Cassai, and D. Di Luca. 1996. Human herpesvirus 8 is present in the lymphoid system of healthy persons and can reactivate in the course of AIDS. J Infect Dis 173:542–9.

39. Birkmann, A., K. Mahr, A. Ensser, S. Yaguboglu, F. Titgemeyer, B. Fleckenstein, and F. Neipel. 2001. Cell surface heparan sulfate is a receptor for human herpesvirus 8 and interacts with envelope glycoprotein K8.1. J Virol 75:11583–93 (2001).

40. Borkovic, S. P., and R. A. Schwartz. 1981. Kaposi's sarcoma presenting in the homosexual man – a new and striking phenomenon! Ariz Med 38:902–4.

41. Boshoff, C., Y. Endo, P. D. Collins, Y. Takeuchi, J. D. Reeves, V. L. Schweickart, M. A. Siani, T. Sasaki, T. J. Williams, P. W. Gray, P. S. Moore, Y. Chang, and R. A. Weiss. 1997. Angiogenic and HIV-inhibitory functions of KSHV-encoded chemokines. Science 278:290–4.

42. Boshoff, C., S. J. Gao, L. E. Healy, S. Matthews, A. J. Thomas, L. Coignet, R. A. Warnke, J. A. Strauchen, E. Matutes, O. W. Kamel, P. S. Moore, R. A. Weiss, and Y. Chang. 1998. Establishing a KSHV+ cell line (BCP-1) from peripheral blood and characterizing its growth in Nod/SCID mice. Blood 91:1671–9.

43. Boshoff, C., T. F. Schulz, M. M. Kennedy, A. K. Graham, C. Fisher, A. Thomas, J. O. McGee, R. A. Weiss, and J. J. O'Leary. 1995. Kaposi's sarcoma-associated herpesvirus infects endothelial and spindle cells. Nat Med 1:1274–8.

44. Boshoff, C., D. Whitby, T. Hatziioannou, C. Fisher, J. van der Walt, A. Hatzakis, R. Weiss, and T. Schulz. 1995. Kaposi's-sarcoma-associated herpesvirus in HIV-negative Kaposi's sarcoma. Lancet **345**:1043–4.

45. Boulanger, E. 1999. [Human herpesvirus 8 (HHV8). II. Pathogenic role and sensitivity to antiviral drugs]. Ann Biol Clin (Paris) **57**:19–28.

46. Bower, M., P. Fox, K. Fife, J. Gill, M. Nelson, and B. Gazzard. 1999. Highly active anti-retroviral therapy (HAART) prolongs time to treatment failure in Kaposi's sarcoma. AIDS **13**:2105–11.

47. Bowser, B. S., S. M. DeWire, and B. Damania. 2002. Transcriptional regulation of the K1 gene product of Kaposi's sarcoma-associated herpesvirus. J Virol **76**:12574–83.

48. Bowser, B. S., S. Morris, M. J. Song, R. Sun, and B. Damania. 2006. Characterization of Kaposi's sarcoma-associated herpesvirus (KSHV) K1 promoter activation by Rta. Virology **348**:309–27.

49. Brander, C., T. Suscovich, Y. Lee, P. T. Nguyen, P. O'Connor, J. Seebach, N. G. Jones, M. van Gorder, B. D. Walker, and D. T. Scadden. 2000. Impaired CTL recognition of cells latently infected with Kaposi's sarcoma-associated herpesvirus. J Immunol **165**:2077–83.

50. Brantjes, H., N. Barker, J. van Es, and H. Clevers. 2002. TCF: Lady Justice casting the final verdict on the outcome of Wnt signaling. Biol Chem **383**:255–61.

51. Bridges, C. C., H. Hu, S. Miyauchi, U. N. Siddaramappa, M. E. Ganapathy, L. Ignatowicz, D. M. Maddox, S. B. Smith, and V. Ganapathy. 2004. Induction of cystine-glutamate transporter xc-by human immunodeficiency virus type 1 transactivator protein tat in retinal pigment epithelium. Invest Ophthalmol Vis Sci **45**:2906–14.

52. Brinkmann, M. M., M. Glenn, L. Rainbow, A. Kieser, C. Henke-Gendo, and T. F. Schulz. 2003. Activation of mitogen-activated protein kinase and NF-kappaB pathways by a Kaposi's sarcoma-associated herpesvirus K15 membrane protein. J Virol **77**:9346–58.

53. Briscoe, J., N. C. Rogers, B. A. Witthuhn, D. Watling, A. G. Harpur, A. F. Wilks, G. R. Stark, J. N. Ihle, and I. M. Kerr. 1996. Kinase-negative mutants of JAK1 can sustain interferon-gamma-inducible gene expression but not an antiviral state. EMBO J **15**:799–809.

54. Brooks, L. A., A. J. Wilson, and T. Crook. 1997. Kaposi's sarcoma-associated herpesvirus (KSHV)/human herpesvirus 8 (HHV8)–a new human tumor virus. J Pathol **182**:262–5.

55. Brown, E. E., D. Whitby, F. Vitale, P. C. Fei, C. Del Carpio, V. Marshall, A. J. Alberg, D. Serraino, A. Messina, L. Gafa, and J. J. Goedert. 2005. Correlates of Human Herpesvirus-8 DNA detection among adults in Italy without Kaposi's sarcoma. Int J Epidemiol **34**:1110–7.

56. Brown, H. J., M. J. Song, H. Deng, T. T. Wu, G. Cheng, and R. Sun. 2003. NF-kappaB inhibits gammaherpesvirus lytic replication. J Virol **77**:8532–40.

57. Brown, L. F., B. J. Dezube, K. Tognazzi, H. F. Dvorak, and G. D. Yancopoulos. 2000. Expression of Tie1, Tie2, and angiopoietins 1, 2, and 4 in Kaposi's sarcoma and cutaneous angiosarcoma. Am J Pathol **156**:2179–83.

58. Browning, P. J., J. M. Sechler, M. Kaplan, R. H. Washington, R. Gendelman, R. Yarchoan, B. Ensoli, and R. C. Gallo. 1994. Identification and culture of Kaposi's sarcoma-like spindle cells from the peripheral blood of human immunodeficiency virus-1-infected individuals and normal controls. Blood **84**:2711–20.

59. Burysek, L., and P. M. Pitha. 2001. Latently expressed human herpesvirus 8-encoded interferon regulatory factor 2 inhibits double-stranded RNA-activated protein kinase. J Virol **75**:2345–52.

60. Burysek, L., W. S. Yeow, B. Lubyova, M. Kellum, S. L. Schafer, Y. Q. Huang, and P. M. Pitha. 1999. Functional analysis of human herpesvirus 8-encoded viral interferon regulatory factor 1 and its association with cellular interferon regulatory factors and p300. J Virol **73**:7334–42.

61. Burysek, L., W. S. Yeow, and P. M. Pitha. 1999. Unique properties of a second human herpesvirus 8-encoded interferon regulatory factor (vIRF-2). J Hum Virol **2**:19–32.

62. Cai, X., and B. R. Cullen. 2006. Transcriptional origin of Kaposi's sarcoma-associated herpesvirus microRNAs. J Virol **80**:2234–42.

63. Cai, X., S. Lu, Z. Zhang, C. M. Gonzalez, B. Damania, and B. R. Cullen. 2005. Kaposi's sarcoma-associated herpesvirus expresses an array of viral microRNAs in latently infected cells. Proc Natl Acad Sci USA **102**:5570–5.

64. Cannon, M., E. Cesarman, and C. Boshoff. 2006. KSHV G protein-coupled receptor inhibits lytic gene transcription in primary-effusion lymphoma cells via p21-mediated inhibition of Cdk2. Blood **107**:277–84.

65. Cannon, M. L., and E. Cesarman. 2004. The KSHV G protein-coupled receptor signals via multiple pathways to induce transcription factor activation in primary effusion lymphoma cells. Oncogene **23**:514–23.

66. Carbone, A., A. M. Cilia, A. Gloghini, V. Canzonieri, C. Pastore, M. Todesco, M. Cozzi, T. Perin, R. Volpe, A. Pinto, and G. Gaidano. 1997. Establishment of HHV-8-positive and HHV-8-negative lymphoma cell lines from primary lymphomatous effusions. Int J Cancer **73**: 562–9.

67. Carbone, A., A. M. Cilia, A. Gloghini, D. Capello, L. Fassone, T. Perin, D. Rossi, V. Canzonieri, P. De Paoli, E. Vaccher, U. Tirelli, R. Volpe, and G. Gaidano. 2000. Characterization of a novel HHV-8-positive cell line reveals implications for the pathogenesis and cell cycle control of primary effusion lymphoma. Leukemia **14**:1301–9.

68. Carbone, A., A. M. Cilia, A. Gloghini, D. Capello, M. Todesco, S. Quattrone, R. Volpe, and G. Gaidano. 1998. Establishment and characterization of EBV-positive and EBV-negative primary effusion lymphoma cell lines harboring human herpesvirus type-8. Br J Haematol **102**:1081–9.

69. Carbone, A., and G. Gaidano. 1997. HHV-8-positive body-cavity-based lymphoma: a novel lymphoma entity. Br J Haematol **97**:515–22.

70. Carmeliet, P. 2005. Angiogenesis in life, disease and medicine. Nature **438**:932–6.

71. Carroll, P. A., E. Brazeau, and M. Lagunoff. 2004. Kaposi's sarcoma-associated herpesvirus infection of blood endothelial cells induces lymphatic differentiation. Virology **328**:7–18.

72. Casalone, R., A. Albini, R. Righi, P. Granata, and A. Toniolo. 2001. Nonrandom chromosome changes in Kaposi's sarcoma: cytogenetic and FISH results in a new cell line (KS-IMM) and literature review. Cancer Genet Cytogenet **124**:16–9.

73. Caselli, E., M. Galvan, E. Cassai, A. Caruso, L. Sighinolfi, and D. Di Luca. 2005. Human herpesvirus 8 enhances human immunodeficiency virus replication in acutely infected cells and induces reactivation in latently infected cells. Blood **106**:2790–7.

74. Caselli, E., M. Galvan, E. Cassai, and D. Di Luca. 2003. Transient expression of human herpesvirus-8 (Kaposi's sarcoma-associated herpesvirus) ORF50 enhances HIV-1 replication. Intervirology **46**:141–9.

75. Caselli, E., M. Galvan, F. Santoni, A. Rotola, A. Caruso, E. Cassai, and D. D. Luca. Human herpesvirus-8 (Kaposi's sarcoma-associated virus) ORF50 increases in vitro cell susceptibility to human immunodeficiency virus type 1 infection. J Gen Virol **84**:1123–31.

76. Caselli, E., P. Menegazzi, A. Bracci, M. Galvan, E. Cassai, and D. Di Luca. 2001. Human herpesvirus-8 (Kaposi's sarcoma-associated herpesvirus) ORF50 interacts synergistically with the tat gene product in transactivating the human immunodeficiency virus type 1 LTR. J Gen Virol **82**:1965–70 (2001).

77. Cathomas, G. 2003. Kaposi's sarcoma-associated herpesvirus (KSHV)/human herpesvirus 8 (HHV-8) as a tumor virus. Herpes **10**:72–7.

78. Cerimele, F., F. Curreli, S. Ely, A. E. Friedman-Kien, E. Cesarman, and O. Flore. 2001. Kaposi's sarcoma-associated herpesvirus can productively infect primary human keratinocytes and alter their growth properties. J Virol **75**:2435–43.

79. Cesarman, E. 2002. The role of Kaposi's sarcoma-associated herpesvirus (KSHV/HHV-8) in lymphoproliferative diseases. Recent Results Cancer Res **159**:27–37.

80. Cesarman, E., Y. Chang, P. S. Moore, J. W. Said, and D. M. Knowles. 1995. Kaposi's sarcoma-associated herpesvirus-like DNA sequences in AIDS-related body-cavity-based lymphomas. N Engl J Med **332**:1186–91.

81. Chan, J., S. Kravcik, and J. B. Angel. 1999. Development of Kaposi's sarcoma despite sustained suppression of HIV plasma viremia. J Acquir Immune Defic Syndr **22**:209–10.

82. Chandran, B., C. Bloomer, S. R. Chan, L. Zhu, E. Goldstein, and R. Horvat. 1998. Human herpesvirus-8 ORF K8.1 gene encodes immunogenic glycoproteins generated by spliced transcripts. Virology **249**:140–9.

83. Chang, H., D. P. Dittmer, S. Y. Chul, Y. Hong, and J. U. Jung. 2005. Role of Notch signal transduction in Kaposi's sarcoma-associated herpesvirus gene expression. J Virol **79**:14371–82

84. Chang, H., Y. Gwack, D. Kingston, J. Souvlis, X. Liang, R. E. Means, E. Cesarman, L. Hutt-Fletcher, and J. U. Jung. 2005. Activation of CD21 and CD23 gene expression by Kaposi's sarcoma-associated herpesvirus RTA. J Virol **79**:4651–63.

85. Chang, H. K., R. Gendelman, J. Lisziewicz, R. C. Gallo, and B. Ensoli. 1994. Block of HIV-1 infection by a combination of antisense tat RNA and TAR decoys: a strategy for control of HIV-1. Gene Ther **1**:208–16.

86. Chang, J., R. Renne, D. Dittmer, and D. Ganem. 2000. Inflammatory cytokines and the reactivation of Kaposi's sarcoma-associated herpesvirus lytic replication. Virology **266**:17–25.

87. Chang, P. J., D. Shedd, L. Gradoville, M. S. Cho, L. W. Chen, J. Chang, and G. Miller. 2002. Open reading frame 50 protein of Kaposi's sarcoma-associated herpesvirus directly activates the viral PAN and K12 genes by binding to related response elements. J Virol **76**:3168–78.

88. Chang, P. J., D. Shedd, and G. Miller. 2005. Two subclasses of Kaposi's sarcoma-associated herpesvirus lytic cycle promoters distinguished by open reading frame 50 mutant proteins that are deficient in binding to DNA. J Virol **79**:8750–63.

89. Chang, Y., E. Cesarman, M. S. Pessin, F. Lee, J. Culpepper, D. M. Knowles, and P. S. Moore. 1994. Identification of herpesvirus-like DNA sequences in AIDS-associated Kaposi's sarcoma. Science **266**:1865–9.

90. Chang, Y., P. S. Moore, S. J. Talbot, C. H. Boshoff, T. Zarkowska, K. Godden, H. Paterson, R. A. Weiss, and S. Mittnacht. 1996. Cyclin encoded by KS herpesvirus. Nature **382**:410.

91. Chang, Y., J. Ziegler, H. Wabinga, E. Katangole-Mbidde, C. Boshoff, T. Schulz, D. Whitby, D. Maddalena, H. W. Jaffe, R. A. Weiss, and P. S. Moore. 1996. Kaposi's sarcoma-associated herpesvirus and Kaposi's sarcoma in Africa. Uganda Kaposi's Sarcoma Study Group. Arch Intern Med **156**:202–4.

92. Chatterjee, M., J. Osborne, G. Bestetti, Y. Chang, and P. S. Moore. 2002. Viral IL-6-induced cell proliferation and immune evasion of interferon activity. Science **298**:1432–5.

93. Chaudhary, P. M., A. Jasmin, M. T. Eby, and L. Hood. 1999. Modulation of the NF-kappa B pathway by virally encoded death effector domains-containing proteins. Oncogene **18**:5738–46.

94. Chehimi, J., Q. Luo, L. Azzoni, L. Shawver, N. Ngoubilly, R. June, G. Jerandi, M. Farabaugh, and L. J. Montaner. 2003. HIV-1 transmission and cytokine-induced expression of DC-SIGN in human monocyte-derived macrophages. J Leukoc Biol **74**:757–63.

95. Chen, J., K. Ueda, S. Sakakibara, T. Okuno, C. Parravicini, M. Corbellino, and K. Yamanishi. 2001. Activation of latent Kaposi's sarcoma-associated herpesvirus by demethylation of the promoter of the lytic transactivator. Proc Natl Acad Sci USA **98**:4119–24.

96. Chen, J., K. Ueda, S. Sakakibara, T. Okuno, and K. Yamanishi. 2000. Transcriptional regulation of the Kaposi's sarcoma-associated herpesvirus viral interferon regulatory factor gene. J Virol **74**:8623–34.

97. Chen, S., K. B. Bacon, L. Li, G. E. Garcia, Y. Xia, D. Lo, D. A. Thompson, M. A. Siani, T. Yamamoto, J. K. Harrison, and L. Feng. 1998. In vivo inhibition of CC and CX3C chemokine-induced leukocyte infiltration and attenuation of glomerulonephritis in Wistar-Kyoto (WKY) rats by vMIP-II. J Exp Med **188**:193–8.

98. Cheng, E. H., J. Nicholas, D. S. Bellows, G. S. Hayward, H. G. Guo, M. S. Reitz, and J. M. Hardwick. 1997. A Bcl-2 homolog encoded by Kaposi's sarcoma-associated virus, human herpesvirus 8, inhibits apoptosis but does not heterodimerize with Bax or Bak. Proc Natl Acad Sci USA **94**:690–4.

99. Cheung, T. W. 2004. AIDS-related cancer in the era of highly active antiretroviral therapy (HAART): a model of the interplay of the immune system, virus, and cancer. "On the offensive–the Trojan Horse is being destroyed"–Part A: Kaposi's sarcoma. Cancer Invest **22**:774–86.

100. Child, E. S., and D. J. Mann. 2001. Novel properties of the cyclin encoded by Human Herpesvirus 8 that facilitate exit from quiescence. Oncogene **20**:3311–22.

101. Chiou, C. J., L. J. Poole, P. S. Kim, D. M. Ciufo, J. S. Cannon, C. M. ap Rhys, D. J. Alcendor, J. C. Zong, R. F. Ambinder, and G. S. Hayward. Patterns of gene expression and a transactivation function exhibited by the vGCR (ORF74) chemokine receptor protein of Kaposi's sarcoma-associated herpesvirus. J Virol **76**:3421–39.

102. Choi, J., R. E. Means, B. Damania, and J. U. Jung. 2001. Molecular piracy of Kaposi's sarcoma associated herpesvirus. Cytokine Growth Factor Rev **12**:245–57.

103. Choi, J. K., B. S. Lee, S. N. Shim, M. Li, and J. U. Jung. 2000. Identification of the novel K15 gene at the rightmost end of the Kaposi's sarcoma-associated herpesvirus genome. J Virol **74**:436–46.

104. Chow, D., X. He, A. L. Snow, S. Rose-John, and K. C. Garcia. 2001. Structure of an extracellular gp130 cytokine receptor signaling complex. Science **291**:2150–5.

105. Chung, Y. H., R. E. Means, J. K. Choi, B. S. Lee, and J. U. Jung. 2002. Kaposi's sarcoma-associated herpesvirus OX2 glycoprotein activates myeloid-lineage cells to induce inflammatory cytokine production. J Virol **76**:4688–98.

106. Ciufo, D. M., J. S. Cannon, L. J. Poole, F. Y. Wu, P. Murray, R. F. Ambinder, and G. S. Hayward. 2001. Spindle cell conversion by Kaposi's sarcoma-associated herpesvirus: formation of colonies and plaques with mixed lytic and latent gene expression in infected primary dermal microvascular endothelial cell cultures. J Virol **75**:5614–26.

107. Cohen, A., C. Brodie, and R. Sarid. 2006. An essential role of ERK signaling in TPA-induced reactivation of Kaposi's sarcoma-associated herpesvirus. J Gen Virol **87**:795–802.

108. Cool, C. D., P. R. Rai, M. E. Yeager, D. Hernandez-Saavedra, A. E. Serls, T. M. Bull, M. W. Geraci, K. K. Brown, J. M. Routes, R. M. Tuder, and N. F. Voelkel. 2003. Expression of human herpesvirus 8 in primary pulmonary hypertension. N Engl J Med **349**:1113–22.

109. Corbellino, M., L. Poirel, J. T. Aubin, M. Paulli, U. Magrini, G. Bestetti, M. Galli, and C. Parravicini. 1996. The role of human herpesvirus 8 and Epstein-Barr virus in the pathogenesis of giant lymph node hyperplasia (Castleman's disease). Clin Infect Dis **22**:1120–1.

110. Coscoy, L., and D. Ganem. 2000. Kaposi's sarcoma-associated herpesvirus encodes two proteins that block cell surface display of MHC class I chains by enhancing their endocytosis. Proc Natl Acad Sci USA **97**:8051–6.

111. Coscoy, L., and D. Ganem. 2001. A viral protein that selectively downregulates ICAM-1 and B7–2 and modulates T cell costimulation. J Clin Invest **107**:1599–606.

112. Cotter, M. A., 2nd, and E. S. Robertson. 1999. The latency-associated nuclear antigen tethers the Kaposi's sarcoma-associated herpesvirus genome to host chromosomes in body cavity-based lymphoma cells. Virology **264**:254–64.

113. Cotter, M. A., 2nd, and E. S. Robertson. 2002. Molecular biology of Kaposi's sarcoma-associated herpesvirus. Front Biosci **7**:d358–75.

114. Couty, J. P., E. Geras-Raaka, B. B. Weksler, and M. C. Gershengorn. 2001. Kaposi's sarcoma-associated herpesvirus G protein-coupled receptor signals through multiple pathways in endothelial cells. J Biol Chem **276**:33805–11.

115. Dagna, L., F. Broccolo, C. T. Paties, M. Ferrarini, L. Sarmati, L. Praderio, M. G. Sabbadini, P. Lusso, and M. S. Malnati. 2005. A relapsing inflammatory syndrome and active human herpesvirus 8 infection. N Engl J Med **353**: 156–63.

116. Damania, B., J. H. Jeong, B. S. Bowser, S. M. DeWire, M. R. Staudt, and D. P. Dittmer. 2004. Comparison of the Rta/Orf50 transactivator proteins of gamma-2-herpesviruses. J Virol **78**:5491–9.

117. Davis, D. A., A. S. Rinderknecht, J. P. Zoeteweij, Y. Aoki, E. L. Read-Connole, G. Tosato, A. Blauvelt, and R. Yarchoan. 2001. Hypoxia induces lytic replication of Kaposi's sarcoma-associated herpesvirus. Blood **97**:3244–50.

118. Dedicoat, M., and R. Newton. 2003. Review of the distribution of Kaposi's sarcoma-associated herpesvirus (KSHV) in Africa in relation to the incidence of Kaposi's sarcoma. Br J Cancer **88**:1–3.

119. Delli Bovi, P., E. Donti, D. M. Knowles, 2nd, A. Friedman-Kien, P. A. Luciw, D. Dina, R. Dalla-Favera, and C. Basilico. 1986. Presence of chromosomal abnormalities and lack of AIDS retrovirus DNA sequences in AIDS-associated Kaposi's sarcoma. Cancer Res **46**:6333–8.

120. Deng, H., J. T. Chu, M. B. Rettig, O. Martinez-Maza, and R. Sun. 2002. Rta of the human herpesvirus 8/Kaposi's sarcoma-associated herpesvirus up-regulates human interleukin-6 gene expression. Blood **100**:1919–21.

121. Deng, H., A. Young, and R. Sun. 2000. Auto-activation of the rta gene of human herpesvirus-8/Kaposi's sarcoma-associated herpesvirus. J Gen Virol **81**:3043–8.

122. Deng, J. H., Y. J. Zhang, X. P. Wang, and S. J. Gao. 2004. Lytic replication-defective Kaposi's sarcoma-associated herpesvirus: potential role in infection and malignant transformation. J Virol **78**:11108–20.

123. Deutsch, E., A. Cohen, G. Kazimirsky, S. Dovrat, H. Rubinfeld, C. Brodie, and R. Sarid. 2004. Role of protein kinase C delta in reactivation of Kaposi's sarcoma-associated herpesvirus. J Virol **78**:10187–92.

124. Dezube, B. J., M. Zambela, D. R. Sage, J. F. Wang, and J. D. Fingeroth. 2002. Characterization of Kaposi's sarcoma-associated herpesvirus/human herpesvirus-8 infection of human vascular endothelial cells: early events. Blood **100**:888–96.

125. Dittmer, D., M. Lagunoff, R. Renne, K. Staskus, A. Haase, and D. Ganem. 1998. A cluster of latently expressed genes in Kaposi's sarcoma-associated herpesvirus. J Virol **72**:8309–15.

126. Dittmer, D., C. Stoddart, R. Renne, V. Linquist-Stepps, M. E. Moreno, C. Bare, J. M. McCune, and D. Ganem. 1999. Experimental transmission of Kaposi's sarcoma-associated herpesvirus (KSHV/HHV-8) to SCID-hu Thy/Liv mice. J Exp Med **190**:1857–68.

127. Dittmer, D. P. 2003. Transcription profile of Kaposi's sarcoma-associated herpesvirus in primary Kaposi's sarcoma lesions as determined by real-time PCR arrays. Cancer Res **63**:2010–5.

128. Djerbi, M., V. Screpanti, A. I. Catrina, B. Bogen, P. Biberfeld, and A. Grandien. 1999. The inhibitor of death receptor signaling, FLICE-inhibitory protein defines a new class of tumor progression factors. J Exp Med **190**:1025–32.

129. Dourmishev, L. A., A. L. Dourmishev, D. Palmeri, R. A. Schwartz, and D. M. Lukac. 2003. Molecular genetics of Kaposi's sarcoma-associated herpesvirus (human herpesvirus-8) epidemiology and pathogenesis. Microbiol Mol Biol Rev **67**:175–212.

130. Dunaief, J. L., B. E. Strober, S. Guha, P. A. Khavari, K. Alin, J. Luban, M. Begemann, G. R. Crabtree, and S. P. Goff. 1994. The retinoblastoma protein and BRG1 form a complex and cooperate to induce cell cycle arrest. Cell **79**:119–30.

131. Dupin, N., T. L. Diss, P. Kellam, M. Tulliez, M. Q. Du, D. Sicard, R. A. Weiss, P. G. Isaacson, and C. Boshoff. 2000. HHV-8 is associated with a plasmablastic variant of Castleman disease that is linked to HHV-8-positive plasmablastic lymphoma. Blood **95**:1406–12.

132. Dupin, N., C. Fisher, P. Kellam, S. Ariad, M. Tulliez, N. Franck, E. van Marck, D. Salmon, I. Gorin, J. P. Escande, R. A. Weiss, K. Alitalo, and C. Boshoff C. 1999. Distribution of human herpesvirus-8 latently infected cells in Kaposi's sarcoma, multicentric Castleman's disease, and primary effusion lymphoma. Proc Natl Acad Sci USA **96**:4546–51.

133. Dupin, N., I. Gorin, J. Deleuze, H. Agut, J. M. Huraux, and J. P. Escande. 1995. Herpes-like DNA sequences, AIDS-related tumors, and Castleman's disease. N Engl J Med **333**:798.

134. Dupin, N., M. Grandadam, V. Calvez, I. Gorin, J. T. Aubin, S. Havard, F. Lamy, M. Leibowitch, J. M. Huraux, J. P. Escande, and et al. 1995. Herpesvirus-like DNA sequences in patients with Mediterranean Kaposi's sarcoma. Lancet **345**:761–2.

135. Duprez, R., E. Kassa-Kelembho, S. Plancoulaine, J. Briere, M. Fossi, L. Kobangue, P. Minsart, M. Huerre, and A. Gessain. 2005. Human herpesvirus 8 serological markers and viral load in patients with AIDS-associated Kaposi's sarcoma in Central African Republic. J Clin Microbiol **43**:4840–3.

136. Ellis, M., Y. P. Chew, L. Fallis, S. Freddersdorf, C. Boshoff, R. A. Weiss, X. Lu, and S. Mittnacht. 1999. Degradation of p27(Kip) cdk inhibitor triggered by Kaposi's sarcoma virus cyclin–cdk6 complex. EMBO J **18**:644–53.

137. Ensoli, B., C. Sgadari, G. Barillari, M. C. Sirianni, M. Sturzl, and P. Monini. 2001. Biology of Kaposi's sarcoma. Eur J Cancer **37**:1251–69.

138. Ensoli, B., and M. C. Sirianni. 1998. Kaposi's sarcoma pathogenesis: a link between immunology and tumor biology. Crit Rev Oncog **9**:107–24.

139. Ensoli, B., and M. Sturzl. 1998. Kaposi's sarcoma: a result of the interplay among inflammatory cytokines, angiogenic factors and viral agents. Cytokine Growth Factor Rev **9**:63–83.

140. Ensoli, B., M. Sturzl, and P. Monini. 2001. Reactivation and role of HHV-8 in Kaposi's sarcoma initiation. Adv Cancer Res **81**:161–200.

141. Fakhari, F. D., J. H. Jeong, Y. Kanan, and D. P. Dittmer. 2006. The latency-associated nuclear antigen of Kaposi's sarcoma-associated herpesvirus induces B cell hyperplasia and lymphoma. J Clin Invest **116**:735–42.

142. Fardet, L., L. Blum, D. Kerob, F. Agbalika, L. Galicier, A. Dupuy, M. Lafaurie, V. Meignin, P. Morel, and C. Lebbe. Human herpesvirus 8-associated hemophagocytic lymphohistiocytosis in human immunodeficiency virus-infected patients. Clin Infect Dis **37**:285–91.

143. Feng, P., J. Park, B. S. Lee, S. H. Lee, R. J. Bram, and J. U. Jung. 2002. Kaposi's sarcoma-associated herpesvirus mitochondrial K7 protein targets a cellular calcium-modulating cyclophilin ligand to modulate intracellular calcium concentration and inhibit apoptosis. J Virol **76**:11491–504.

144. Fickenscher, H., and B. Fleckenstein. 2001. Herpesvirus saimiri. Philos Trans R Soc Lond B Biol Sci **356**:545–67.

145. Field, N., W. Low, M. Daniels, S. Howell, L. Daviet, C. Boshoff, and M. Collins. 2003. KSHV vFLIP binds to IKK-gamma to activate IKK. J Cell Sci **116**:3721–8.

146. Fife, K., M. R. Howard, F. Gracie, R. H. Phillips, and M. Bower. 1998. Activity of thalidomide in AIDS-related Kaposi's sarcoma and correlation with HHV8 titer. Int J STD AIDS **9**:751–5.

147. Flore, O., S. Rafii, S. Ely, J. J. O'Leary, E. M. Hyjek, and E. Cesarman. 1998. Transformation of primary human endothelial cells by Kaposi's sarcoma-associated herpesvirus. Nature **394**:588–92.

148. Flowers, C. C., S. P. Flowers, and G. J. Nabel. 1998. Kaposi's sarcoma-associated herpesvirus viral interferon regulatory factor confers resistance to the antiproliferative effect of interferon-alpha. Mol Med **4**:402–12.

149. Foglieni, C., S. Scabini, D. Belloni, F. Broccolo, P. Lusso, M. S. Malnati, and E. Ferrero. 2005. Productive infection of HUVEC by HHV-8 is associated with changes compatible with angiogenic transformations. Eur J Histochem **49**:273–84.

150. Folkman, J. 2006. Angiogenesis. Annu Rev Med **57**:1–18.

151. Foreman, K. E., J. Friborg, B. Chandran, H. Katano, T. Sata, M. Mercader, G. J. Nabel, and B. J. Nickoloff. 2001. Injection of human herpesvirus-8 in human skin engrafted on SCID mice induces Kaposi's sarcoma-like lesions. J Dermatol Sci **26**:182–93.

152. Foreman, K. E., J. Friborg, Jr., W. P. Kong, C. Woffendin, P. J. Polverini, B. J. Nickoloff, and G. J. Nabel. 1997. Propagation of a human herpesvirus from AIDS-associated Kaposi's sarcoma. N Engl J Med **336**:163–71.

153. Foussat, A., J. Wijdenes, L. Bouchet, G. Gaidano, F. Neipel, K. Balabanian, P. Galanaud, J. Couderc, and D. Emilie, D. 1999. Human interleukin-6 is in vivo an autocrine growth factor for human herpesvirus-8-infected malignant B lymphocytes. Eur Cytokine Netw **10**:501–8.

154. Friborg, J., Jr., W. Kong, M. O. Hottiger, and G. J. Nabel. 1999. p53 inhibition by the LANA protein of KSHV protects against cell death. Nature **402**:889–94.

155. Friedman-Kien, A. E. 1981. Disseminated Kaposi's sarcoma syndrome in young homosexual men. J Am Acad Dermatol **5**:468–71.

156. Fujimuro, M., and S. D. Hayward. 2003. The latency-associated nuclear antigen of Kaposi's sarcoma-associated herpesvirus manipulates the activity of glycogen synthase kinase-3beta. J Virol **77**:8019–30.

157. Fujimuro, M., J. Liu, J. Zhu, H. Yokosawa, and S. D. Hayward. 2005. Regulation of the interaction between glycogen synthase kinase 3 and the Kaposi's sarcoma-associated herpesvirus latency-associated nuclear antigen. J Virol **79**:10429–41.

158. Fujimuro, M., F. Y. Wu, C. ApRhys, H. Kajumbula, D. B. Young, G. S. Hayward, and S. D. Hayward. 2003. A novel viral mechanism for dysregulation of beta-catenin in Kaposi's sarcoma-associated herpesvirus latency. Nat Med **9**:300–6.

159. Gage, J. R., E. C. Breen, A. Echeverri, L. Magpantay, T. Kishimoto, S. Miles, and O. Martinez-Maza. 1999. Human herpesvirus 8-encoded interleukin 6 activates HIV-1 in the U1 monocytic cell line. AIDS **13**:1851–5.

160. Gaidano, G., D. Capello, C. Pastore, A. Antinori, A. Gloghini, A. Carbone, L. M. Larocca, and G. Saglio. 1997. Analysis of human herpesvirus type 8 infection in AIDS-related and AIDS-unrelated primary central nervous system lymphoma. J Infect Dis **175**:1193–7.

161. Gaidano, G., and A. Carbone. 2001. Primary effusion lymphoma: a liquid phase lymphoma of fluid-filled body cavities. Adv Cancer Res **80**:115–46.

162. Gaidano, G., K. Cechova, Y. Chang, P. S. Moore, D. M. Knowles, and R. Dalla-Favera. 1996. Establishment of AIDS-related lymphoma cell lines from lymphomatous effusions. Leukemia **10**:1237–40.

163. Gaidano, G., C. Pastore, A. Gloghini, D. Capello, U. Tirelli, G. Saglio, and A. Carbone. 1997. Microsatellite instability in KSHV/HHV-8 positive body-cavity-based lymphoma. Hum Pathol **28**:748–50.

164. Ganem, D. 1998. Human herpesvirus 8 and its role in the genesis of Kaposi's sarcoma. Curr Clin Top Infect Dis **18**:237–51.

165. Gao, S. J., C. Boshoff, S. Jayachandra, R. A. Weiss, Y. Chang, and P. S. Moore. 1997. KSHV ORF K9 (vIRF) is an oncogene which inhibits the interferon signaling pathway. Oncogene **15**:1979–85.

166. Gao, S. J., J. H. Deng, and F. C. Zhou. 2003. Productive lytic replication of a recombinant Kaposi's sarcoma-associated herpesvirus in efficient primary infection of primary human endothelial cells. J Virol **77**:9738–49.

167. Gao, S. J., L. Kingsley, D. R. Hoover, T. J. Spira, C. R. Rinaldo, A. Saah, J. Phair, R. Detels, P. Parry, Y. Chang, and P. S. Moore. 1996. Seroconversion to antibodies against Kaposi's sarcoma-associated herpesvirus-related latent nuclear antigens before the development of Kaposi's sarcoma. N Engl J Med **335**:233–41.

168. Gao, S. J., L. Kingsley, M. Li, W. Zheng, C. Parravicini, J. Ziegler, R. Newton, C. R. Rinaldo, A. Saah, J. Phair, R. Detels, Y. Chang, and P. S. Moore. 1996. KSHV antibodies among Americans, Italians and Ugandans with and without Kaposi's sarcoma. Nat Med **2**:925–8.

169. Gao, S. J., Y. J. Zhang, J. H. Deng, C. S. Rabkin, O. Flore, and H. B. Jenson. 1999. Molecular polymorphism of Kaposi's sarcoma-associated herpesvirus (Human herpesvirus 8) latent nuclear antigen: evidence for a large repertoire of viral genotypes and dual infection with different viral genotypes. J Infect Dis **180**:1466–76.

170. Garber, A. C., M. A. Shu, J. Hu, and R. Renne. 2001. DNA binding and modulation of gene expression by the latency-associated nuclear antigen of Kaposi's sarcoma-associated herpesvirus. J Virol **75**:7882–92.

171. Geijtenbeek, T. B., and Y. van Kooyk. 2003. Pathogens target DC-SIGN to influence their fate DC-SIGN functions as a pathogen receptor with broad specificity. Apmis **111**:698–714.

172. Giancotti, F. G., and E. Ruoslahti. 1999. Integrin signaling. Science **285**:1028–32.

173. Giraldo, G., E. Beth, and F. Haguenau. 1972. Herpes-type virus particles in tissue culture of Kaposi's sarcoma from different geographic regions. J Natl Cancer Inst **49**:1509–26.

174. Glenn, M., L. Rainbow, F. Aurade, A. Davison, and T. F. Schulz. 1999. Identification of a spliced gene from Kaposi's sarcoma-associated herpesvirus encoding a protein with similarities to latent membrane proteins 1 and 2A of Epstein-Barr virus. J Virol **73**:6953–63.

175. Godden-Kent, D., S. J. Talbot, C. Boshoff, Y. Chang, P. Moore, R. A. Weiss, and S. Mittnacht. 1997. The cyclin encoded by Kaposi's sarcoma-associated herpesvirus stimulates cdk6 to phosphorylate the retinoblastoma protein and histone H1. J Virol **71**:4193–8.

176. Goedert, J. J. 2000. The epidemiology of acquired immunodeficiency syndrome malignancies. Semin Oncol **27**:390–401.

177. Goedert, J. J., T. R. Cote, P. Virgo, S. M. Scoppa, D. W. Kingma, M. H. Gail, E. S. Jaffe, and R. J. Biggar. 1998. Spectrum of AIDS-associated malignant disorders. Lancet **351**:1833–9.

178. Gonzalez, C. M., E. L. Wong, B. S. Bowser, G. K. Hong, S. Kenney, and B. Damania. 2006. Identification and characterization of the Orf49 protein of Kaposi's sarcoma-associated herpesvirus. J Virol **80**:3062–70.

179. Goodwin, D. J., M. S. Walters, P. G. Smith, M. Thurau, H. Fickenscher, and A. Whitehouse. 2001. Herpesvirus saimiri open reading frame 50 (Rta) protein reactivates the lytic replication cycle in a persistently infected A549 cell line. J Virol **75**:4008–13.

180. Gottlieb, G. J., A. Ragaz, J. V. Vogel, A. Friedman-Kien, A. M. Rywlin, E. A. Weiner, and A. B. Ackerman. 1981. A preliminary communication on extensively disseminated Kaposi's sarcoma in young homosexual men. Am J Dermatopathol **3**:111–4.

181. Gradoville, L., J. Gerlach, E. Grogan, D. Shedd, S. Nikiforow, C. Metroka, and G. Miller. 2000. Kaposi's sarcoma-associated herpesvirus open reading frame 50/Rta protein activates the entire viral lytic cycle in the HH-B2 primary effusion lymphoma cell line. J Virol **74**:6207–12.

182. Grisotto, M. G., A. Garin, A. P. Martin, K. K. Jensen, P. Chan, S. C. Sealfon, and S. A. Lira. 2006. The human herpesvirus 8 chemokine receptor vGPCR triggers autonomous proliferation of endothelial cells. J Clin Invest **116**:1264–73.

183. Grisotto, M. G., A. Garin, A. P. Martin, K. K. Jensen, P. Chan, S. C. Sealfon, and S. A. Lira. 2006. The human herpesvirus 8 chemokine receptor vGPCR triggers autonomous proliferation of endothelial cells. J Clin Invest **116**:1264–73.

184. Grundhoff, A., and D. Ganem. 2003. The latency-associated nuclear antigen of Kaposi's sarcoma-associated herpesvirus permits replication of terminal repeat-containing plasmids. J Virol **77**:2779–83.

185. Grundhoff, A., and D. Ganem. 2001. Mechanisms governing expression of the v-FLIP gene of Kaposi's sarcoma-associated herpesvirus. J Virol **75**:1857–63.

186. Grundhoff, A., C. S. Sullivan, and D. Ganem. 2006. A combined computational and microarray-based approach identifies novel microRNAs encoded by human gamma-herpesviruses. RNA **12**:733–50.

187. Guasparri, I., S. A. Keller, and E. Cesarman. KSHV vFLIP is essential for the survival of infected lymphoma cells. J Exp Med **199**:993–1003.

188. Guasparri, I., H. Wu, and E. Cesarman. 2006. The KSHV oncoprotein vFLIP contains a TRAF-interacting motif and requires TRAF2 and TRAF3 for signaling. EMBO Rep **7**:114–9.

189. Guo, H. G., S. Pati, M. Sadowska, M. Charurat, and M. Reitz. 2004. Tumorigenesis by human herpesvirus 8 vGPCR is accelerated by human immunodeficiency virus type 1 Tat. J Virol **78**:9336–42.

190. Gwack, Y., H. J. Baek, H. Nakamura, S. H. Lee, M. Meisterernst, R. G. Roeder, and J. U. Jung. 2003. Principal role of TRAP/mediator and SWI/SNF complexes in Kaposi's sarcoma-associated herpesvirus RTA-mediated lytic reactivation. Mol Cell Biol **23**:2055–67.

191. Gwack, Y., H. Byun, S. Hwang, C. Lim, and J. Choe. 2001. CREB-binding protein and histone deacetylase regulate the transcriptional activity of Kaposi's sarcoma-associated herpesvirus open reading frame 50. J Virol **75**:1909–17.

192. Gwack, Y., S. Hwang, H. Byun, C. Lim, J. W. Kim, E. J. Choi, and J. Choe. 2001. Kaposi's sarcoma-associated herpesvirus open reading frame 50 represses p53-induced transcriptional activity and apoptosis. J Virol **75**:6245–8.

193. Gwack, Y., H. Nakamura, S. H. Lee, J. Souvlis, J. T. Yustein, S. Gygi, H. J. Kung, and J. U. Jung. 2003. Poly(ADP-ribose) polymerase 1 and Ste20-like kinase hKFC act as transcriptional repressors for gamma-2 herpesvirus lytic replication. Mol Cell Biol **23**:8282–94.

194. Han, Z., and S. Swaminathan. 2006. Kaposi's sarcoma-associated herpesvirus lytic gene ORF57 is essential for infectious virion production. J Virol **80**:5251–60.

195. Haque, M., J. Chen, K. Ueda, Y. Mori, K. Nakano, Y. Hirata, S. Kanamori, Y. Uchiyama, R. Inagi, T. Okuno, and K. Yamanishi. 2000. Identification and analysis of the K5 gene of Kaposi's sarcoma-associated herpesvirus. J Virol **74**:2867–75.

196. Harada, H., T. Fujita, M. Miyamoto, Y. Kimura, M. Maruyama, A. Furia, T. Miyata, and T. Taniguchi. 1989. Structurally similar but functionally distinct factors, IRF-1 and IRF-2, bind to the same regulatory elements of IFN and IFN-inducible genes. Cell **58**:729–39.

197. Harada, H., M. Kitagawa, N. Tanaka, H. Yamamoto, K. Harada, M. Ishihara, and T. Taniguchi. 1993. Anti-oncogenic and oncogenic potentials of interferon regulatory factors-1 and -2. Science **259**:971–4.

198. Harrington, W., Jr., L. Sieczkowski, C. Sosa, S. Chan-a-Sue, J. P. Cai, L. Cabral, and C. Wood. 1997. Activation of HHV-8 by HIV-1 tat. Lancet **349**:774–5.

199. Hayward, S. D. 2004. Viral interactions with the Notch pathway. Semin Cancer Biol **14**:387–96.

200. Helin, K., E. Harlow, and A. Fattaey. 1993. Inhibition of E2F-1 transactivation by direct binding of the retinoblastoma protein. Mol Cell Biol **13**:6501–8.

201. Hengge, U. R., T. Ruzicka, S. K. Tyring, M. Stuschke, M. Roggendorf, R. A. Schwartz, and S. Seeber. 2002. Update on Kaposi's sarcoma and other HHV8 associated diseases. Part 1: epidemiology, environmental predispositions, clinical manifestations, and therapy. Lancet Infect Dis **2**:281–92.

202. Hideshima, T., D. Chauhan, G. Teoh., N. Raje, S. P. Treon, Y. T. Tai, Y. Shima, and K. C. Anderson. 2000. Characterization of signaling cascades triggered by human interleukin-6 versus Kaposi's sarcoma-associated herpes virus-encoded viral interleukin 6. Clin Cancer Res **6**:1180–9.

203. Hiscott, J., N. Grandvaux, S. Sharma, B. R. Tenoever, M. J. Servant, and R. Lin. 2003. Convergence of the NF-kappaB and interferon signaling pathways in the regulation of antiviral defense and apoptosis. Ann N Y Acad Sci **1010**:237–48.

204. Holkova, B., K. Takeshita, D. M. Cheng, M. Volm, C. Wasserheit, R. Demopoulos, and A. Chanan-Khan. 2001. Effect of highly active antiretroviral therapy on survival in patients with AIDS-associated pulmonary Kaposi's sarcoma treated with chemotherapy. J Clin Oncol **19**:3848–51.

205. Hu, J., A. C. Garber, and R. Renne. 2002. The latency-associated nuclear antigen of Kaposi's sarcoma-associated herpesvirus supports latent DNA replication in dividing cells. J Virol **76**:11677–87.

206. Huang, L. M., M. F. Chao, M.Y. Chen, H. Shih,Y. P. Chiang, C.Y. Chuang, and C.Y. Lee. 2001. Reciprocal regulatory interaction between human herpesvirus 8 and human immunodeficiency virus type 1. J Biol Chem **276**:13427–32.

207. Huang,Y. Q., A. E. Friedman-Kien, J. J. Li, and B. J. Nickoloff. 1993. Cultured Kaposi's sarcoma cell lines express factor XIIIa, CD14, and VCAM-1, but not factor VIII or ELAM-1. Arch Dermatol **129**:1291–6.

208. Hunter, T., and J. Pines. 1991. Cyclins and cancer. Cell **66**:1071–4.

209. Hwang, S.,Y. Gwack, H. Byun, C. Lim, and J. Choe. 2001. The Kaposi's sarcoma-associated herpesvirus K8 protein interacts with CREB-binding protein (CBP) and represses CBP-mediated transcription. J Virol **75**:9509–16.

210. Hymes, K. B., T. Cheung, J. B. Greene, N. S. Prose, A. Marcus, H. Ballard, D. C. William, and L. J. Laubenstein. 1981. Kaposi's sarcoma in homosexual men-a report of eight cases. Lancet **2**:598–600.

211. Hyun, T. S., C. Subramanian, M. A. Cotter, 2nd, R. A. Thomas, and E. S. Robertson. 2001. Latency-associated nuclear antigen encoded by Kaposi's sarcoma-associated herpesvirus interacts with Tat and activates the long terminal repeat of human immunodeficiency virus type 1 in human cells. J Virol **75**:8761–71.

212. Inoue, N., J. Winter, R. B. Lal, M. K. Offermann, and S. Koyano. 2003. Characterization of entry mechanisms of human herpesvirus 8 by using an Rta-dependent reporter cell line. J Virol **77**:8147–52.

213. Ishido, S., J. K. Choi, B. S. Lee, C. Wang, M. DeMaria, R. P. Johnson, G. B. Cohen, and J. U. Jung. 2000. Inhibition of natural killer cell-mediated cytotoxicity by Kaposi's sarcoma-associated herpesvirus K5 protein. Immunity **13**:365–74.

214. Ishido, S., C. Wang, B. S. Lee, G. B. Cohen, and J. U. Jung. 2000. Downregulation of major histocompatibility complex class I molecules by Kaposi's sarcoma-associated herpesvirus K3 and K5 proteins. J Virol **74**:5300–9.

215. Jacobson, L. P., F. J. Jenkins, G. Springer, A. Munoz, K.V. Shah, J. Phair, Z. Zhang, and H. Armenian. 2000. Interaction of human immunodeficiency virus type 1 and human herpesvirus type 8 infections on the incidence of Kaposi's sarcoma. J Infect Dis **181**:1940–9.

216. Jarviluoma, A., S. Koopal, S. Rasanen, T. P. Makela, and P. M. Ojala. 2004. KSHV viral cyclin binds to p27KIP1 in primary effusion lymphomas. Blood **104**:3349–54.

217. Jenner, R. G., M. M. Alba, C. Boshoff, and P. Kellam. 2001. Kaposi's sarcoma-associated herpesvirus latent and lytic gene expression as revealed by DNA arrays. J Virol **75**:891–902.

218. Jensen, K. K., D. J. Manfra, M. G. Grisotto, A. P. Martin, G. Vassileva, K. Kelley, T. W. Schwartz, and S. A. Lira. 2005. The human herpes virus 8-encoded chemokine receptor is required for angioproliferation in a murine model of Kaposi's sarcoma. J Immunol **174**:3686–94.

219. Jeong, J. H., J. Orvis, J. W. Kim, C. P. McMurtrey, R. Renne, and D. P. Dittmer. 2004. Regulation and autoregulation of the promoter for the latency-associated nuclear antigen of Kaposi's sarcoma-associated herpesvirus. J Biol Chem **279**:16822–31.

220. Johansson, S., G. Svineng, K. Wennerberg, A. Armulik, and L. Lohikangas. 1997. Fibronectin-integrin interactions. Front Biosci **2**:d126–46.

221. Jones, J. L., D. L. Hanson, S.Y. Chu, J. W. Ward, and H. W. Jaffe. 1995. AIDS-associated Kaposi's sarcoma. Science **267**:1078–9.

222. Judde, J. G.,V. Lacoste, J. Briere, E. Kassa-Kelembho, E. Clyti, P. Couppie, C. Buchrieser, M. Tulliez, J. Morvan, and A. Gessain. 2000. Monoclonality or oligoclonality of human herpesvirus 8 terminal repeat sequences in Kaposi's sarcoma and other diseases. J Natl Cancer Inst **92**:729–36.

223. Juhasz, I., S. M. Albelda, D. E. Elder, G. F. Murphy, K. Adachi, D. Herlyn, I. T. Valyi-Nagy, and M. Herlyn. 1993. Growth and invasion of human melanomas in human skin grafted to immunodeficient mice. Am J Pathol **143**:528–37.

224. Kaldis, P., P. M. Ojala, L. Tong, T. P. Makela, and M. J. Solomon. 2001. CAK-independent activation of CDK6 by a viral cyclin. Mol Biol Cell **12**:3987–99.

225. Kaleeba, J. A., and E. A. Berger. 2006. Kaposi's sarcoma-associated herpesvirus fusion-entry receptor: cystine transporter xCT. Science **311**:1921–4.

226. Karlin, S., E. S. Mocarski, and G. A. Schachtel. 1994. Molecular evolution of herpesviruses: genomic and protein sequence comparisons. J Virol **68**:1886–902.
227. Katano, H., Y. Sato, T. Kurata, S. Mori, and T. Sata. 2000. Expression and localization of human herpesvirus 8-encoded proteins in primary effusion lymphoma, Kaposi's sarcoma, and multicentric Castleman's disease. Virology **269**:335–44.
228. Katano, H., Y. Sato, and T. Sata. 2001. Expression of p53 and human herpesvirus-8 (HHV-8)-encoded latency-associated nuclear antigen with inhibition of apoptosis in HHV-8-associated malignancies. Cancer **92**:3076–84.
229. Kedes, D. H., E. Operskalski, M. Busch, R. Kohn, J. Flood, and D. Ganem. 1996. The seroepidemiology of human herpesvirus 8 (Kaposi's sarcoma-associated herpesvirus): distribution of infection in KS risk groups and evidence for sexual transmission. Nat Med **2**:918–24.
230. Keller, S. A., D. Hernandez-Hopkins, J. Vider, V. Ponomarev, E. Hyjek, E. J. Schattner, and E. Cesarman. 2006. NF-kappaB is essential for the progression of KSHV- and EBV-infected lymphomas in vivo. Blood **107**:3295–302.
231. Keller, S. A., E. J. Schattner, and E. Cesarman. 2000. Inhibition of NF-kappaB induces apoptosis of KSHV-infected primary effusion lymphoma cells. Blood **96**:2537–42.
232. Kirchhausen, T. 2000. Three ways to make a vesicle. Nat Rev Mol Cell Biol **1**:187–98.
233. Kirchhoff, S., T. Sebens, S. Baumann, A. Krueger, R. Zawatzky, M. Li-Weber, E. Meinl, F. Neipel, B. Fleckenstein, and P. H. Krammer. 2002. Viral IFN-regulatory factors inhibit activation-induced cell death via two positive regulatory IFN-regulatory factor 1-dependent domains in the CD95 ligand promoter. J Immunol **168**:1226–34.
234. Kledal, T. N., M. M. Rosenkilde, F. Coulin, G. Simmons, A. H. Johnsen, S. Alouani, C. A. Power, H. R. Luttichau, J. Gerstoft, P. R. Clapham, I. Clark-Lewis, T. N. Wells, and T. W. Schwartz. 1997. A broad-spectrum chemokine antagonist encoded by Kaposi's sarcoma-associated herpesvirus. Science **277**:1656–9.
235. Kliche, S., W. Nagel, E. Kremmer, C. Atzler, A. Ege, T. Knorr, U. Koszinowski, W. Kolanus, and J. Haas. 2001. Signaling by human herpesvirus 8 kaposin A through direct membrane recruitment of cytohesin-1. Mol Cell **7**:833–43.
236. Klouche, M., N. Brockmeyer, C. Knabbe, and S. Rose-John. 2002. Human herpesvirus 8-derived viral IL-6 induces PTX3 expression in Kaposi's sarcoma cells. AIDS **16**:F9–18.
237. Knipe, D. M., Howley, Peter, M., Fields Virology. Lippincott Williams and Wilkins. 2001.
238. Komatsu, T., M. E. Ballestas, A. J. Barbera, B. Kelley-Clarke, and K. M. Kaye. 2004. KSHV LANA1 binds DNA as an oligomer and residues N-terminal to the oligomerization domain are essential for DNA binding, replication, and episome persistence. Virology **319**:225–36.
239. Kouzarides, T. 1999. Histone acetylases and deacetylases in cell proliferation. Curr Opin Genet Dev **9**:40–8.
240. Koyano, S., E. C. Mar, F. R. Stamey, and N. Inoue. Glycoproteins M and N of human herpesvirus 8 form a complex and inhibit cell fusion. J Gen Virol **84**:1485–91.
241. Krishnan, H. H., P. P. Naranatt, M. S. Smith, L. Zeng, C. Bloomer, and B. Chandran. 2004. Concurrent expression of latent and a limited number of lytic genes with immune modulation and antiapoptotic function by Kaposi's sarcoma-associated herpesvirus early during infection of primary endothelial and fibroblast cells and subsequent decline of lytic gene expression. J Virol **78**:3601–20.
242. Krishnan, H. H., N. Sharma-Walia, L. Zeng, S. J. Gao, and B. Chandran. 2005. Envelope glycoprotein gB of Kaposi's sarcoma-associated herpesvirus is essential for egress from infected cells. J Virol **79**:10952–67.
243. Krithivas, A., M. Fujimuro, M. Weidner, D. B. Young, and S. D. Hayward. 2002. Protein interactions targeting the latency-associated nuclear antigen of Kaposi's sarcoma-associated herpesvirus to cell chromosomes. J Virol **76**:11596–604.
244. Krithivas, A., D. B. Young, G. Liao, D. Greene, and S. D. Hayward. 2000. Human herpesvirus 8 LANA interacts with proteins of the mSin3 corepressor complex and negatively regulates Epstein-Barr virus gene expression in dually infected PEL cells. J Virol **74**:9637–45.
245. Lacoste, V., C. de la Fuente, F. Kashanchi, and A. Pumfery. 2004. Kaposi's sarcoma-associated herpesvirus immediate early gene activity. Front Biosci **9**:2245–72.
246. Lagunoff, M., J. Bechtel, E. Venetsanakos, A. M. Roy, N. Abbey, B. Herndier, M. McMahon, and D. Ganem. 2002. De novo infection and serial transmission of Kaposi's sarcoma-associated herpesvirus in cultured endothelial cells. J Virol **76**:2440–8.
247. Lan, K., D. A. Kuppers, and E. S. Robertson. 2005. Kaposi's sarcoma-associated herpesvirus reactivation is regulated by interaction of latency-associated nuclear antigen with recombination signal sequence-binding protein Jkappa, the major downstream effector of the Notch signaling pathway. J Virol **79**:3468–78.

248. Lan, K., D. A. Kuppers, S. C. Verma, and E. S. Robertson. 2004. Kaposi's sarcoma-associated herpesvirus-encoded latency-associated nuclear antigen inhibits lytic replication by targeting Rta: a potential mechanism for virus-mediated control of latency. J Virol **78**:6585–94.
249. Lan, K., D. A. Kuppers, S. C. Verma, N. Sharma, M. Murakami, and E. S. Robertson. 2005. Induction of Kaposi's sarcoma-associated herpesvirus latency-associated nuclear antigen by the lytic transactivator RTA: a novel mechanism for establishment of latency. J Virol **79**:7453–65.
250. Laney, A. S., T. De Marco, J. S. Peters, M. Malloy, C. Teehankee, P. S. Moore, and Y. Chang. 2005. Kaposi's sarcoma-associated herpesvirus and primary and secondary pulmonary hypertension. Chest **127**:762–7.
251. Lang, M. E., C. Lottersberger, B. Roth, G. Bock, H. Recheis, R. Sgonc, M. Sturzl, A. Albini, E. Tschachler, R. Zangerle, S. Donini, H. Feichtinger, and S. Schwarz. 1997. Induction of apoptosis in Kaposi's sarcoma spindle cell cultures by the subunits of human chorionic gonadotropin. AIDS **11**:1333–40.
252. Lazzi, S., C. Bellan, T. Amato, N. Palummo, C. Cardone, A. D'Amuri, F. De Luca, M. Beyanga, F. Facchetti, P. Tosi, and L. Leoncini. 2006. Kaposi's sarcoma-associated herpesvirus/human herpesvirus 8 infection in reactive lymphoid tissues: a model for KSHV/HHV-8-related lymphomas? Hum Pathol **37**:23–31.
253. Lebbe, C., P. de Cremoux, M. Rybojad, C. Costa da Cunha, P. Morel, and F. Calvo. 1995. Kaposi's sarcoma and new herpesvirus. Lancet **345**:1180.
254. Lee, B. S., S. H. Lee, P. Feng, H. Chang, N. H. Cho, and J. U. Jung. 2005. Characterization of the Kaposi's sarcoma-associated herpesvirus K1 signalosome. J Virol **79**:12173–84.
255. Lee, H., R. Veazey, K. Williams, M. Li, J. Guo, F. Neipel, B. Fleckenstein, A. Lackner, R. C. Desrosiers, and J. U. Jung. 1998. Deregulation of cell growth by the K1 gene of Kaposi's sarcoma-associated herpesvirus. Nat Med **4**:435–40.
256. Lennette, E. T., M. P. Busch, F. M. Hecht, J. A. and Levy. Potential herpesvirus interaction during HIV type 1 primary infection. AIDS Res Hum Retroviruses **21**:869–75.
257. Li, H., T. Komatsu, B. J. Dezube, and K. M. Kaye. 2002. The Kaposi's sarcoma-associated herpesvirus K12 transcript from a primary effusion lymphoma contains complex repeat elements, is spliced, and initiates from a novel promoter. J Virol **76**:11880–8.
258. Li, M., B. Damania, X. Alvarez, V. Ogryzko, K. Ozato, and J. U. Jung. 2000. Inhibition of p300 histone acetyltransferase by viral interferon regulatory factor. Mol Cell Biol **20**:8254–63.
259. Li, M., H. Lee, J. Guo, F. Neipel, B. Fleckenstein, K. Ozato, and J. U. Jung. 1998. Kaposi's sarcoma-associated herpesvirus viral interferon regulatory factor. J Virol **72**:5433–40
260. Li, M., H. Lee, D. W. Yoon, J. C. Albrecht, B. Fleckenstein, F. Neipel, and J. U. Jung. 1997. Kaposi's sarcoma-associated herpesvirus encodes a functional cyclin. J Virol **71**:1984–91.
261. Li, M., J. MacKey, S. C. Czajak, R. C. Desrosiers, A. A. Lackner, and J. U. Jung. 1999. Identification and characterization of Kaposi's sarcoma-associated herpesvirus K8.1 virion glycoprotein. J Virol **73**:1341–9.
262. Liang, Y., J. Chang, S. J. Lynch, D. M. Lukac, and D. Ganem. 2002. The lytic switch protein of KSHV activates gene expression via functional interaction with RBP-Jkappa (CSL), the target of the Notch signaling pathway. Genes Dev **16**:1977–89.
263. Liang, Y., and D. Ganem. 2003. Lytic but not latent infection by Kaposi's sarcoma-associated herpesvirus requires host CSL protein, the mediator of Notch signaling. Proc Natl Acad Sci USA **100**: 8490–5.
264. Liao, W., Y. Tang, Y. L. Kuo, B. Y. Liu, C. J. Xu, and C. Z. Giam. 2003. Kaposi's sarcoma-associated herpesvirus/human herpesvirus 8 transcriptional activator Rta is an oligomeric DNA-binding protein that interacts with tandem arrays of phased A/T-trinucleotide motifs. J Virol **77**:9399–411.
265. Lim, C., Y. Gwack, S. Hwang, S. Kim, and J. Choe. 2001. The transcriptional activity of cAMP response element-binding protein-binding protein is modulated by the latency associated nuclear antigen of Kaposi's sarcoma-associated herpesvirus. J Biol Chem **276**:31016–22.
266. Lim, C., H. Sohn, Y. Gwack, and J. Choe. 2000. Latency-associated nuclear antigen of Kaposi's sarcoma-associated herpesvirus (human herpesvirus-8) binds ATF4/CREB2 and inhibits its transcriptional activation activity. J Gen Virol **81**:2645–52.
267. Lin, R., P. Genin, Y. Mamane, M. Sgarbanti, A. Battistini, W. J. Harrington Jr., G. N. Barber, and J. Hiscott. 2001. HHV-8 encoded vIRF-1 represses the interferon antiviral response by blocking IRF-3 recruitment of the CBP/p300 coactivators. Oncogene **20**:800–11.
268. Lin, S. F., D. R. Robinson, J. Oh, J. U. Jung, P. A. Luciw, and H. J. Kung. 2002. Identification of the bZIP and Rta homologues in the genome of rhesus monkey rhadinovirus. Virology **298**:181–8.
269. Liu, C., Y. Okruzhnov, H. Li, and J. Nicholas. 2001. Human herpesvirus 8 (HHV-8)-encoded cytokines induce expression of and autocrine signaling by vascular endothelial growth factor (VEGF) in HHV-8-infected primary-effusion lymphoma cell lines and mediate VEGF-independent antiapoptotic effects. J Virol **75**:10933–40.

270. Liu, L., M. T. Eby, N. Rathore, S. K. Sinha, A. Kumar, and P. M. Chaudhary. 2002. The human herpes virus 8-encoded viral FLICE inhibitory protein physically associates with and persistently activates the Ikappa B kinase complex. J Biol Chem **277**:13745–51.
271. Locati, M., and P. M. Murphy. 1999. Chemokines and chemokine receptors: biology and clinical relevance in inflammation and AIDS. Annu Rev Med **50**:425–40.
272. Low, W., M. Harries, H. Ye, M. Q. Du, C. Boshoff, and M. Collins. 2001. Internal ribosome entry site regulates translation of Kaposi's sarcoma-associated herpesvirus FLICE inhibitory protein. J Virol **75**:2938–45.
273. Lu, F., J. Zhou, A. Wiedmer, K. Madden, Y. Yuan, and P. M. Lieberman. 2003. Chromatin remodeling of the Kaposi's sarcoma-associated herpesvirus ORF50 promoter correlates with reactivation from latency. J Virol **77**:11425–35.
274. Lubyova, B., and P. M. Pitha. 2000. Characterization of a novel human herpesvirus 8-encoded protein, vIRF-3, that shows homology to viral and cellular interferon regulatory factors. J Virol **74**:8194–201.
275. Lukac, D. M., J. R. Kirshner, and D. Ganem. 1999. Transcriptional activation by the product of open reading frame 50 of Kaposi's sarcoma-associated herpesvirus is required for lytic viral reactivation in B cells. J Virol **73**:9348–61.
276. Lukac, D. M., R. Renne, J. R. Kirshner, and D. Ganem. 1998. Reactivation of Kaposi's sarcoma-associated herpesvirus infection from latency by expression of the ORF 50 transactivator, a homolog of the EBV R protein. Virology **252**:304–12.
277. Luna, R. E., F. Zhou, A. Baghian, V. Chouljenko, B. Forghani, S. J. Gao, and K. G. Kousoulas. 2004. Kaposi's sarcoma-associated herpesvirus glycoprotein K8.1 is dispensable for virus entry. J Virol **78**:6389–98.
278. Malmgaard, L. 2004. Induction and regulation of IFNs during viral infections. J Interferon Cytokine Res **24**:439–54.
279. Mann, D. J., E. S. Child, C. Swanton, H. Laman, and N. Jones. 1999. Modulation of p27(Kip1) levels by the cyclin encoded by Kaposi's sarcoma-associated herpesvirus. EMBO J **18**:654–63.
280. Martin, R. W., 3rd, A. F. Hood, and E. R. Farmer. 1993. Kaposi's sarcoma. Medicine (Baltimore) **72**:245–61.
281. Masood, R., J. Cai, T. Zheng, D. L. Smith, Y. Naidu, and P. S. Gill. 1997. Vascular endothelial growth factor/vascular permeability factor is an autocrine growth factor for AIDS-Kaposi's sarcoma. Proc Natl Acad Sci USA **94**:979–84.
282. Masood, R., E. Cesarman, D. L. Smith, P. S. Gill, and O. Flore. 2002. Human herpesvirus-8-transformed endothelial cells have functionally activated vascular endothelial growth factor/vascular endothelial growth factor receptor. Am J Pathol **160**:23–9.
283. Masood, R., G. Xia, D. L. Smith, P. Scalia, J. G. Still, A. Tulpule, and P. S. Gill. 2005. Ephrin B2 expression in Kaposi's sarcoma is induced by human herpesvirus type 8: phenotype switch from venous to arterial endothelium. Blood **105**:1310–8.
284. Matta, H., and P. M. Chaudhary. 2004. Activation of alternative NF-kappa B pathway by human herpes virus 8-encoded Fas-associated death domain-like IL-1 beta-converting enzyme inhibitory protein (vFLIP). Proc Natl Acad Sci USA **101**:9399–404.
285. Mattsson, K., C. Kiss, G. M. Platt, G. R. Simpson, E. Kashuba, G. Klein, T. F. Schulz, and L. Szekely. 2002. Latent nuclear antigen of Kaposi's sarcoma herpesvirus/human herpesvirus-8 induces and relocates RING3 to nuclear heterochromatin regions. J Gen Virol **83**:179–88.
286. McAllister, S. C., S. G. Hansen, R. A. Ruhl, C. M. Raggo, V. R. DeFilippis, D. Greenspan, K. Fruh, and A. V. Moses. 2004. Kaposi's sarcoma-associated herpesvirus (KSHV) induces heme oxygenase-1 expression and activity in KSHV-infected endothelial cells. Blood **103**:3465–73.
287. McCormick, C., and D. Ganem. 2005. The kaposin B protein of KSHV activates the p38/MK2 pathway and stabilizes cytokine mRNAs. Science **307**:739–41.
288. McGeoch, D. J., and A. J. Davison. 1999. The descent of human herpesvirus 8. Semin Cancer Biol **9**:201–9.
289. McPherson, P. S., B. K. Kay, and N. K. Hussain. 2001. Signaling on the endocytic pathway. Traffic **2**:375–84.
290. Meads, M. B., and P. G. Medveczky. 2004. Kaposi's sarcoma-associated herpesvirus-encoded viral interleukin-6 is secreted and modified differently than human interleukin-6: evidence for a unique autocrine signaling mechanism. J Biol Chem **279**:51793–803.
291. Melbye, M., P. M. Cook, H. Hjalgrim, K. Begtrup, G. R. Simpson, R. J. Biggar, P. Ebbesen, and T. F. Schulz. 1998. Risk factors for Kaposi's-sarcoma-associated herpesvirus (KSHV/HHV-8) seropositivity in a cohort of homosexual men, 1981–1996. Int J Cancer **77**:543–8.

292. Merat, R., A. Amara, C. Lebbe, H. de The, P. Morel, and A. Saib. 2002. HIV-1 infection of primary effusion lymphoma cell line triggers Kaposi's sarcoma-associated herpesvirus (KSHV) reactivation. Int J Cancer **97**:791–5.

293. Mercader, M., B. J. Nickoloff, and K. E. Foreman. Induction of human immunodeficiency virus 1 replication by human herpesvirus 8. Arch Pathol Lab Med **125**:785–9.

294. Mercader, M., B. Taddeo, J. R. Panella, B. Chandran, B. J. Nickoloff, and K. E. Foreman. 2000. Induction of HHV-8 lytic cycle replication by inflammatory cytokines produced by HIV-1-infected T cells. Am J Pathol **156**:1961–71.

295. Miller, G., L. Heston, E. Grogan, L. Gradoville, M. Rigsby, R. Sun, D. Shedd, V. M. Kushnaryov, S. Grossberg, and Y. Chang. 1997. Selective switch between latency and lytic replication of Kaposi's sarcoma herpesvirus and Epstein-Barr virus in dually infected body cavity lymphoma cells. J Virol **71**:314–24.

296. Mocroft, A., M. Youle, B. Gazzard, J. Morcinek, R. Halai, and A. N. Phillips. Anti-herpesvirus treatment and risk of Kaposi's sarcoma in HIV infection. Royal Free/Chelsea and Westminster Hospitals Collaborative Group. AIDS **10**:1101–5.

297. Molden, J., Y. Chang, Y. You, P. S. Moore, and M. A. Goldsmith. 1997. A Kaposi's sarcoma-associated herpesvirus-encoded cytokine homolog (vIL-6) activates signaling through the shared gp130 receptor subunit. J Biol Chem **272**:19625–31.

298. Montaner, S., A. Sodhi, S. Pece, E. A. Mesri, and J. S. Gutkind. 2001. The Kaposi's sarcoma-associated herpesvirus G protein-coupled receptor promotes endothelial cell survival through the activation of Akt/protein kinase B. Cancer Res **61**:2641–8.

299. Montaner, S., A. Sodhi, J. M. Servitja, A. K. Ramsdell, A. Barac, E. T. Sawai, and J. S. Gutkind. 2004. The small GTPase Rac1 links the Kaposi's sarcoma-associated herpesvirus vGPCR to cytokine secretion and paracrine neoplasia. Blood **104**:2903–11.

300. Moore, P. S., C. Boshoff, R. A. Weiss, and Y. Chang. 1996. Molecular mimicry of human cytokine and cytokine response pathway genes by KSHV. Science **274**:1739–44.

301. Moore, P. S., and Y. Chang. 1995. Detection of herpesvirus-like DNA sequences in Kaposi's sarcoma in patients with and without HIV infection. N Engl J Med **332**:1181–5.

302. Moore, P. S., and Y. Chang. 2001. Molecular virology of Kaposi's sarcoma-associated herpesvirus. Philos Trans R Soc Lond B Biol Sci **356**:499–516.

303. Moore, P. S., S. J. Gao, G. Dominguez, E. Cesarman, O. Lungu, D. M. Knowles, R. Garber, P. E. Pellett, D. J. McGeoch, and Y. Chang. 1996. Primary characterization of a herpesvirus agent associated with Kaposi's sarcoma. J Virol **70**:549–58.

304. Moore, P. S., L. A. Kingsley, S. D. Holmberg, T. Spira, P. Gupta, D. R. Hoover, J. P. Parry, L. J. Conley, H. W. Jaffe, and Y. Chang. 1996. Kaposi's sarcoma-associated herpesvirus infection prior to onset of Kaposi's sarcoma. AIDS **10**:175–80.

305. Mori, Y., N. Nishimoto, M. Ohno, R. Inagi, P. Dhepakson, K. Amou, K. Yoshizaki, and K. Yamanishi. 2000. Human herpesvirus 8-encoded interleukin-6 homologue (viral IL-6) induces endogenous human IL-6 secretion. J Med Virol **61**:332–5.

306. Moses, A. V., K. N. Fish, R. Ruhl, P. P. Smith, J. G. Strussenberg, L. Zhu, B. Chandran, and J. A. Nelson. 1999. Long-term infection and transformation of dermal microvascular endothelial cells by human herpesvirus 8. J Virol **73**:6892–902.

307. Moynagh, P. N. 2005. TLR signaling and activation of IRFs: revisiting old friends from the NF-kappaB pathway. Trends Immunol **26**:469–76.

308. Mullberg, J., T. Geib, T. Jostock, S. H. Hoischen, P. Vollmer, N. Voltz, D. Heinz, P. R. Galle, M. Klouche, and S. Rose-John. 2000. IL-6 receptor independent stimulation of human gp130 by viral IL-6. J Immunol **164**:4672–7.

309. Mullick, J., J. Bernet, A. K. Singh, J. D. Lambris, and A. Sahu. 2003. Kaposi's sarcoma-associated herpesvirus (human herpesvirus 8) open reading frame 4 protein (kaposica) is a functional homolog of complement control proteins. J Virol **77**:3878–81.

310. Munshi, N., R. K. Ganju, S. Avraham, E. A. Mesri, and J. E. Groopman. 1999. Kaposi's sarcoma-associated herpesvirus-encoded G protein-coupled receptor activation of c-jun amino-terminal kinase/stress-activated protein kinase and lyn kinase is mediated by related adhesion focal tyrosine kinase/proline-rich tyrosine kinase 2. J Biol Chem **274**:31863–7.

311. Muralidhar, S., A. M. Pumfery, M. Hassani, M. R. Sadaie, M. Kishishita, J. N. Brady, J. Doniger, P. Medveczky, and L. J. Rosenthal. 1998. Identification of kaposin (open reading frame K12) as a human herpesvirus 8 (Kaposi's sarcoma-associated herpesvirus) transforming gene. J Virol **72**:4980–8.

312. Muralidhar, S., G. Veytsmann, B. Chandran, D. Ablashi, J. Doniger, and L. J. Rosenthal. 2000. Characterization of the human herpesvirus 8 (Kaposi's sarcoma-associated herpesvirus) oncogene, kaposin (ORF K12). J Clin Virol **16**:203–13.

313. Murdoch, C., and A. Finn. 2000. Chemokine receptors and their role in inflammation and infectious diseases. Blood **95**:3032–43.
314. Murdoch, C., and A. Finn. 2000. Chemokine receptors and their role in vascular biology. J Vasc Res **37**:1–7.
315. Muromoto, R., K. Okabe, M. Fujimuro, K. Sugiyama, H. Yokosawa, T. Seya, and T. Matsuda. 2006. Physical and functional interactions between STAT3 and Kaposi's sarcoma-associated herpesvirus-encoded LANA. FEBS Lett **580**:93–8.
316. Nair, V., and M. Zavolan. 2006. Virus-encoded microRNAs: novel regulators of gene expression. Trends Microbiol **14**:169–75.
317. Nakamura, H., M. Li, J. Zarycki, and J. U. Jung. 2001. Inhibition of p53 tumor suppressor by viral interferon regulatory factor. J Virol **75**:7572–82.
318. Nakamura, H., M. Lu, Y. Gwack, J. Souvlis, S. L. Zeichner, and J. U. Jung. 2003. Global changes in Kaposi's sarcoma-associated virus gene expression patterns following expression of a tetracycline-inducible Rta transactivator. J Virol **77**:4205–20.
319. Nakano, H., M. Shindo, S. Sakon, S. Nishinaka, M. Mihara, H. Yagita, and K. Okumura. 1998. Differential regulation of IkappaB kinase alpha and beta by two upstream kinases, NF-kappaB-inducing kinase and mitogen-activated protein kinase/ERK kinase kinase-1. Proc Natl Acad Sci USA **95**:3537–42.
320. Naranatt, P. P., S. M. Akula, C. A. Zien, H. H. Krishnan, and B. Chandran. 2003. Kaposi's sarcoma-associated herpesvirus induces the phosphatidylinositol 3-kinase-PKC-zeta-MEK-ERK signaling pathway in target cells early during infection: implications for infectivity. J Virol **77**:1524–39.
321. Naranatt, P. P., H. H. Krishnan, M. S. Smith, and B. Chandran. 2005. Kaposi's sarcoma-associated herpesvirus modulates microtubule dynamics via RhoA-GTP-diaphanous 2 signaling and utilizes the dynein motors to deliver its DNA to the nucleus. J Virol **79**:1191–206.
322. Naranatt, P. P., H. H. Krishnan, S. R. Svojanovsky, C. Bloomer, S. Mathur, and B. Chandran. 2004. Host gene induction and transcriptional reprogramming in Kaposi's sarcoma-associated herpesvirus (KSHV/HHV-8)-infected endothelial, fibroblast, and B cells: insights into modulation events early during infection. Cancer Res **64**:72–84.
323. Nealon, K., W. W. Newcomb, T. R. Pray, C. S. Craik, J. C. Brown, and D. H. Kedes. 2001. Lytic replication of Kaposi's sarcoma-associated herpesvirus results in the formation of multiple capsid species: isolation and molecular characterization of A, B, and C capsids from a gammaherpesvirus. J Virol **75**:2866–78.
324. Nicholas, J., V. R. Ruvolo, W. H. Burns, G. Sandford, X. Wan, D. Ciufo, S. B. Hendrickson, H. G. Guo, G. S. Hayward, and M. S. Reitz. 1997. Kaposi's sarcoma-associated human herpesvirus-8 encodes homologues of macrophage inflammatory protein-1 and interleukin-6. Nat Med **3**:287–92.
325. Nocera, A., M. Corbellino, U. Valente, S. Barocci, F. Torre, R. De Palma, A. Sementa, G. B. Traverso, A. Icardi, I. Fontana, V. Arcuri, F. Poli, P. Cagetti, P. Moore, and C. Parravicini. 1998. Posttransplant human herpes virus 8 infection and seroconversion in a Kaposi's sarcoma affected kidney recipient transplanted from a human herpes virus 8 positive living related donor. Transplant Proc **30**:2095–6.
326. O'Brien, T. R., D. Kedes, D. Ganem, D. R. Macrae, P. S. Rosenberg, J. Molden, and J. J. Goedert. Evidence for concurrent epidemics of human herpesvirus 8 and human immunodeficiency virus type 1 in US homosexual men: rates, risk factors, and relationship to Kaposi's sarcoma. J Infect Dis **180**:1010–7.
327. O'Leary, J. J., M. Kennedy, K. Luttich, V. Uhlmann, I. Silva, J. Russell, O. Sheils, M. Ring, M. Sweeney, C. Kenny, N. Bermingham, C. Martin, M. O'Donovan, D. Howells, S. Picton, and S. B. Lucas. 2000. Localization of HHV-8 in AIDS related lymphadenopathy. Mol Pathol **53**:43–7.
328. Oettle, A. G. 1962. Geographical and racial differences in the frequency of Kaposi's sarcoma as evidence of environmental or genetic causes. Acta Unio Int Contra Cancrum **18**:330–63.
329. Offermann, M. K. 1999. Consideration of host-viral interactions in the pathogenesis of Kaposi's sarcoma. J Acquir Immune Defic Syndr **21**(Suppl 1):S58–65.
330. Ojala, P. M., M. Tiainen, P. Salven, T. Veikkola, E. Castanos-Velez, R. Sarid, P. Biberfeld, and T. P. Makela. 1999. Kaposi's sarcoma-associated herpesvirus-encoded v-cyclin triggers apoptosis in cells with high levels of cyclin-dependent kinase 6. Cancer Res **59**:4984–9.
331. Ojala, P. M., K. Yamamoto, E. Castanos-Velez, P. Biberfeld S. J., Korsmeyer, and T. P. Makela. 2000. The apoptotic v-cyclin–CDK6 complex phosphorylates and inactivates Bcl-2. Nat Cell Biol **2**:819–25.
332. Osborne, J., P. S. Moore, and Y. Chang. 1999. KSHV-encoded viral IL-6 activates multiple human IL-6 signaling pathways. Hum Immunol **60**:921–7.
333. Pan, H., J. Xie, F. Ye, S-J. Gao. Modulation of Kaposi's sarcoma -associated herpesvirus infection and replication by MEK/ERK, JNK, and p38 multiple mitogen-activated protein kinase pathways during primary infection. J Virol **80**:5371–82.

334. Pan, H., F. Zhou, and S. J. Gao. 2004. Kaposi's sarcoma-associated herpesvirus induction of chromosome instability in primary human endothelial cells. Cancer Res **64**:4064–8.

335. Pan, H. Y., Y. J. Zhang, X. P. Wang, J. H. Deng, F. C. Zhou, and S. J. Gao. 2003. Identification of a novel cellular transcriptional repressor interacting with the latent nuclear antigen of Kaposi's sarcoma-associated herpesvirus. J Virol **77**:9758–68.

336. Parravicini, C., B. Chandran, M. Corbellino, E. Berti, M. Paulli, P. S. Moore, and Y. Chang. 2000. Differential viral protein expression in Kaposi's sarcoma-associated herpesvirus-infected diseases: Kaposi's sarcoma, primary effusion lymphoma, and multicentric Castleman's disease. Am J Pathol **156**: 743–9.

337. Parravicini, C., S. J. Olsen, M. Capra, F. Poli, G. Sirchia, S. J. Gao, E. Berti, A. Nocera, E. Rossi, G. Bestetti, M. Pizzuto, M. Galli, M. Moroni, P. S. Moore, and M. Corbellino. 1997. Risk of Kaposi's sarcoma-associated herpes virus transmission from donor allografts among Italian posttransplant Kaposi's sarcoma patients. Blood **90**:2826–9.

338. Parsons, C. H., B. Szomju, and D. H. Kedes. 2004. Susceptibility of human fetal mesenchymal stem cells to Kaposi's sarcoma-associated herpesvirus. Blood **104**:2736–8.

339. Pati, S., M. Cavrois, H. G. Guo, J. S. Foulke, Jr., J. Kim, R. A. Feldman, and M. Reitz. 2001. Activation of NF-kappaB by the human herpesvirus 8 chemokine receptor ORF74: evidence for a paracrine model of Kaposi's sarcoma pathogenesis. J Virol **75**:8660–73.

340. Pearce, M., S. Matsumura, and A. C. Wilson. 2005. Transcripts encoding K12, v-FLIP, v-cyclin, and the microRNA cluster of Kaposi's sarcoma-associated herpesvirus originate from a common promoter. J Virol **79**:14457–64.

341. Pertel, P. E. 2002. Human herpesvirus 8 glycoprotein B (gB), gH, and gL can mediate cell fusion. J Virol **76**:4390–400.

342. Pfeffer, S., A. Sewer, M. Lagos-Quintana, R. Sheridan, C. Sander, F. A. Grasser, L. F. van Dyk, C. K. Ho, S. Shuman, M. Chien, J. J. Russo, J. Ju, G. Randall, B. D. Lindenbach, C. M. Rice, V. Simon, D. D.Ho, M. Zavolan, and T. Tuschl. 2005. Identification of microRNAs of the herpesvirus family. Nat Methods **2**:269–76.

343. Picchio, G. R. E. Sabbe, R. J. Gulizia, M. McGrath, B. G. Herndier, and D. E. Mosier. 1997. The KSHV/HHV8-infected BCBL-1 lymphoma line causes tumors in SCID mice but fails to transmit virus to a human peripheral blood mononuclear cell graft. Virology **238**:22–9.

344. Pine, R., T. Decker, D. S. Kessler, D. E. Levy, and J. E. Darnell, Jr. 1990. Purification and cloning of interferon-stimulated gene factor 2 (ISGF2): ISGF2 (IRF-1) can bind to the promoters of both beta interferon- and interferon-stimulated genes but is not a primary transcriptional activator of either. Mol Cell Biol **10**:2448–57.

345. Piolot, T., M. Tramier, M. Coppey, J. C. Nicolas, and V. Marechal. 2001. Close but distinct regions of human herpesvirus 8 latency-associated nuclear antigen 1 are responsible for nuclear targeting and binding to human mitotic chromosomes. J Virol **75**:3948–59.

346. Plow, E. F., T. A. Haas, L. Zhang, J. Loftus, and J. W. Smith. 2000. Ligand binding to integrins. J Biol Chem **275**:21785–8.

347. Polstra, A. M., M. Cornelissen, J. Goudsmit, and A. C. van der Kuyl. 2004. Retrospective, longitudinal analysis of serum human herpesvirus-8 viral DNA load in AIDS-related Kaposi's sarcoma patients before and after diagnosis. J Med Virol **74**:390–6.

348. Popescu, N. C., D. B. Zimonjic, S. Leventon-Kriss, J. L. Bryant, Y. Lunardi-Iskandar, and R. C. Gallo. 1996. Deletion and translocation involving chromosome 3 (p14) in two tumorigenic Kaposi's sarcoma cell lines. J Natl Cancer Inst **88**:450–5.

349. Prakash, O., O. R. Swamy, X. Peng, Z. Y. Tang, L. Li, J. E. Larson, J. C. Cohen, J. Gill, G. Farr, S. Wang, and F. Samaniego. 2005. Activation of Src kinase Lyn by the Kaposi's sarcoma-associated herpesvirus K1 protein: implications for lymphomagenesis. Blood **105**:3987–94.

350. Pyakurel, P., F. Pak, A. R. Mwakigonja, E. Kaaya, T. Heiden, and P. Biberfeld. 2006. Lymphatic and vascular origin of Kaposi's sarcoma spindle cells during tumor development. Int J Cancer **119**:1262–7.

351. Qiang, W., J. M. Cahill, J. Liu, X. Kuang, N. Liu, V. L. Scofield, J. R. Voorhees, A. J. Reid, M. Yan, W. S. Lynn, and P. K. Wong. 2004. Activation of transcription factor Nrf-2 and its downstream targets in response to Maloney murine leukemia virus ts1-induced thiol depletion and oxidative stress in astrocytes. J Virol **78**:11926–38.

352. Radkov, S. A., P. Kellam, and C. Boshoff. The latent nuclear antigen of Kaposi's sarcoma-associated herpesvirus targets the retinoblastoma-E2F pathway and with the oncogene Hras transforms primary rat cells. Nat Med **6**:1121–7.

353. Rainbow, L., G. M. Platt, G. R. Simpson, R. Sarid, S. J. Gao, H. Stoiber, C. S. Herrington, P. S. Moore, and T. F. Schulz. 1997. The 222- to 234-kilodalton latent nuclear protein (LNA) of Kaposi's

sarcoma-associated herpesvirus (human herpesvirus 8) is encoded by orf73 and is a component of the latency-associated nuclear antigen. J Virol 71:5915–21.

354. Rappocciolo, G., F. J. Jenkins, H. R. Hensler, P. Piazza, M. Jais, L. Borowski, S. C. Watkins, and C. R. Rinaldo, Jr. 2006. DC-SIGN is a receptor for human herpesvirus 8 on dendritic cells and macrophages. J Immunol 176:1741–9.

355. Regamey, N., M. Tamm, M. Wernli, A. Witschi, G. Thiel, G. Cathomas, and P. Erb. 1998. Transmission of human herpesvirus 8 infection from renal-transplant donors to recipients. N Engl J Med 339:1358–63.

356. Relloso, M., A. Puig-Kroger, O. M. Pello, J. L. Rodriguez-Fernandez, G. de la Rosa, N. Longo, J. Navarro, M. A. Munoz-Fernandez, P. Sanchez-Mateos, and A. L. Corbi. 2002. DC-SIGN (CD209) expression is IL-4 dependent and is negatively regulated by IFN, TGF-beta, and anti-inflammatory agents. J Immunol 168:2634–43.

357. Renne, R., D. Blackbourn, D. Whitby, J. Levy, and D. Ganem. Limited transmission of Kaposi's sarcoma-associated herpesvirus in cultured cells. J Virol 72:5182–8.

358. Renne, R., D. Dittmer, D. Kedes, K. Schmidt, R. C. Desrosiers, P. A. Luciw, and D. Ganem. 2004. Experimental transmission of Kaposi's sarcoma-associated herpesvirus (KSHV/HHV-8) to SIV-positive and SIV-negative rhesus macaques. J Med Primatol 33:1–9.

359. Renne, R., W. Zhong, B. Herndier, M. McGrath, N. Abbey, D. Kedes, and D. Ganem. 1996. Lytic growth of Kaposi's sarcoma-associated herpesvirus (human herpesvirus 8) in culture. Nat Med 2:342–6.

360. Renwick, N., T. Halaby, G. J. Weverling, N. H. Dukers, G. R. Simpson, R. A. Coutinho, J. M. Lange, T. F. Schulz, and J. Goudsmit. Seroconversion for human herpesvirus 8 during HIV infection is highly predictive of Kaposi's sarcoma. AIDS 12:2481–8.

361. Rimessi, P., A. Bonaccorsi, M. Sturzl, M. Fabris, E. Brocca-Cofano, A. Caputo, G. Melucci-Vigo, M. Falchi, A. Cafaro, E. Cassai, B. Ensoli, and P. Monini. 2001. Transcription pattern of human herpesvirus 8 open reading frame K3 in primary effusion lymphoma and Kaposi's sarcoma. J Virol 75:7161–74.

362. Rivas, C., A. E. Thlick, C. Parravicini, P. S. Moore, and Y. Chang. 2001. Kaposi's sarcoma-associated herpesvirus LANA2 is a B-cell-specific latent viral protein that inhibits p53. J Virol 75:429–38.

363. Roan, F., N. Inoue, and M. K. Offermann. 2002. Activation of cellular and heterologous promoters by the human herpesvirus 8 replication and transcription activator. Virology 301:293–3040.

364. Roan, F., J. C. Zimring, S. Goodbourn., and M. K. Offermann. 1999. Transcriptional activation by the human herpesvirus-8-encoded interferon regulatory factor. J Gen Virol 80(Pt 8):2205–9.

365. Robles, R., D. Lugo, L. Gee, and M. A. Jacobson. 1999. Effect of antiviral drugs used to treat cytomegalovirus end-organ disease on subsequent course of previously diagnosed Kaposi's sarcoma in patients with AIDS. J Acquir Immune Defic Syndr Hum Retrovirol 20:34–8.

366. Rose, T. M., K. B. Strand, E. R. Schultz, G. Schaefer, G. W. Rankin, Jr., M. E. Thouless, C. C. Tsai, and M. L. Bosch. 1997. Identification of two homologs of the Kaposi's sarcoma-associated herpesvirus (human herpesvirus 8) in retroperitoneal fibromatosis of different macaque species. J Virol 71:4138–44.

367. Russo, J. J., R. A. Bohenzky, M. C. Chien, J. Chen, M. Yan, D. Maddalena, J. P. Parry, D. Peruzzi, I. S. Edelman, Y. Chang, and P. S. Moore. 1996. Nucleotide sequence of the Kaposi's sarcoma-associated herpesvirus (HHV8). Proc Natl Acad Sci USA 93:14862–7.

368. Sadler, R., L. Wu, B. Forghani, R. Renne, W. Zhong, B. Herndier, and D. Ganem. 1999. A complex translational program generates multiple novel proteins from the latently expressed kaposin (K12) locus of Kaposi's sarcoma-associated herpesvirus. J Virol 73:5722–30.

369. Sailer, A., K. Nagata, D. Naf, M. Aebi, and C. Weissmann. 1992. Interferon regulatory factor-1 (IRF-1) activates the synthetic IRF-1-responsive sequence (GAAAGT)4 in Saccharomyces cerevisiae. Gene Expr 2:329–37.

370. Sakakibara, S., K. Ueda, J. Chen, T. Okuno, and K. Yamanishi. 2001. Octamer-binding sequence is a key element for the autoregulation of Kaposi's sarcoma-associated herpesvirus ORF50/Lyta gene expression. J Virol 75:6894–900.

371. Sakakibara, S., K. Ueda, K. Nishimura, E. Do, E. Ohsaki, T. Okuno, and K. Yamanishi. 2004. Accumulation of heterochromatin components on the terminal repeat sequence of Kaposi's sarcoma-associated herpesvirus mediated by the latency-associated nuclear antigen. J Virol 78:7299–310.

372. Samaniego, F., P. D. Markham, R. C. Gallo, and B. Ensoli. 1995. Inflammatory cytokines induce AIDS-Kaposi's sarcoma-derived spindle cells to produce and release basic fibroblast growth factor and enhance Kaposi's sarcoma-like lesion formation in nude mice. J Immunol 154:3582–92.

373. Samaniego, F., D. Young, C. Grimes, V. Prospero, M. Christofidou-Solomidou, H. M. DeLisser, O. Prakash, A. A. Sahin, and S. Wang. 2002. Vascular endothelial growth factor and Kaposi's sarcoma cells in human skin grafts. Cell Growth Differ 13:387–95.

374. Samols, M. A., J. Hu, R. L. Skalsky, and R. Renne. 2005. Cloning and identification of a microRNA cluster within the latency-associated region of Kaposi's sarcoma-associated herpesvirus. J Virol 79:9301–5.

375. Sanchez, D. J., J. E. Gumperz, and D. Ganem. 2005. Regulation of CD1d expression and function by a herpesvirus infection. J Clin Invest 115:1369–78.
376. Sarek, G., A. Jarviluoma, and P. M. Ojala. 2006. KSHV viral cyclin inactivates p27KIP1 through Ser10 and Thr187 phosphorylation in proliferating primary effusion lymphomas. Blood 107:725–32.
377. Sarid, R., O. Flore, R. A. Bohenzky, Y. Chang, and P. S. Moore. 1998. Transcription mapping of the Kaposi's sarcoma-associated herpesvirus (human herpesvirus 8) genome in a body cavity-based lymphoma cell line (BC-1). J Virol 72:1005–12.
378. Sarid, R., T. Sato, R. A. Bohenzky, J. J. Russo, and Y. Chang. Kaposi's sarcoma-associated herpesvirus encodes a functional bcl-2 homologue. Nat Med 3:293–8.
379. Sasaki, H., H. Sato, K. Kuriyama-Matsumura, K. Sato, K. Maebara, H. Wang, M. Tamba, K. Itoh, M. Yamamoto, and S. Bannai. 2002. Electrophile response element-mediated induction of the cystine/glutamate exchange transporter gene expression. J Biol Chem 277:44765–71.
380. Sastry, S. K., and K. Burridge. 2000. Focal adhesions: a nexus for intracellular signaling and cytoskeletal dynamics. Exp Cell Res 261:25–36.
381. Scadden, D. T. 2003. AIDS-related malignancies. Annu Rev Med 54:285–303.
382. Schaller, M. D. 2001. Biochemical signals and biological responses elicited by the focal adhesion kinase. Biochim Biophys Acta 1540:1–21.
383. Schalling, M., M. Ekman, E. E. Kaaya, A. Linde, and P. Biberfeld. 1995. A role for a new herpes virus (KSHV) in different forms of Kaposi's sarcoma. Nat Med 1:707–8.
384. Schulz, T. F. 2006. The pleiotropic effects of Kaposi's sarcoma herpesvirus. J Pathol 208:187–98.
385. Schwam, D. R., R. L. Luciano, S. S. Mahajan, L. Wong, and A. C. Wilson. 2000. Carboxy terminus of human herpesvirus 8 latency-associated nuclear antigen mediates dimerization, transcriptional repression, and targeting to nuclear bodies. J Virol 74:8532–40.
386. Schwarz, M., and P. M. Murphy. 2001. Kaposi's sarcoma-associated herpesvirus G protein-coupled receptor constitutively activates NF-kappa B and induces proinflammatory cytokine and chemokine production via a C-terminal signaling determinant. J Immunol 167:505–13.
387. Seaman, W. T., D. Ye, R. X. Wang, E. E. Hale, M. Weisse, and E. B. Quinlivan. 1999. Gene expression from the ORF50/K8 region of Kaposi's sarcoma-associated herpesvirus. Virology 263:436–49.
388. Searles, R. P., E. P. Bergquam, M. K. Axthelm, and S. W. Wong. 1999. Sequence and genomic analysis of a Rhesus macaque rhadinovirus with similarity to Kaposi's sarcoma-associated herpesvirus/human herpesvirus 8. J Virol 73:3040–53.
389. Seo, T., D. Lee, B. Lee, J. H. Chung, and J. Choe. 2000. Viral interferon regulatory factor 1 of Kaposi's sarcoma-associated herpesvirus (human herpesvirus 8) binds to, and inhibits transactivation of, CREB-binding protein. Biochem Biophys Res Commun 270:23–7.
390. Seo, T., D. Lee, Y. S. Shim, J. E. Angell, N. V. Chidambaram, D. V. Kalvakolanu, and J. Choe. 2002. Viral interferon regulatory factor 1 of Kaposi's sarcoma-associated herpesvirus interacts with a cell death regulator, GRIM19, and inhibits interferon/retinoic acid-induced cell death. J Virol 76:8797–807.
391. Seo, T., J. Park, D. Lee, S. G. Hwang, and J. Choe. 2001. Viral interferon regulatory factor 1 of Kaposi's sarcoma-associated herpesvirus binds to p53 and represses p53-dependent transcription and apoptosis. J Virol 75:6193–8.
392. Sgarbanti, M., M. Arguello. B. R. tenOever, A. Battistini, R. Lin, and J. Hiscott. A requirement for NF-kappaB induction in the production of replication-competent HHV-8 virions. Oncogene 23:5770–81.
393. Sharma-Walia, N., H. H. Krishnan, P. P. Naranatt, L. Zeng, M. S. Smith, and B. Chandran. 2005. RK1/2 and MEK1/2 induced by Kaposi's sarcoma-associated herpesvirus (human herpesvirus 8) early during infection of target cells are essential for expression of viral genes and for establishment of infection. J Virol 79:10308–29.
394. Sharma-Walia, N., P. P. Naranatt, H. H. Krishnan, L. Zeng, and B. Chandran. 2004. Kaposi's sarcoma-associated herpesvirus/human herpesvirus 8 envelope glycoprotein gB induces the integrin-dependent focal adhesion kinase-Src-phosphatidylinositol 3-kinase-rho GTPase signal pathways and cytoskeletal rearrangements. J Virol 78:4207–23.
395. Sharp, T. V., H. W. Wang, A. Koumi, D. Hollyman, Y. Endo, H. Ye, M. Q. Du, and C. Boshoff. 2002. K15 protein of Kaposi's sarcoma-associated herpesvirus is latently expressed and binds to HAX-1, a protein with antiapoptotic function. J Virol 76:802–16.
396. Shepard, L. W., M. Yang, P. Xie, D. D. Browning, T. Voyno-Yasenetskaya, T. Kozasa, R. D. and Ye. 2001. Constitutive activation of NF-kappa B and secretion of interleukin-8 induced by the G protein-coupled receptor of Kaposi's sarcoma-associated herpesvirus involve G alpha[13] and RhoA. J Biol Chem 276:45979–87.
397. Shin, Y. C., H. Nakamura, X. Liang, P. Feng, H. Chang, T. F. Kowalik, and J. U. Jung. 2006. Inhibition of the ATM/p53 signal transduction pathway by Kaposi's sarcoma-associated herpesvirus interferon regulatory factor 1. J Virol 80:2257–66.

398. Shinohara, H., M. Fukushi, M. Higuchi, M. Oie, O. Hoshi, T. Ushiki, J. Hayashi, and M. Fujii. 2002. Chromosome binding site of latency-associated nuclear antigen of Kaposi's sarcoma-associated herpesvirus is essential for persistent episome maintenance and is functionally replaced by histone H1. J Virol **76**:12917–24.

399. Si, H., and E. S. Robertson. 2006. Kaposi's sarcoma-associated herpesvirus-encoded latency-associated nuclear antigen induces chromosomal instability through inhibition of p53 function. J Virol **80**:697–709.

400. Sieczkarski, S. B., and G. R. Whittaker. 2002. Dissecting virus entry via endocytosis. J Gen Virol **83**:1535–45.

401. Siegal, F. P., C. Lopez, G. S. Hammer, A. E. Brown, S. J. Kornfeld, J. Gold, J. Hassett, S. Z. Hirschman, C. Cunningham-Rundles, B. R. Adelsberg, and et al. 1981. Severe acquired immunodeficiency in male homosexuals, manifested by chronic perianal ulcerative herpes simplex lesions. N Engl J Med **305**:1439–44.

402. Simonart, T., C. Degraef, M. Heenen, P. Hermans, J. P. Van Vooren, and J. C. Noel. 2002. Expression of the fibroblast/macrophage marker 1B10 by spindle cells in Kaposi's sarcoma lesions and by Kaposi's sarcoma-derived tumor cells. J Cutan Pathol **29**:72–8.

403. Simpson, G. R., T. F. Schulz, D. Whitby, P. M. Cook, C. Boshoff, L. Rainbow, M. R. Howard, S. J. Gao, R. A. Bohenzky, P. Simmonds, C. Lee, A. de Ruiter, A. Hatzakis, R. S. Tedder, I. V. Weller, R. A. Weiss, and P. S. Moore. 1996. Prevalence of Kaposi's sarcoma associated herpesvirus infection measured by antibodies to recombinant capsid protein and latent immunofluorescence antigen. Lancet **348**:1133–8.

404. Smit, M. J., D. Verzijl, P. Casarosa, M. Navis, H. Timmerman, and R. Leurs. 2002. Kaposi's sarcoma-associated herpesvirus-encoded G protein-coupled receptor ORF74 constitutively activates p44/p42 MAPK and Akt via G(i) and phospholipase C-dependent signaling pathways. J Virol **76**:1744–52.

405. Sodhi, A., S. Montaner, and J. S. Gutkind. 2004. Does dysregulated expression of a deregulated viral GPCR trigger Kaposi's sarcomagenesis? Faseb J **18**:422–7.

406. Sodhi, A., S. Montaner, V. Patel, J. J. Gomez-Roman, Y. Li, E. A. Sausville, E. T. Sawai, and J. S. Gutkind. 2004. Akt plays a central role in sarcomagenesis induced by Kaposi's sarcoma herpesvirus-encoded G protein-coupled receptor. Proc Natl Acad Sci USA **101**:4821–6.

407. Sodhi, A., S. Montaner, V. Patel, M. Zohar, C. Bais, E. A. Mesri, and J. S. Gutkind. 2000. The Kaposi's sarcoma-associated herpes virus G protein-coupled receptor up-regulates vascular endothelial growth factor expression and secretion through mitogen-activated protein kinase and p38 pathways acting on hypoxia-inducible factor 1alpha. Cancer Res **60**:4873–80.

408. Soilleux, E. J., L. S. Morris, B. Lee, S. Pohlmann, J. Trowsdale, R. W. Doms, and N. Coleman. 2001. Placental expression of DC-SIGN may mediate intrauterine vertical transmission of HIV. J Pathol **195**:586–92.

409. Soilleux, E. J., L. S. Morris, G. Leslie, J. Chehimi, Q. Luo, E. Levroney, J. Trowsdale, L. J. Montaner, R. W. Doms, D. Weissman, N. Coleman, and B. Lee. 2002. Constitutive and induced expression of DC-SIGN on dendritic cell and macrophage subpopulations in situ and in vitro. J Leukoc Biol **71**:445–57.

410. Soilleux, E. J., L. S. Morris, S. Rushbrook, B. Lee, and N. Coleman. 2002. Expression of human immunodeficiency virus (HIV)-binding lectin DC-SIGNR: Consequences for HIV infection and immunity. Hum Pathol **33**:652–9.

411. Soilleux, E. J., L. S. Morris, J. Trowsdale, N. Coleman, and J. J. Boyle. 2002. Human atherosclerotic plaques express DC-SIGN, a novel protein found on dendritic cells and macrophages. J Pathol **198**:511–6.

412. Song, J., T. Ohkura, M. Sugimoto, Y. Mori, R. Inagi, K. Yamanishi, K. Yoshizaki, and N. Nishimoto. 2002. Human interleukin-6 induces human herpesvirus-8 replication in a body cavity-based lymphoma cell line. J Med Virol **68**:404–11.

413. Song, M. J., H. Deng, and R. Sun. 2003. Comparative study of regulation of RTA-responsive genes in Kaposi's sarcoma-associated herpesvirus/human herpesvirus 8. J Virol **77**:9451–62.

414. Song, M. J., X. Li, H. J. Brown, and R. Sun. 2002. Characterization of interactions between RTA and the promoter of polyadenylated nuclear RNA in Kaposi's sarcoma-associated herpesvirus/human herpesvirus 8. J Virol **76**:5000–13.

415. Soulier, J., L. Grollet, E. Oksenhendler, P. Cacoub, D. Cazals-Hatem, P. Babinet, M. F. d'Agay, J. P. Clauvel, M. Raphael, L. Degos, and et al. 1995. Kaposi's sarcoma-associated herpesvirus-like DNA sequences in multicentric Castleman's disease. Blood **86**:1276–80.

416. Spear, P. G., and R. Longnecker. 2003. Herpesvirus entry: an update. J Virol **77**:10179–85.

417. Spiller, O. B., D. J. Blackbourn, L. Mark, D. G. Proctor, and A. M. Blom. 2003. Functional activity of the complement regulator encoded by Kaposi's sarcoma-associated herpesvirus. J Biol Chem **278**:9283–9.

418. Spiller, O. B., L. Mark, C. E. Blue, D. G. Proctor, J. A. Aitken, A. M. Blom, and D. J. Blackbourn. 2006. Dissecting the regions of virion-associated Kaposi's sarcoma-associated herpesvirus complement control protein required for complement regulation and cell binding. J Virol **80**:4068–78.

419. Spiller, O. B., M. Robinson, E. O'Donnell, S. Milligan, B. P. Morgan, A. J. Davison, and D. J. Blackbourn. 2003. Complement regulation by Kaposi's sarcoma-associated herpesvirus ORF4 protein. J Virol **77**:592–9.

420. Staskus, K. A., W. Zhong, K. Gebhard, B. Herndier, H. Wang, R. Renne, J. Beneke, J. Pudney, D. J. Anderson, D. Ganem, and A. T. Haase. 1997. Kaposi's sarcoma-associated herpesvirus gene expression in endothelial (spindle) tumor cells. J Virol **71**:715–9.

421. Staudt, M. R., and D. P. Dittmer. 2006. Promoter switching allows simultaneous transcription of LANA and K14/vGPCR of Kaposi's sarcoma-associated herpesvirus. Virology **350**:192–205.

422. Staudt, M. R., and D. P. Dittmer. 2003. Viral latent proteins as targets for Kaposi's sarcoma and Kaposi's sarcoma-associated herpesvirus (KSHV/HHV-8) induced lymphoma. Curr Drug Targets Infect Disord **3**:129–35.

423. Stedman, W., Z. Deng, F. Lu, and P. M. Lieberman. 2004. ORC, MCM, and histone hyperacetylation at the Kaposi's sarcoma-associated herpesvirus latent replication origin. J Virol **78**:12566–75.

424. Sternbach, G., and J. Varon. 1995. Moritz Kaposi: idiopathic pigmented sarcoma of the skin. J Emerg Med **13**:671–4.

425. Stine, J. T., C. Wood, M. Hill, A. Epp, C. J. Raport, V. L. Schweickart, Y. Endo, T. Sasaki, G. Simmons, C. Boshoff, P. Clapham, Y. Chang, P. Moore, P. W. Gray, and D. Chantry, D. 2000. KSHV-encoded CC chemokine vMIP-III is a CCR4 agonist, stimulates angiogenesis, and selectively chemoattracts TH2 cells. Blood **95**:1151–7.

426. Sturzl, M., C. Hohenadl, C. Zietz, E. Castanos-Velez, A. Wunderlich, G. Ascherl, P. Biberfeld, P., Monini, J. Browning, and B. Ensoli. 1999. Expression of K13/v-FLIP gene of human herpesvirus 8 and apoptosis in Kaposi's sarcoma spindle cells. J Natl Cancer Inst **91**:1725–33.

427. Sturzl, M., A. Wunderlich, G. Ascherl, C. Hohenadl, P. Monini, C. Zietz, P. J. Browning, F. Neipel, P. Biberfeld, and B. Ensoli. 1999. Human herpesvirus-8 (HHV-8) gene expression in Kaposi's sarcoma (KS) primary lesions: an in situ hybridization study. Leukemia **13**(Suppl 1):S110–2.

428. Sturzl, M., C. Zietz, P. Monini, and B. Ensoli. 2001. Human herpesvirus-8 and Kaposi's sarcoma: relationship with the multistep concept of tumorigenesis. Adv Cancer Res **81**:125–59.

429. Sun, Q., S. Zachariah, and P. M. Chaudhary. 2003. The human herpes virus 8-encoded viral FLICE-inhibitory protein induces cellular transformation via NF-kappaB activation. J Biol Chem **278**:52437–45.

430. Sun, R., S. F. Lin, L. Gradoville, Y. Yuan, F. Zhu, and G. Miller. 1998. A viral gene that activates lytic cycle expression of Kaposi's sarcoma-associated herpesvirus. Proc Natl Acad Sci USA **95**:10866–71.

431. Sun, R., S. F. Lin, K. Staskus, L. Gradoville, E. Grogan, A. Haase, and G. Miller. 1999. Kinetics of Kaposi's sarcoma-associated herpesvirus gene expression. J Virol **73**:2232–42.

432. Swanton, C., D. J. Mann, B. Fleckenstein, F. Neipel, G. Peters, and N. Jones. Herpes viral cyclin/Cdk6 complexes evade inhibition by CDK inhibitor proteins. Nature **390**:184–7.

433. Szekely, L., C. Kiss, K. Mattsson, E. Kashuba, K. Pokrovskaja, A. Juhasz, P. Holmvall, and G. Klein. 1999. Human herpesvirus-8-encoded LNA-1 accumulates in heterochromatin-associated nuclear bodies. J Gen Virol **80**(Pt 11):2889–900.

434. Talbot, S. J., R. A. Weiss, P. Kellam, and C. Boshoff. 1999. Transcriptional analysis of human herpesvirus-8 open reading frames 71, 72, 73, K14, and 74 in a primary effusion lymphoma cell line. Virology **257**:84–94.

435. Tamura, T., M. Ishihara, M. S. Lamphier, N. Tanaka, I. Oishi, S. Aizawa, T. Matsuyama, T. W. Mak, S. Taki, and T. Taniguchi. 1995. An IRF-1-dependent pathway of DNA damage-induced apoptosis in mitogen-activated T lymphocytes. Nature **376**:596–9.

436. Tanaka, N., M. Ishihara, M. S. Lamphier, H. Nozawa, T. Matsuyama, T. W. Mak, S. Aizawa, T. Tokino, M. Oren, and T. Taniguchi. 1996. Cooperation of the tumor suppressors IRF-1 and p53 in response to DNA damage. Nature **382**:816–8.

437. Tanaka, N., M. Ishihara, and T. Taniguchi. 1994. Suppression of c-myc or fosB-induced cell transformation by the transcription factor IRF-1. Cancer Lett **83**:191–6.

438. Tang, J., G. M. Gordon, M. G. Muller, M. Dahiya, and K. E. Foreman. 2003. Kaposi's sarcoma-associated herpesvirus latency-associated nuclear antigen induces expression of the helix-loop-helix protein Id-1 in human endothelial cells. J Virol **77**:5975–84.

439. Tang, S., and Z. M. Zheng. 2002. Kaposi's sarcoma-associated herpesvirus K8 exon 3 contains three 5′-splice sites and harbors a K8.1 transcription start site. J Biol Chem **277**:14547–56.

440. Taylor, K. R., and R. L. Gallo. 2006. Glycosaminoglycans and their proteoglycans: host-associated molecular patterns for initiation and modulation of inflammation. Faseb J **20**:9–22.

441. Tetsuka, T., M. Higuchi, M. Fukushi, A. Watanabe, S. Takizawa, M. Oie, F. Gejyo, and M. Fujii. 2004. Visualization of a functional KSHV episome-maintenance protein LANA in living cells. Virus Genes **29**:175–82.

442. Tomkowicz, B., S. P. Singh, M. Cartas, and A. Srinivasan. 2002. Human herpesvirus-8 encoded Kaposin: subcellular localization using immunofluorescence and biochemical approaches. DNA Cell Biol **21**:151–62.
443. Turville, S. G., P. U. Cameron, A. Handley, G. Lin, S. Pohlmann, R. W. Doms, and A. L. Cunningham. 2002. Diversity of receptors binding HIV on dendritic cell subsets. Nat Immunol **3**:975–83.
444. Uccini, S., L. P. Ruco, F. Monardo, A. Stoppacciaro, E. Dejana, I. L. La Parola, D. Cerimele, and C. D. Baroni. 1994. Co-expression of endothelial cell and macrophage antigens in Kaposi's sarcoma cells. J Pathol **173**:23–31.
445. Ueda, K., K. Ishikawa, K. Nishimura, S. Sakakibara, E. Do, and K. Yamanishi. 2002. Kaposi's sarcoma-associated herpesvirus (human herpesvirus 8) replication and transcription factor activates the K9 (vIRF) gene through two distinct cis elements by a non-DNA-binding mechanism. J Virol **76**:12044–54.
446. Verma, S. C., S. Borah, and E. S. Robertson. 2004. Latency-associated nuclear antigen of Kaposi's sarcoma-associated herpesvirus up-regulates transcription of human telomerase reverse transcriptase promoter through interaction with transcription factor Sp1. J Virol **78**:10348–59.
447. Verma, S. C., T. Choudhuri, R. Kaul, and E. S. Robertson. 2006. Latency-associated nuclear antigen (LANA) of Kaposi's sarcoma-associated herpesvirus interacts with origin recognition complexes at the LANA binding sequence within the terminal repeats. J Virol **80**:2243–56.
448. Verma, S. C., K. Lan, T. Choudhuri, and E. S. Robertson. 2006. Kaposi's sarcoma-associated herpesvirus-encoded latency-associated nuclear antigen modulates K1 expression through its cis-acting elements within the terminal repeats. J Virol **80**:3445–58.
449. Verschuren, E. W., N. Jones, and G. I. Evan. 2004. The cell cycle and how it is steered by Kaposi's sarcoma-associated herpesvirus cyclin. J Gen Virol **85**:1347–61.
450. Verschuren, E. W., J. Klefstrom, G. I. Evan, and N. Jones. 2002. The oncogenic potential of Kaposi's sarcoma-associated herpesvirus cyclin is exposed by p53 loss in vitro and in vivo. Cancer Cell **2**:229–41.
451. Vieira, J., P. O'Hearn, L. Kimball, B. Chandran, and L. Corey. 2001. Activation of Kaposi's sarcoma-associated herpesvirus (human herpesvirus 8) lytic replication by human cytomegalovirus. J Virol **75**:1378–86.
452. Viejo-Borbolla, A., M. Ottinger, E. Bruning, A. Burger, R. Konig, E. Kati, J. A. Sheldon, and T. F. Schulz. 2005. Brd2/RING3 interacts with a chromatin-binding domain in the Kaposi's sarcoma-associated herpesvirus latency-associated nuclear antigen 1 (LANA-1) that is required for multiple functions of LANA-1. J Virol **79**:13618–29.
453. Wan, X., H. Wang, and J. Nicholas. 1999. Human herpesvirus 8 interleukin-6 (vIL-6) signals through gp130 but has structural and receptor-binding properties distinct from those of human IL-6. J Virol **73**:8268–78.
454. Wang, F. Z., S. M. Akula, N. P. Pramod, L. Zeng, and B. Chandran. Human herpesvirus 8 envelope glycoprotein K8.1A interaction with the target cells involves heparan sulfate. J Virol **75**:7517–27.
455. Wang, F. Z., S. M. Akula, N. Sharma-Walia, L. Zeng, and B. Chandran. 2003. Human herpesvirus 8 envelope glycoprotein B mediates cell adhesion via its RGD sequence. J Virol **77**:3131–47.
456. Wang, H. W., T. V. Sharp, A. Koumi, G. Koentges, and C. Boshoff. 2002. Characterization of an anti-apoptotic glycoprotein encoded by Kaposi's sarcoma-associated herpesvirus which resembles a spliced variant of human survivin. EMBO J **21**:2602–15.
457. Wang, H. W., M. W. Trotter, D. Lagos, D. Bourboulia, S. Henderson, T. Makinen, S. Elliman, A. M. Flanagan, K. Alitalo, and C. Boshoff. 2004. Kaposi's sarcoma herpesvirus-induced cellular reprogramming contributes to the lymphatic endothelial gene expression in Kaposi's sarcoma. Nat Genet **36**:687–93.
458. Wang, J., J. Zhang, L. Zhang, W. Harrington, Jr., J. T. West, and C. Wood. 2005. Modulation of human herpesvirus 8/Kaposi's sarcoma-associated herpesvirus replication and transcription activator transactivation by interferon regulatory factor 7. J Virol **79**:2420–31.
459. Wang, J. F., X. Zhang, and J. E. Groopman. 2004. Activation of vascular endothelial growth factor receptor-3 and its downstream signaling promote cell survival under oxidative stress. J Biol Chem **279**:27088–97.
460. Wang, L., D. P. Dittmer, C. C. Tomlinson, F. D. Fakhari, and B. Damania. 2006. Immortalization of primary endothelial cells by the K1 protein of Kaposi's sarcoma-associated herpesvirus. Cancer Res **66**:3658–66.
461. Wang, L., N. Wakisaka, C. C. Tomlinson, S. M. DeWire, S. Krall, J. S. Pagano, and B. Damania. The Kaposi's sarcoma-associated herpesvirus (KSHV/HHV-8) K1 protein induces expression of angiogenic and invasion factors. Cancer Res **64**:2774–81.
462. Wang, S., S. Liu, M. H. Wu, Y. Geng, and C. Wood. 2001. Identification of a cellular protein that interacts and synergizes with the RTA (ORF50) protein of Kaposi's sarcoma-associated herpesvirus in transcriptional activation. J Virol **75**:11961–73.

463. Wang, S. E., F. Y. Wu, H. Chen, M. Shamay, Q. Zheng, and G. S. Hayward. 2004. Early activation of the Kaposi's sarcoma-associated herpesvirus RTA, RAP, and MTA promoters by the tetradecanoyl phorbol acetate-induced AP1 pathway. J Virol 78:4248–67.

464. Wang, S. E., F. Y. Wu, M. Fujimuro, J. Zong, S. D. Hayward, and G. S. Hayward. 2003. Role of CCAAT/enhancer-binding protein alpha (C/EBPalpha) in activation of the Kaposi's sarcoma-associated herpesvirus (KSHV) lytic-cycle replication-associated protein (RAP) promoter in cooperation with the KSHV replication and transcription activator (RTA) and RAP. J Virol 77:600–23.

465. Wang, X. P., and S. J. Gao. 2003. Auto-activation of the transforming viral interferon regulatory factor encoded by Kaposi's sarcoma-associated herpesvirus (human herpesvirus-8). J Gen Virol 84:329–36.

466. Wang, X. P., Y. J. Zhang, J. H. Deng, H. Y. Pan, F. C. Zhou, and S. J. Gao. 2002. Transcriptional regulation of Kaposi's sarcoma-associated herpesvirus-encoded oncogene viral interferon regulatory factor by a novel transcriptional silencer, Tis. J Biol Chem 277:12023–31.

467. Wang, X. P., Y. J. Zhang, J. H. Deng, H. Y. Pan, F. C. Zhou, E. A. Montalvo, and S. J. Gao. 2001. Characterization of the promoter region of the viral interferon regulatory factor encoded by Kaposi's sarcoma-associated herpesvirus. Oncogene 20:523–30.

468. Watanabe, T., M. Sugaya, A. M. Atkins, E. A. Aquilino, A. Yang, D. L. Borris, J. Brady, and A. Blauvelt. 2003. Kaposi's sarcoma-associated herpesvirus latency-associated nuclear antigen prolongs the life span of primary human umbilical vein endothelial cells. J Virol 77:6188–96.

469. Weindel, K., D. Marme, and H. A. Weich. 1992. AIDS-associated Kaposi's sarcoma cells in culture express vascular endothelial growth factor. Biochem Biophys Res Commun 183:1167–74.

470. Weisz, A., S. Kirchhoff, and B. Z. Levi. 1994. IFN consensus sequence binding protein (ICSBP) is a conditional repressor of IFN inducible promoters. Int Immunol 6:1125–31.

471. Weisz, A., P. Marx, R. Sharf, E. Appella, P. H. Driggers, K. Ozato, and B. Z. Levi. 1992. Human interferon consensus sequence binding protein is a negative regulator of enhancer elements common to interferon-inducible genes. J Biol Chem 267:25589–96.

472. Weninger, W., T. A. Partanen, S. Breiteneder-Geleff, C. Mayer, H. Kowalski, M. Mildner, J. Pammer, M. Sturzl, D. Kerjaschki, K. Alitalo, and E. Tschachler. 1999. Expression of vascular endothelial growth factor receptor-3 and podoplanin suggests a lymphatic endothelial cell origin of Kaposi's sarcoma tumor cells. Lab Invest 79:243–51.

473. Weninger, W., M. Rendl, J. Pammer, M. Mildner, W. Tschugguel, C. Schneeberger, M. Sturzl, and E. Tschachler. 1998. Nitric oxide synthases in Kaposi's sarcoma are expressed predominantly by vessels and tissue macrophages. Lab Invest 78:949–55.

474. West, J. T., and C. Wood. 2003. The role of Kaposi's sarcoma-associated herpesvirus/human herpesvirus-8 regulator of transcription activation (RTA) in control of gene expression. Oncogene 22:5150–63.

475. Widmer, I., M. Wernli, F. Bachmann, G. Gudat, G. Cathomas, and P. Erb. 2002. Differential expression of viral Bcl-2 encoded by Kaposi's sarcoma-associated herpesvirus and human Bcl-2 in primary effusion lymphoma cells and Kaposi's sarcoma lesions. J Virol 76:2551–6.

476. Wong, E. L., and B. Damania, B. 2006. Transcriptional regulation of the Kaposi's sarcoma-associated herpesvirus K15 gene. J Virol 80:1385–92.

477. Wong, S. W., E. P. Bergquam, R. M. Swanson, F. W. Lee, S. M. Shiigi, N. A. Avery, J. W Fanton, and M. K. Axthelm. 1999. Induction of B cell hyperplasia in simian immunodeficiency virus-infected rhesus macaques with the simian homologue of Kaposi's sarcoma-associated herpesvirus. J Exp Med 190:827–40.

478. Wood, C., and W. Harrington Jr. 2005. AIDS and associated malignancies. Cell Res 15:947–52.

479. Wu, C., A. E. Chung, and J. A. McDonald. 1995. A novel role for alpha 3 beta 1 integrins in extracellular matrix assembly. J Cell Sci 108(Pt 6):2511–23.

480. Wu, F. Y., Q. Q. Tang, H. Chen, C. ApRhys, C. Farrell, J. Chen, M. Fujimuro, M. D. Lane, and G. S. Hayward. 2002. Lytic replication-associated protein (RAP) encoded by Kaposi's sarcoma-associated herpesvirus causes p21CIP-1-mediated G1 cell cycle arrest through CCAAT/enhancer-binding protein-alpha. Proc Natl Acad Sci USA 99:10683–8.

481. Wu, L., P. Lo, X. Yu, J. K. Stoops, B. Forghani, and Z. H. Zhou. 2000. Three-dimensional structure of the human herpesvirus 8 capsid. J Virol 74:9646–54.

482. Wu, T. T., E. J. Usherwood, J. P. Stewart, A. A. Nash, and R. Sun. 2000. Rta of murine gammaherpesvirus 68 reactivates the complete lytic cycle from latency. J Virol 74:3659–67.

483. Wu, W., J. Vieira, N. Fiore, P. Banerjee, M. Sieburg, R. Rochford, W. Harrington Jr., and G. Feuer. 2006. KSHV/HHV-8 infection of human hematopoietic progenitor (CD34+) cells: persistence of infection during hematopoiesis in vitro and in vivo. Blood.

484. Xie, J., H. Pan, S. Yoo, and S. J. Gao. 2005. Kaposi's sarcoma-associated herpesvirus induction of AP-1 and interleukin 6 during primary infection mediated by multiple mitogen-activated protein kinase pathways. J Virol 79:15027–37.

485. Xu, Y., D. P. AuCoin, A. R. Huete, S. A. Cei, L. J. Hanson, and G. S. Pari. 2005. A Kaposi's sarcoma-associated herpesvirus/human herpesvirus 8 ORF50 deletion mutant is defective for reactivation of latent virus and DNA replication. J Virol **79**:3479–87.

486. Yang, T. Y., S. C. Chen, M. W. Leach, D. Manfra, B. Homey, M. Wiekowski, L. Sullivan, C. H. Jenh, S. K. Narula, S. W. Chensue, and S. A. Lira. 2000. Transgenic expression of the chemokine receptor encoded by human herpesvirus 8 induces an angioproliferative disease resembling Kaposi's sarcoma. J Exp Med **191**:445–54.

487. Ye, F. C., F. C. Zhou, S. M. Yoo, J. P. Xie, P. J. Browning, and S. J. Gao. Disruption of Kaposi's sarcoma-associated herpesvirus latent nuclear antigen leads to abortive episome persistence. J Virol **78**:11121–9.

488. Ye, J., D. Shedd, and G. Miller. 2005. An Sp1 response element in the Kaposi's sarcoma-associated herpesvirus open reading frame 50 promoter mediates lytic cycle induction by butyrate. J Virol **79**:1397–408.

489. Yoo, S. M., F. C. Zhou, F. C. Ye, H. Y. Pan, and S. J. Gao. 2005. Early and sustained expression of latent and host modulating genes in coordinated transcriptional program of KSHV productive primary infection of human primary endothelial cells. Virology **343**:47–64.

490. You, R. I., M. C. Chen, H. W. Wang, Y. C. Chou, C. H. Lin, and S. L. Hsieh. 2006. Inhibition of lymphotoxin-beta receptor-mediated cell death by survivin-DeltaEx3. Cancer Res **66**:3051–61.

491. Yu, Y., S. E. Wang, and G. S. Hayward. 2005. The KSHV immediate-early transcription factor RTA encodes ubiquitin E3 ligase activity that targets IRF7 for proteosome-mediated degradation. Immunity **22**:59–70.

492. Zala, C., C. Ochoa, A. Krolewiecki, P. Patterson, P. Cahn, R. I. Crawford, and J. S. Montaner. 2000. Highly active antiretroviral therapy does not protect against Kaposi's sarcoma in HIV-infected individuals. AIDS **14**:2217–8.

493. Zhang, J., J. Wang, C. Wood, D. Xu, and L. Zhang. 2005. Kaposi's sarcoma-associated herpesvirus/human herpesvirus 8 replication and transcription activator regulates viral and cellular genes via interferon-stimulated response elements. J Virol **79**:5640–52.

494. Zhang, X., J. F. Wang, B. Chandran, K. Persaud, B. Pytowski, J. Fingeroth, and J. E. Groopman. 2005. Kaposi's sarcoma-associated herpesvirus activation of vascular endothelial growth factor receptor 3 alters endothelial function and enhances infection. J Biol Chem **280**:26216–24.

495. Zhang, Y. J., J. H. Deng, C. Rabkin, and S. J. Gao. 2000. Hot-spot variations of Kaposi's sarcoma-associated herpesvirus latent nuclear antigen and application in genotyping by PCR-RFLP. J Gen Virol **81**:2049–58.

496. Zhang, Y. J., H. Y. Pan, and S. J. Gao. 2001. Reverse transcription slippage over the mRNA secondary structure of the LIP1 gene. Biotechniques **31**:1286, 1288, 1290.

497. Zhong, W., H. Wang, B. Herndier, and D. Ganem. 1996. Restricted expression of Kaposi's sarcoma-associated herpesvirus (human herpesvirus 8) genes in Kaposi's sarcoma. Proc Natl Acad Sci USA **93**:6641–6.

498. Zhou, F. C., Y. J. Zhang, J. H. Deng, X. P. Wang, H. Y. Pan, E. Hettler, and S. J. Gao. 2002. Efficient infection by a recombinant Kaposi's sarcoma-associated herpesvirus cloned in a bacterial artificial chromosome: application for genetic analysis. J Virol **76**:6185–96.

499. Zhu, F. X., J. M. Chong, L. Wu, and Y. Yuan. 2005. Virion proteins of Kaposi's sarcoma-associated herpesvirus. J Virol 79, 800–11 (2005).

500. Zhu, F. X., T. Cusano, and Y. Yuan. 1999. Identification of the immediate-early transcripts of Kaposi's sarcoma-associated herpesvirus. J Virol **73**:5556–67.

501. Zhu, F. X., S. M. King, E. J. Smith, D. E. Levy, and Y. Yuan. 2002. A Kaposi's sarcoma-associated herpesviral protein inhibits virus-mediated induction of type I interferon by blocking IRF-7 phosphorylation and nuclear accumulation. Proc Natl Acad Sci USA **99**:5573–8.

502. Zhu, F. X., and Y. Yuan. 2003. The ORF45 protein of Kaposi's sarcoma-associated herpesvirus is associated with purified virions. J Virol **77**:4221–30.

503. Zimring, J. C., S. Goodbourn, and M. K. Offermann. 1998. Human herpesvirus 8 encodes an interferon regulatory factor (IRF) homolog that represses IRF-1-mediated transcription. J Virol **72**:701–7.

504. Zoeteweij, J. P., A. V. Moses, A. S. Rinderknecht, D. A. Davis, W. W. Overwijk, R. Yarchoan, J. M. Orenstein, and A. Blauvelt. 2001. Targeted inhibition of calcineurin signaling blocks calcium-dependent reactivation of Kaposi's sarcoma-associated herpesvirus. Blood **97**:2374–80.

4. KSHV-ASSOCIATED DISEASE IN THE AIDS PATIENT

DIRK P. DITTMER AND BLOSSOM DAMANIA

Department of Microbiology & Immunology & Lineberger Comprehensive Cancer Center, University of North Carolina, Chapel Hill, NC

INTRODUCTION

Twenty five to thirty percent of all human cancers are etiologically linked to an infectious agent, such as a virus or bacterium. These microbes are normally kept in check by the host immune system. However, in individuals that are immunodeficient, such as acquired immunodeficiency syndrome (AIDS) patients or those receiving immunosuppressive drugs following organ transplantation, this checkpoint fails and there is a correspondingly higher risk for the development of cancers associated with infectious agents. Viruses contribute to the development of neoplasia either cell autonomously through the activities of viral oncogenes, or through paracrine mechanisms that modulate the transformed cell as well as the supporting microenvironment.

Prior to the onset of the AIDS epidemic and the emergence of HIV, Kaposi's sarcoma (KS) was described in 1872 by Moritz Kaposi, the then head of the Vienna Dermatology clinic, as *"idiopathisches multiples Pigmentsarkom,"* a rare angiosarcoma in elderly men of Mediterranean descent.[1] In the early and mid-1980s, the HIV epidemic lead to a dramatic increase in the incidence of KS. KS remains the most common neoplasm seen in individuals with AIDS today. In 1994, KSHV (also known as human herpesvirus 8; HHV-8) was initially identified in KS lesions of AIDS patients by Chang et al.[2] using representational difference analysis.

KSHV AND THE DEVELOPMENT OF KS

KS is divided into four subtypes with distinct clinical manifestations: classic, endemic, epidemic or AIDS associated, and iatrogenic. Classic KS is a disease of elderly Mediterranean and Eastern European men, while endemic KS is found in parts of equatorial Africa such as Uganda and Zambia[3] where it is responsible for an estimated 1% of all tumors. Hence, in these regions, transmission of KSHV is thought to occur vertically. Endemic KS tends to be more aggressive than classic KS.

Widespread HIV-1 infection has resulted in an epidemic of KS. Prevalence levels for KSHV antibodies reach 30% in black South African HIV patients, and childhood KS has become the most common neoplasm in many regions of sub-Saharan Africa that are ravaged by HIV infection. In 1981, KS was recognized as a defining pathology for the diagnosis of AIDS. Highly active antiretroviral therapy (HAART) has led to a substantial decline of AIDS-related KS in the United States. However, even in the current post-HAART era, standardized incidence rates for KS are higher than that of any other AIDS-defining or AIDS-associated cancers.[4] This suggests that KS will remain a permanent health problem for years to come.

Iatrogenic KS occurs after solid-organ transplantation in patients receiving immunosuppressive therapy[5] and KS comprises an estimated 3% of all tumors associated with transplantation.[6] This is seen particularly in regions of high KSHV prevalence, such as Southern Italy, Turkey, and Saudi Arabia. KSHV present in the recipient may be acquired during iatrogenically induced immunodeficiency after transplantation or may be transmitted through the graft itself.[7] The frequency of KS in AIDS patients is 20,000 times higher than in the general population and the frequency of KS in transplant recipients is 500 times higher than in healthy individuals.[8] It is important to note that 95% of all KS lesions, regardless of clinical type or HIV status, contain KSHV viral DNA thus strongly linking KS to KSHV infection.

In the mid-1980s, incidence rates for KS showed a greater than exponential increase. At that time KS was only observed in AIDS patients with a history of men who had sex with men, but not in individuals who became HIV infected through blood transfusion.[9] In AIDS-associated KS, incidence rates correlated significantly with the lifetime number of male sexual partners,[3] which established KSHV as a sexually transmitted agent responsible for the development of this cancer. Today, more women have become infected with HIV and consequently KS has now also been reported in this group. African KS also affects both genders; while classic (Mediterranean) KS affects predominantly elder men. The reason for the gender bias in classic KS is unknown. In the US, KS incidence rates per age group follow a bimodal distribution that peaks at ages 30–36 and again at ages >70. Since incidence rates for most spontaneously occurring cancers increase exponentially with age due to the accumulation of spontaneous mutations in tumor suppressor genes and oncogenes, the bimodal distribution of KS posits an infectious agent with oncogenic potential as the cause of the disease.

The KS lesion itself is highly angiogenic and comprises spindle-shaped cells, slit-like endothelium-lined vasculature, and infiltrating blood cells. The spindle cells form the majority of the cell population, and are thought to arise from lymphatic endothelial cells.[10] In fact experimental KSHV infection can reprogram the endothelial gene expression profile into that of the lymphatic endothelium.[11,12]

KS lesions are classified as plaque, patched, or nodular. As the KS tumor progresses clinically, the number of KSHV-infected cells increase and the endothelial cell population within the lesion expands. KS lesions range from patches or plaques, to nodules and there is evidence for both polyclonality and monoclonality of the lesions.[13,14] It is thought that KS probably initiates as a polyclonal hyperplasia and develops into a clonal neoplasia. KS not only affects the skin but also involves multiple organs such as the liver, lung, spleen, and gastrointestinal tract. Very aggressive types of KS can lead to foci formation in the visceral organs and ultimately result in hemorrhage and death.

KSHV viral load in PBMC rises in the one-to-six months that precede lesion formation.[15] A rise in viral load predicts imminent clinical lesions independent of HIV or immune status[16] (also Dittmer and Martin, unpublished). KSHV is found in circulating B cells as well as macrophages and endothelial cells.[17–19] The presence of anti-KSHV antibodies documents prior exposure but does not allow a prediction of KS development, since in HIV-positive individuals the median time from seroconversion to disease is 7 years or greater.[3,9]

KS-tumor explants lose the virus after serial passage in tissue culture over time. This suggests that ex vivo passage selects for cells that no longer depend on the virus for survival and that the cells previously infected with KSHV have undergone epigenetic mutations or changes that allow the cells to persist without the virus. KSHV-infected endothelial cell preparations in culture generally also lose the virus over time,[20,21] though KSHV-positive tumor cell lines have recently been derived.[22]

KSHV AND THE DEVELOPMENT OF LYMPHOMAS

In addition to KS, KSHV is also found in B lymphoproliferative diseases; primary effusion lymphoma (PEL) and multicentric Castleman's disease (MCD). In fact, the first association of KS and a B-cell lymphoproliferative disorder, MCD, was reported in a patient who presented with both diseases.[23] Greater than 50% of KSHV-positive transplant recipients develop lymphoproliferative disease[24] and KSHV transmission can occur from organ donor to organ recipient[7].

The association between KSHV and MCD and PEL has been firmly established.[25,26] MCD is a B-cell lymphoproliferative disorder that is sometimes referred to as multicentric angiofollicular hyperplasia. As the name implies, patients usually present with diffuse lymphadenopathy and a series of constitutional symptoms. The disease is characterized by vascular proliferation of the germinal centers of the lymph node. There are two forms of MCD: (1) a plasmablastic variant form that is associated with lymphadenopathy and immune dysregulation and (2) a hyaline vascular form, which presents as a solid mass. Nearly 100% of AIDS-associated MCD is associated with KSHV, while approximately 50% of non–AIDS-associated MCD contains KSHV DNA. AIDS-associated MCD is usually accompanied by the development of KS in the affected individual.

MCD is a polyclonal tumor and is highly dependent on cytokines such as inter-leukin-6 (IL-6). KSHV itself encodes a viral IL-6 that is also expressed in these lesions.[27] Viral antigens can be detected in the immunoblastic B cells in the mantle zone of the lymph node. The plasmablasts in MCD express monotypic IgM light chains[28] yet MCD is a polyclonal disorder with MCD patients frequently developing cytopenia, autoimmune disease, and other malignancies such as KS and non-Hodgkin's lymphoma.[29]

PELs, sometimes referred to as body cavity based lymphomas, represent a specific subset of non-Hodgkin's B-cell lymphomas that involve body cavities and form a distinct clinicopathologic group.[30] Most PELs are KSHV-positive, and are often coinfected with EBV as well. PELs may involve the peritoneal, pleural, or pericardial cavities. These tumors are typically large-cell immunoblastic or anaplastic large-cell lymphomas that express CD45, clonal immunoglobulin gene rearrangements, and lack c-myc, bcl-2, ras, and p53 gene alterations.[29,30]

PELs have the characteristics of a preterminal stage of B-cell differentiation. Since PELs have mutations in their immunoglobulin genes, they are thought to arise from post-germinal center B cells. Most PELs express CD138/syndecan-1 antigen, which is normally expressed by a subset of plasma cells, but they do not express immunoglobulins.

Although KSHV has been associated with PEL and MCD in HIV patients, other reports describe cases of KSHV-positive lymphomas that do not fit the classic PEL phenotypes. For instance, KSHV has been linked to cases of ger-minotropic lymphoproliferative disease (GLD).[31] This disease also involves plasma-blasts but unlike plasmablastic lymphomas, the GLD lymphomas contain polyclonal immunoglobulin receptors. Another report suggests a high incidence of KSHV infection in solid HIV-associated immunoblastic/plasmablastic non-Hodgkin's lymphomas that developed in patients lacking PEL and MCD[32] and yet others have found KSHV associated with solid lymphomas, which resemble PEL cell morphology but do not present as effusions.[33] This suggests a model in which KSHV infects an early germinal center B cell that can still differentiate into multiple lymphoma phenotypes dependent on secondary mutations to the cellular genome.

The evidence linking KSHV to KS, PEL, and MCD is overwhelming and has been confirmed by multiple laboratories. KSHV DNA has also been detected in multiple myeloma, primary pulmonary hypertension, angiosarcomas, as well as malignant skin tumors in post-transplant patients such as Bowen's disease, squamous cell carcinomas, actinic keratosis, and extramammary Paget's disease. However, these disease associations are at present controversial.[29]

PREVALENCE OF VIRAL INFECTION

Several serology studies have suggested that KSHV infection is widespread in Africa with 30–60% of people being KSHV-positive, but is uncommon in the United States and Western Europe with seropositivity ranging from 3 to 10% in these areas.[34] Regions such as Italy and Greece show a higher prevalence of KSHV at about 4–35%,[35] which correlates with correspondingly higher incidence rates for classical or transplant-associated KS. Transmission routes include sexual transmission, mother-to-child transmission, as well as salivary transmission.[3,36,37]

THE KSHV GENOME

Herpesviruses are a diverse group of DNA viruses that differ in their biology and disease induction. A hallmark of herpesviruses is their ability to establish a latent infection for the lifetime of their host. Pathogenesis caused by these viruses is usually seen in the context of host immune suppression. All herpesviruses share a common evolutionary origin, which is highly evident from the homology seen among a substantial number of herpesviral genes.[38] Based on biological characteristics and genomic organization, herpesviruses are classified into three subfamilies: alpha, beta, and gamma. The gamma herpesviruses are lymphotropic and some are capable of undergoing lytic replication in epithelial, endothelial, or fibroblast cells. The gammaherpesvirinae are grouped into two classes: lymphocryptoviruses (gamma-1) and rhadinoviruses (gamma-2). Epstein-Barr virus (EBV) or human herpesvirus 4 (HHV4) is a lymphocryptovirus while KSHV (HHV8) is a rhadinovirus.

During latent infection, viral gene expression is highly attenuated and the viral genome remains stably associated with the cell. In the lytic phase of infection, viral gene expression and DNA replication ensue, leading to the production of progeny virions and eventual lysis of the cell. The KSHV viral genome comprises a ~140 kb long unique region flanked by multiple terminal repeat sequences with the total genomic size being ~160–170 kb. KSHV has at least 80 open-reading frames (ORFs) that encode for proteins greater than 100 amino acids.[39] The viral genes encoded by KSHV can be divided into three classes: (1) genes common to all herpesviruses, (2) genes unique to KSHV (these are generally given a "K" designation followed by the number of the open reading frame (ORF), and (3) KSHV-encoded genes that are homologous to cellular genes (these may be unique to KSHV or shared with other herpesviruses), and are likely to have been usurped from the host genome during the course of evolution. It is likely that several viral genes contribute to the neoplastic process[38] (Wong, 2005).

MOLECULAR BIOLOGY OF KSHV-ASSOCIATED DISEASE

A relatively recent concept in understanding KSHV gene expression in human KS, PEL, and MCD disease has involved the use of microarrays to profile viral gene expression. Since the KSHV genome is orders of magnitude smaller than the human genome, it has been feasible to develop whole genome arrays based upon real-time quantitative RT-PCR for all individual viral genes and to analyze primary KS biopsy samples and KSHV-infected lymphomas.[40,41] Conventional microarray-based viral gene expression in KSHV-infected lymphomas has also been performed.[42,43] These new techniques generate a viral signature for each disease state and offer a chance to classify KS beyond Moritz Kaposi's observational diagnosis. High-throughput genomic profiling offers the chance to accelerate our investigations into KSHV-associated cancers as much as it has benefited research into nonviral cancers. Microarray analyses of host cell transcription[44–46] proved that KSHV-positive PEL differ from other types of B-cell lymphomas. This is consistent with the idea that KSHV reprograms the tumor cell. Recently it was shown that KSHV infection reprograms endothelial cells toward a specialized cell fate; that of lymphatic endothelium, which expresses characteristic lineage markers, such as LYVE-1.[11,12] Several studies have ascertained the host transcription profile in tissue culture models of KSHV infection.[47–50] Potentially interesting drug targets emerged in each of these studies but a consensus among the different models has yet to be found. KS will almost certainly have a cellular transcription signature that is distinct from other cancers and tied to the unique pathology of this disease, as an angio-proliferative, cytokine driven disease. For instance, c-Kit and other growth factor receptors in microarray studies of KSHV-infected endothelial cells led to a successful pilot study using the kinase inhibitor gleevec (imatinib).[51] Another recent phase II study found a significant response rate of KS to a matrix metalloproteinase inhibitor.[52]

Every KS tumor transcribed high levels of the canonical KSHV latency genes LANA, vFLIP, vCyclin, and Kaposin. These genes are under control of the same promoter and are expressed in every KS tumor cell.[53,54] Kaposin is located immediately downstream of these three genes and, in addition to the common promoter, can be regulated by a promoter located between LANA and cyclin[55] and during lytic reactivation yet another, ORF-proximal promoter.[56] Like LANA, Kaposin too is expressed in every tumor cell[18] and has recently been shown to stabilize cellular cytokine mRNAs.[57] In addition to these latent proteins, many KS tumors as well as PEL engrafts[58,59] express an extended set of proteins that were initially classified as lytic viral genes, but in the context of the tumor may be the result of abortive or incomplete viral reactivation. These include the KSHV interferon regulatory factor (vIRF-1) and G-coupled receptor (vGPCR) homologs[41] and the K1 signaling protein,[59,60] which suggests that a subset of KS phenotypes may be attributable to these genes and the paracrine mechanisms that they invoke.[61–63] Interestingly, vIRF-3, a duplicated KSHV IRF homolog, is constitutively transcribed in KSHV-infected PEL,[64] but not KS. Thus we speculate that KSHV has to interfere with the host cell's innate interferon response in every infected cell regardless of cell lineage or mode of infection and has thus placed multiple copies of the vIRFs, all of which interfere with normal interferon signaling, under different control elements, e.g., vIRF-3 is specific for B cells while vIRF-1 is specific for endothelial cells. Thus, both latent and select lytic genes can be considered tumor-specific therapy targets for KS.

THERAPIES TO TREAT KS, PEL, AND MCD

Currently, treatment modalities for KS include observation, local therapy, or systemic chemotherapy such as paclitaxel and liopsomal anthracycline,[65] depending on the severity of the disease. Interferon alpha is also used to treat KS. KS is a highly angiogenic tumor but a clinical trial targeting the angiogenic nature of KS using IM862 proved ineffective in obliterating KS.[66] However, a clinical trial involving daily doses of imatinib mesylate (Gleevec), which targets c-kit and platelet derived growth factor receptor signaling, resulted in clinical and histologic regression of cutaneous KS,[51] as did a trial of a matrix metalloproteinase inhibitor.[52]

A recent report suggested that recipients of organ transplants, who were suscepti-ble to KS due to immunosuppressive therapy, benefited from treatment with sirolimus (also called rapamycin) since sirolimus displayed both immunosuppressive and antineoplastic properties.[8] Sirolimus likely acts via an antiangiogenic mechanism ultimately reducing the levels of VEGF and of VEGF receptor on endothelial cells.

Currently many clinical investigations are underway for treating KS. These include (1) bevacizumab, a recombinant human anti-VEGF antibody, (2) valproic acid, which can activate KSHV lytic replication in vitro leading to lysis of the infected cells,[67] (3) halofuginone, an inhibitor of matrix metalloproteinases, and (4) IL-12 therapy, with and without liposomal doxorubicin.[68]

Risk for KS and virally associated lymphomas increases rapidly as the CD4$^+$ cell counts of HIV-infected individuals diminish,[69] and the risk of developing AIDS-associated cancers is lower for individuals who are less severely immune suppressed. Since the prevalence of KS in AIDS patients is very high, and HIV coinfection is thought to be an important factor in the development of KS, attempts to control KS by improving the immune system of HIV-infected individuals through HAART is recommended. Indeed, the incidence of KS has declined considerably following the introduction of HAART therapy and often HAART alone will lead to KS regression in AIDS patients. However, it is important to note that even in the face of HAART therapy, the likelihood of an HIV-positive individual developing KS is still 20 times higher than uninfected individuals.[69]

Current treatments for MCD, PEL, and other AIDS lymphomas include standard chemotherapy such as CHOP, which contains four drugs – prednisone, vincristine, cyclophosphamide, and doxorubicin, or EPOCH, which in addition contains etoposide. These can be given simultaneously with HAART.[70,71] Case reports in the literature also suggest that rituximab (rituxan) is effective against PEL and MCD. Rituximab is an anti-CD20 antibody, but because rituximab targets normal B cells as well, it can be associated with an increased risk of infection when used in AIDS patients.[72] Scott et al. have reported on two MCD patients that went into sustained remission with just oral etoposide.[71] Another line of thinking has lead to exploratory studies using antiherpesviral drugs that inhibit herpesviral replication such as ganciclovir or AZT[73–75] in patients, suggesting a mechanism of action that suppresses viral reactivation and dissemination rather than direct tumor toxicity. Cidofovir, another herpesvirus polymerase inhibitor, did not show a clinical bene-fit.[76] HAART therapy has resulted in varying degrees of success with respect to decline in the incidence of non-Hodgkin lymphoma. It is estimated that HAART therapy decreases the incidence of non-Hodgkin lymphoma anywhere in the range of 40–76%. Moreover, there is emerging, but as of yet controversial evidence that protease inhibitors such as indinavir, which also inhibit matrix metalloproteinase may have direct anti-KS activity in addition to HAART-associated reconstitution of the immune system.[77]

More information on current trials that are underway to treat KS, PEL, and MCD can be gleaned by visiting the National Cancer Institute (NCI) website: http://www-dcs.nci.nih.gov/branches/aidstrials/adlist.html or http://www.amcoperations.com.

CONCLUSIONS

As a consequence of HAART therapy, the life expectancy of HIV-infected individuals has increased tremendously. It is likely that as these HIV-infected patients continue to age, there will be a corresponding increase in the incidence of AIDS-defining as well as non–AIDS defining cancers. Most of the current therapies with the exception of antiherpesviral drugs do not take advantage of the unique viral etiology of KSHV-associated cancers, and antiherpesviral drugs themselves are not effective against latent virus. Thus it will be important to show that "traditional" anticancer therapies are safe in the context of HAART and HIV infection, and to develop future therapies that directly impact upon, and obliterate, the function of viral genes.

ACKNOWLEDGMENTS

Due to space restrictions we regret that we had to omit many important references. We thank W. Harrington for critical reading. This work was supported by NIH grant CA096500 and HL083469 to BD, and NIH grant CA109232, a translational award from the Leukemia & Lymphoma Society and funding from the AIDS-associated clinical trials consortium (AMC) to DD. BD is a Leukemia & Lymphoma Society Scholar.

REFERENCES

1. Kaposi, M. 1872. Idiopatisches multiples Pigmentsarkom der Haut. Arch Dermatol Syphillis **4**:265–73.
2. Chang, Y., E. Cesarman, M. S. Pessin, et al. 1994. Identification of herpesvirus-like DNA sequences in AIDS-associated Kaposi's sarcoma. Science **266**(5192):1865–9.
3. Martin, J. N., D. E. Ganem, D. H. Osmond, et al. 1998. Sexual transmission and the natural history of human herpesvirus 8 infection. N Engl J Med **338**(14):948–54.
4. Mbulaiteye, S. M., R. J. Biggar, J. J. Goedert, et al. 2003. Immune deficiency and risk for malignancy among persons with AIDS. J Acquir Immune Defic Syndr **32**(5):527–33.
5. Civati, G., G. Busnach, B. Brando, et al. 1988. Occurrence of Kaposi's sarcoma in renal transplant recipients treated with low doses of cyclosporine. Transplant Proc **20**(Suppl 3):924–8.
6. Mendez, J. C., G. W. Procop, M. J. Espy, et al. 1999. Relationship of HHV8 replication and Kaposi's sarcoma after solid organ trasplantation. Transplantation **67**(8):1200–1.
7. Barozzi, P., M. Luppi, F. Facchetti, et al. 2003. Post-transplant Kaposi sarcoma originates from the seeding of donor-derived progenitors. Nat Med **9**(5):554–61.
8. Stallone, G., A. Schena, B. Infante, et al. 2005. Sirolimus for Kaposi's sarcoma in renal-transplant recipients. N Engl J Med **352**(13):1317–23.
9. Gao, S. J., L. Kingsley, D. R. Hoover, et al. 1996. Seroconversion to antibodies against Kaposi's sarcoma-associated herpesvirus-related latent nuclear antigens before the development of Kaposi's sarcoma. N Engl J Med **335**(4):233–41.

10. Dupin, N., M. Grandadam, V. Calvez, et al. 1995. Herpesvirus-like DNA sequences in patients with Mediterranean Kaposi's sarcoma. Lancet **345**(8952):761–2.
11. Hong, Y. K., K. Foreman, J. W. Shin, et al. 2004. Lymphatic reprogramming of blood vascular endothelium by Kaposi sarcoma-associated herpesvirus. Nat Genet **36**(7):683–5.
12. Wang, H. W., M. W. Trotter, D. Lagos, et al. 2004. Kaposi sarcoma herpesvirus-induced cellular reprogramming contributes to the lymphatic endothelial gene expression in Kaposi sarcoma. Nat Genet **36**(7):687–93.
13. Kaaya, S. F., M. T. Leshabari, and J. K. Mbwambo. 1998. Risk behaviors and vulnerability to HIV infection among Tanzanian youth. J Health Popul Dev Ctries **1**(2):51–60.
14. Rabkin, C. S., S. Janz, A. Lash, et al. 1997. Monoclonal origin of multicentric Kaposi's sarcoma lesions. N Engl J Med **336**(14):988–93.
15. Whitby, D., M. R. Howard, M. Tenant-Flowers, et al. 1995. Detection of Kaposi sarcoma associated herpesvirus in peripheral blood of HIV-infected individuals and progression to Kaposi's sarcoma. Lancet **346**(8978):799–802.
16. Pellet, C., D. Kerob, A. Dupuy, et al. 2006. Kaposi's sarcoma-associated herpesvirus viremia is associated with the progression of classic and endemic Kaposi's sarcoma. J Invest Dermatol **126**(3):621–7.
17. Decker, L. L., P. Shankar, G. Khan, et al. 1996. The Kaposi sarcoma-associated herpesvirus (KSHV) is present as an intact latent genome in KS tissue but replicates in the peripheral blood mononuclear cells of KS patients. J Exp Med **184**(1):283–8.
18. Staskus, K. A., W. Zhong, K. Gebhard, et al. 1997. Kaposi's sarcoma-associated herpesvirus gene expression in endothelial (spindle) tumor cells. J Virol **71**(1):715–9.
19. Rappocciolo, G., F. J. Jenkins, H. R. Hensler, et al. 2006. DC-SIGN is a receptor for human herpesvirus 8 on dendritic cells and macrophages. J Immunol **176**(3):1741–9.
20. Lagunoff, M., J. Bechtel, E. Venetsanakos, et al. 2002. De novo infection and serial transmission of Kaposi's sarcoma-associated herpesvirus in cultured endothelial cells. J Virol **76**(5):2440–8.
21. Grundhoff, A., and D. Ganem. 2004. Inefficient establishment of KSHV latency suggests an additional role for continued lytic replication in Kaposi sarcoma pathogenesis. J Clin Invest **113**(1):124–36.
22. An, F. Q., H. M. Folarin, N. Compitello, et al. 2006. Long-term-infected telomerase-immortalized endothelial cells: A model for Kaposi's sarcoma-associated herpesvirus latency in vitro and in vivo. J Virol **80**(10):4833–46.
23. De Rosa, G., E. Barra, M. Guarino, et al. 1989. Multicentric Castleman's disease in association with Kaposi's sarcoma. Appl Pathol **7**(2):105–10.
24. Farge, D., C. Lebbe, Z. Marjanovic, et al. 1999. Human herpes virus-8 and other risk factors for Kaposi's sarcoma in kidney transplant recipients. Groupe Cooperatif de Transplantation d' Ile de France (GCIF). Transplantation **67**(9):1236–42.
25. Cesarman, E., Y. Chang, P. S. Moore, et al. 1995. Kaposi's sarcoma-associated herpesvirus-like DNA sequences in AIDS-related body-cavity-based lymphomas. N Engl J Med **332**(18):1186–91.
26. Soulier, J., L. Grollet, E. Oksenhendler, et al. 1995. Kaposi's sarcoma-associated herpesvirus-like DNA sequences in multicentric Castleman's disease. Blood **86**(4):1276–80.
27. Parravicini, C., B. Chandran, M. Corbellino, et al. 2000. Differential viral protein expression in Kaposi's sarcoma-associated herpesvirus-infected diseases: Kaposi's sarcoma, primary effusion lymphoma, and multicentric Castleman's disease. Am J Pathol **156**(3):743–9.
28. Du, M. Q., H. Liu, T. C. Diss, et al. 2001. Kaposi sarcoma-associated herpesvirus infects monotypic (IgM lambda) but polyclonal naive B cells in Castleman disease and associated lymphoproliferative disorders. Blood **97**(7):2130–6.
29. Ablashi, D. V., L. G. Chatlynne, J. E. Whitman, Jr., et al. 2002. Spectrum of Kaposi's sarcoma-associated herpesvirus, or human herpesvirus 8, diseases. Clin Microbiol Rev **15**(3):439–64.
30. Nador, R. G., E. Cesarman, A. Chadburn, et al. 1996. Primary effusion lymphoma: a distinct clinico-pathologic entity associated with the Kaposi's sarcoma-associated herpes virus. Blood **88**(2):645–56.
31. Du, M. Q., T. C. Diss, H. Liu, et al. 2002. KSHV- and EBV-associated germinotropic lymphoproliferative disorder. Blood **100**(9):3415–8.
32. Deloose, S. T., L. A. Smit, F. T. Pals, et al. 2005. High incidence of Kaposi sarcoma-associated herpesvirus infection in HIV-related solid immunoblastic/plasmablastic diffuse large B-cell lymphoma. Leukemia **19**(5):851–5.
33. Carbone, A., A. Gloghini, E. Vaccher, et al. 2005. Kaposi's sarcoma-associated herpesvirus/human herpesvirus type 8-positive solid lymphomas: a tissue-based variant of primary effusion lymphoma. J Mol Diagn **7**(1):17–27.
34. Kedes, D. H., E. Operskalski, M. Busch, et al. 1996. The seroepidemiology of human herpesvirus 8 (Kaposi's sarcoma-associated herpesvirus): distribution of infection in KS risk groups and evidence for sexual transmission. Nat Med **2**(8):918–24.

35. Whitby, D., M. Luppi, P. Barozzi, et al. 1998. Human herpesvirus 8 seroprevalence in blood donors and lymphoma patients from different regions of Italy. J Natl Cancer Inst **90**(5):395–7.
36. Vieira, J., M. L. Huang, D. M. Koelle, et al. 1997. Transmissible Kaposi's sarcoma-associated herpesvirus (human herpesvirus 8) in saliva of men with a history of Kaposi's sarcoma. J Virol **71**(9):7083–7.
37. Brayfield, B. P., C. Kankasa, J. T. West, et al. 2004. Distribution of Kaposi sarcoma-associated herpesvirus/human herpesvirus 8 in maternal saliva and breast milk in Zambia: implications for transmission. J Infect Dis **189**(12):2260–70.
38. Damania, B. 2004. Oncogenic gamma-herpesviruses: comparison of viral proteins involved in tumorigenesis. Nat Rev Microbiol **2**(8):656–68.
39. Russo, J. J., R. A. Bohenzky, M. C. Chien, et al. 1996. Nucleotide sequence of the Kaposi sarcoma-associated herpesvirus (HHV8). Proc Natl Acad Sci USA **93**(25):14862–7.
40. Fakhari, F. D., and D. P. Dittmer. 2002. Charting latency transcripts in Kaposi's sarcoma-assocaited herpesvirus by whole-genome real-time quantitative reverse transcription-PCR. J Virol **76**(12):6213–23.
41. Dittmer, D. P. 2003. Transcription profile of Kaposi's sarcoma-associated herpesvirus in primary Kaposi's sarcoma lesions as determined by real-time PCR arrays. Cancer Res **63**(9):2010–5.
42. Jenner, R. G., M. M. Alba, C. Boshoff, et al. 2001. Kaposi's sarcoma-associated herpesvirus latent and lytic gene expression as revealed by DNA arrays. J Virol **75**(2):891–902.
43. Paulose-Murphy, M., N. K. Ha, C. Xiang, et al. 2001. Transcription program of human herpesvirus 8 (kaposi's sarcoma-associated herpesvirus). J Virol **75**(10):4843–53.
44. Jenner, R. G., K. Maillard, N. Cattini, et al. 2003. Kaposi's sarcoma-associated herpesvirus-infected primary effusion lymphoma has a plasma cell gene expression profile. Proc Natl Acad Sci USA **100**(18): 10399–404.
45. Klein, U., A. Gloghini, G. Gaidano, et al. 2003. Gene expression profile analysis of AIDS-related primary effusion lymphoma (PEL) suggests a plasmablastic derivation and identifies PEL-specific transcripts. Blood **101**(10):4115–21.
46. Fan, W., D. Bubman, A. Chadburn, et al. 2005. Distinct subsets of primary effusion lymphoma can be identified based on their cellular gene expression profile and viral association. J Virol **79**(2):1244–51.
47. Poole, L. J., Y. Yu, P. S. Kim, et al. 2002. Altered patterns of cellular gene expression in dermal microvascular endothelial cells infected with Kaposi's sarcoma-associated herpesvirus. J Virol **76**(7):3395–420.
48. Naranatt, P. P., H. H. Krishnan, S. R. Svojanovsky, et al. 2004. Host gene induction and transcriptional reprogramming in Kaposi's sarcoma-associated herpesvirus (KSHV/HHV-8)-infected endothelial, fibroblast, and B cells: insights into modulation events early during infection. Cancer Res **64**(1):72–84.
49. Moses, A. V., M. A. Jarvis, C. Raggo, et al. 2002. Kaposi's sarcoma-associated herpesvirus-induced upregulation of the c-kit proto-oncogene, as identified by gene expression profiling, is essential for the transformation of endothelial cells. J Virol **76**(16):8383–99.
50. Renne, R., C. Barry, D. Dittmer, et al. 2001. Modulation of cellular and viral gene expression by the latency-associated nuclear antigen of Kaposi's sarcoma-associated herpesvirus [In Process Citation]. J Virol **75**(1):458–68.
51. Koon, H. B., G. J. Bubley, L. Pantanowitz, et al. 2005. Imatinib-induced regression of AIDS-related Kaposi's sarcoma. J Clin Oncol **23**(5):982–9.
52. Dezube, B. J., S. E. Krown, J. Y. Lee, et al. 2006. Randomized phase II trial of matrix metalloproteinase inhibitor COL-3 in AIDS-related Kaposi's sarcoma: an AIDS Malignancy Consortium Study. J Clin Oncol **24**(9):1389–94.
53. Dittmer, D., M. Lagunoff, R. Renne, et al. 1998. A cluster of latently expressed genes in Kaposi's sarcoma-associated herpesvirus. J Virol **72**(10):8309–15.
54. Dupin, N., C. Fisher, P. Kellam, et al. 1999. Distribution of human herpesvirus-8 latently infected cells in Kaposi's sarcoma, multicentric Castleman's disease, and primary effusion lymphoma. Proc Natl Acad Sci USA **96**(8):4546–51.
55. Li, H., T. Komatsu, B. J. Dezube, et al. 2002. The Kaposi's sarcoma-associated herpesvirus K12 transcript from a primary effusion lymphoma contains complex repeat elements, is spliced, and initiates from a novel promoter. J Virol **76**(23):11880–8.
56. Sadler, R., L. Wu, B. Forghani, et al. 1999. A complex translational program generates multiple novel proteins from the latently expressed kaposin (K12) locus of Kaposi's sarcoma-associated herpesvirus. J Virol **73**(7):5722–30.
57. McCormick, C., and D. Ganem. 2005. The kaposin B protein of KSHV activates the p38/MK2 pathway and stabilizes cytokine mRNAs. Science **307**(5710):739–41.
58. Staudt, M. R., Y. Kanan, J. H. Jeong, et al. 2004. The tumor microenvironment controls primary effusion lymphoma growth in vivo. Cancer Res **64**(14):4790–9.
59. Wang, L., D. P. Dittmer, C. C. Tomlinson, et al. 2006. Immortalization of primary endothelial cells by the K1 protein of Kaposi's sarcoma-associated herpesvirus. Cancer Res **66**(7):3658–66.

60. Wang, L., N. Wakisaka, C. C. Tomlinson, S. DeWire, S. Krall, J. S. Pagano, and B. Damania. 2004. The Kaposi's sarcoma-associated herpesvirus (KSHV/HHV8) K1 protein induces expression of angiogenic and invasion factors. Cancer Res **64**:2774–81.
61. Moore, P. S., L. A. Kingsley, S. D. Holmberg, et al. 1996. Kaposi's sarcoma-associated herpesvirus infection prior to onset of Kaposi's sarcoma. Aids **10**(2):175–80.
62. Bais, C., A. Van Geelen, P. Eroles, et al. 2003. Kaposi's sarcoma associated herpesvirus G protein-coupled receptor immortalizes human endothelial cells by activation of the VEGF receptor-2/KDR. Cancer Cell **3**(2):131–43.
63. Montaner, S., A. Sodhi, A. Molinolo, et al. 2003. Endothelial infection with KSHV genes in vivo reveals that vGPCR initiates Kaposi's sarcomagenesis and can promote the tumorigenic potential of viral latent genes. Cancer Cell **3**(1):23–36.
64. Rivas, C., A. E. Thlick, C. Parravicini, et al. 2001. Kaposi's sarcoma-associated herpesvirus LANA2 is a B-cell-specific latent viral protein that inhibits p53 [In Process Citation]. J Virol **75**(1):429–38.
65. Northfelt, D. W., B. J. Dezube, J. A. Thommes, et al. 1998. Pegylated-liposomal doxorubicin versus doxorubicin, bleomycin, and vincristine in the treatment of AIDS-related Kaposi's sarcoma: results of a randomized phase III clinical trial. J Clin Oncol **16**(7):2445–51.
66. Noy, A., D. T. Scadden, J. Lee, et al. 2005. Angiogenesis inhibitor IM862 is ineffective against AIDS-Kaposi's sarcoma in a phase III trial, but demonstrates sustained, potent effect of highly active antiretroviral therapy: from the AIDS Malignancy Consortium and IM862 Study Team. J Clin Oncol **23**(5):990–8.
67. Klass, C. M., L. T. Krug, V. P. Pozharskaya, et al. 2005. The targeting of primary effusion lymphoma cells for apoptosis by inducing lytic replication of human herpesvirus 8 while blocking virus production. Blood **105**(10):4028–34.
68. Little, R. F., J. M. Pluda, K. M. Wyvill, et al. 2006. Activity of subcutaneous interleukin-12 in AIDS-related Kaposi's sarcoma. Blood **107**(12):4650–7.
69. Clifford, G. M., J. Polesel, M. Rickenbach, et al. 2005. Cancer risk in the Swiss HIV Cohort Study: associations with immunodeficiency, smoking, and highly active antiretroviral therapy. J Natl Cancer Inst **97**(6):425–32.
70. Ratner, L., J. Lee, S. Tang, et al. 2001. Chemotherapy for human immunodeficiency virus-associated non-Hodgkin's lymphoma in combination with highly active antiretroviral therapy. J Clin Oncol **19**(8):2171–8.
71. Scott, D., L. Cabral, W. J. Harrington, Jr. 2001. Treatment of HIV-associated multicentric Castleman's disease with oral etoposide. Am J Hematol **66**(2):148–50.
72. Kaplan, L. D., J. Y. Lee, R. F. Ambinder, et al. 2005. Rituximab does not improve clinical outcome in a randomized phase 3 trial of CHOP with or without rituximab in patients with HIV-associated non-Hodgkin lymphoma: AIDS-Malignancies Consortium Trial 010. Blood **106**(5):1538–43.
73. Ghosh, S. K., C. Wood, L. H. Boise, et al. 2003. Potentiation of TRAIL-induced apoptosis in primary effusion lymphoma through azidothymidine-mediated inhibition of NF-kappa B. Blood **101**(6):2321–7.
74. Casper, C., W. G. Nichols, M. L. Huang, et al. 2004. Remission of HHV-8 and HIV-associated multicentric Castleman disease with ganciclovir treatment. Blood **103**(5):1632–4.
75. Martin, D. F., B. D. Kuppermann, R. A. Wolitz, et al. 1999. Oral ganciclovir for patients with cytomegalovirus retinitis treated with a ganciclovir implant. Roche Ganciclovir Study Group. N Engl J Med **340**(14):1063–70.
76. Little, R. F., F. Merced-Galindez, K. Staskus, et al. 2003. A pilot study of cidofovir in patients with kaposi sarcoma. J Infect Dis **187**(1):149–53.
77. Sgadari, C., G. Barillari, E. Toschi, et al. 2002. HIV protease inhibitors are potent anti-angiogenic molecules and promote regression of Kaposi sarcoma. Nat Med **8**(3):225–32.
78. Wong E. L., B. Damania. 2005. Linking KSHV to human cancer. Curr Oncol Rep **7**(5):349–356.

5. MOLECULAR BIOLOGY OF EBV IN RELATIONSHIP TO AIDS-ASSOCIATED ONCOGENESIS

BHARAT G. BAJAJ, MASANAO MURAKAMI, AND ERLE S. ROBERTSON

Department of Microbiology and the Tumor Virology Program, Abramson Comprehensive Cancer Center, University of Pennsylvania Medical School, Philadelphia, PA

INTRODUCTION

Epstein–Barr virus (EBV) was first isolated from cultures stemming from a biopsy of a Burkitt's lymphoma in 1958.[12] It was identified as a large DNA virus belonging to the gammaherpesvirus family.[26] Soon afterward, it was shown to possess the ability to transform primary B lymphocytes in vitro, leading to blast formation and uncontrolled proliferation.[38] This in vitro transformation ability supported an oncogenic potential associated with the virus. Since then, EBV has been shown to be associated with a number of human malignant disorders.

EBV has all the hallmarks of an extremely successful virus. Current estimates indicate that EBV infects approximately 95% of the world's adult population.[133] In most healthy individuals EBV infection does not lead to any overt pathogenesis, thus ensuring survival of both the host and the virus. EBV infection is able to persist, and escape clearance, for the lifetime of the host. Primary infection is followed by long periods of latency, during which the viral DNA is replicated along with the host cell, but little or no virus particles are produced.[93] This is in large part responsible for successful immune evasion by EBV. These long periods of latency are interspersed by occasional short periods of reactivation.[29] It is during these periods that fresh virus particles are produced and these find new host cells either in the same or in a different individual. Thus, many millennia of coevolution along

Table 1. EBV-associated diseases. EBV-associated malignancies and other contributing factors are tabulated

Disease	AIDS association	Other association
Burkitt's lymphoma	Yes	Immunosuppression, endemic (parts of Africa)
B-lymphoproliferative disease	Yes	Post-transplant
Infectious mononucleosis	No	Immunosuppression
Nasopharyngeal carcinoma	No	Endemic (parts of Asia)
Primary effusion lymphoma	Yes	Kaposi's sarcoma-associated herpes virus
Natural killer cell lymphoma	Yes	
Hodgkin's disease	No	
X-linked lymphoproliferative disease	No	Human XLP gene
Post transplant lymphoproliferative disease	No	Immunosuppression after organ transplant

with humans have ensured that EBV exists along a fine line between pathogenesis accompanied by possible death of the individual and clearance by the immune system. Primary or acquired immunodeficiency on the part of humans disturbs this equilibrium and leads to the development of EBV-associated diseases (Table 1). Ironically, this is not likely to be in the best interest of the virus.

During the past decade, substantial gaps in our understanding of the biology of EBV in the healthy host, the molecular mechanisms involved in viral proliferation and in oncogenesis, have been somewhat narrowed. This progress has provided the building blocks for development of novel therapeutic approaches that will eventually make a significant impact on clinical management of EBV-associated lymphoid malignancies. In this chapter, we discuss the molecular biological aspect of EBV-associated malignancies, and focus on recent advances in our understanding of the functions of specific latent viral antigens and their contribution to disease pathogenesis.

THE EPSTEIN-BARR VIRUS LIFECYCLE

As mentioned earlier, EBV belongs to the herpes virus family. The viral genome is ~184 kb and 85 open reading frames (ORFs) have thus far been identified.[7] Most of these genes have cellular homologs and play a part in either viral replication or immune modulation. Like all herpes viruses, EBV has a biphasic lifecycle.[82] Infection can occur by exchange of bodily fluids including but not limited to saliva.[32] EBV predominantly infects human primary B lymphocytes and certain epithelial cells of the nasopharynx. Primary infection is followed by a brief period of virus replication at the site of infection. In a healthy individual, this is usually modulated by the immune system and is believed to be as short as 24 h. This replication period allows for the systemic spread of the virus and allows establishment of latency in memory B cells, which are infected in the lymphoid organs.[29] Latent infection is characterized by minimal viral gene expression and no lytic replication of the virus occurs during this period.[93] The viral genome is maintained episomally and replicates at the same time, using the same machinery, as host cells DNA replication.[2] Typically, 11 of the 85 identified ORFs are expressed during virus latency program III.[93] This low profile likely helps the virus evade immune surveillance, a key requirement for lifetime persistence. Differentiation of the memory B cells can cause reactivation of the latent virus into lytic replication.[67] This is important for shedding and transmission of the virus. Onset of lytic replication is characterized by a preprogrammed pattern of virus gene expression.[77] Genes expressed during this phase have been divided into three subgroups, immediate-early, early, and late genes, depending on the timeframe of their expression.[77] Immediate early genes are mostly transcriptional inducers of early gene expression, early genes are involved in facilitating host cell independent virus DNA replication and late genes facilitate production of viral structural genes, virus assembly and egress.[77] Thus, while the latent program ensures immune evasion and host survival, the occasional lytic reactivation of a small fraction of the latently infected pool ensures virus shedding and transmission (Fig. 1).

There are three distinct types of the EBV latent programs. Each type is characterized by a distinct viral gene expression pattern and is associated with different disease states (Table 2). It is generally believed that the first latency program established

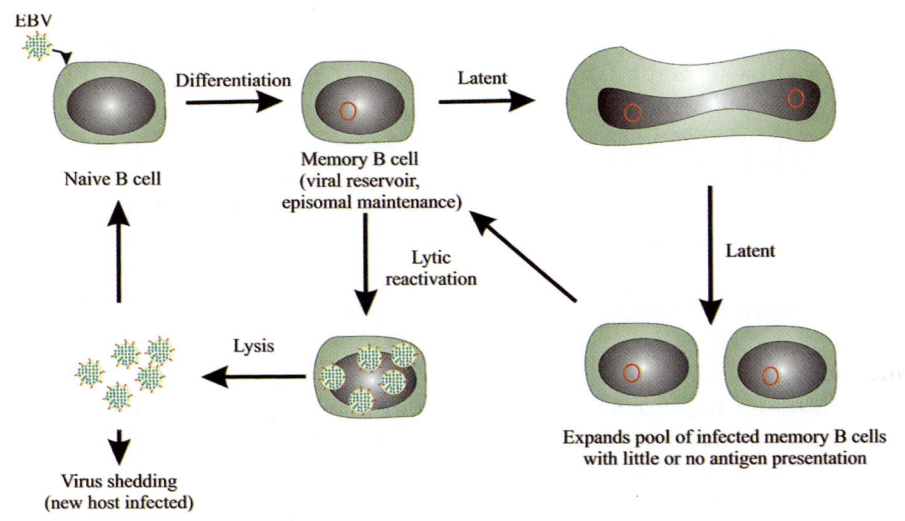

Figure 1. Schematic of the biphasic lifecycle of Epstein-Barr virus (EBV). EBV, like most herpes viruses, follows a biphasic life cycle.[82] The virus infects a naïve B cell, which then differentiates into a B memory cell. In these long-lived memory cells, the virus establishes a latent program where the virus genome replicates at the same time as the host cell DNA, using the host cell DNA replication machinery, and is efficiently portioned into daughter cells during mitosis. This expands the pool of cells carrying the viral genome with minimal expression of viral antigens. Certain stimuli, like B-cell receptor signaling, can lead to lytic reactivation of the virus in these infected B memory cells. Triggering of the lytic cascade leads to productive replication of the virus and eventual lysis of the cell. These newly formed particles can infect fresh cells within the same individual or be transmitted by exchange of body fluids to an uninfected individual leading to the spread of virus (adapted from Laichalk et al., J. Virol 2005).

Table 2. EBV protein expression in AIDS-associated malignancies. EBV antigen expression and the type of latency program

Disease	Type of latency	EBV proteins expressed
Burkitt's lymphoma	Type I	EBNA1, EBERs, BARTs
B-lymphoproliferative disease	Type III	EBNAs (EBNA1, 2, 3A, 3B, 3C, LP), LMP1, LMP2, EBERs, BARTs
Primary effusion lymphoma	Type III	
Natural killer cell lymphoma	Type II	EBNA1, LMP1, LMP2, EBERs, BARTs

by EBV is latency type III, also known as the growth program.[93] This is characterized by the expression of six EBV nuclear antigens (EBNAs; EBNA1, EBNA2, EBNA3A, EBNA3B, EBNA3C, and EBNA-LP), two membrane proteins (latent membrane protein (LMP); LMP1 and LMP2), two small viral RNAs (EBERs), and the BamA rightward transcripts (BARTs). These proteins drive the infected B cells to proliferate. Since the viral genome is replicated at the same time as the host cell DNA, this type of latency thus expands the pool of infected B cells, and hence the name growth program. However, this large gene expression profile engenders a robust T-cell response and clearance of cells expressing this type of latency. This leaves some B cells in which gene expression has been further tempered to minimize a possible T-cell response. These B cells are in either latency type II (also known as the default program) or in latency I (or the latency program). Type II latency is characterized by the expression of EBNA1, LMP1, LMP2, EBERs, and BARTs. In latency type I, only EBNA1, EBERs, and BARTs are expressed.[134]

EBV-RELATED DISEASES

EBV infection in healthy individuals is characterized by a rapid IgM viral capsid antigen (VCA) antibody response. This usually wanes within 2–3 months postinfection and is replaced by an IgG VCA response which persists for the lifetime of the infected individual. At the same time, antibodies against the EBNA are produced, which also persist for life. In healthy individuals, the number of infected B cells is tightly controlled by EBV specific cytotoxic T lymphocytes. These cells are responsible for the shift from the growth program, established immediately postinfection, to either the default or the latency program, and thus prevent uncontrolled proliferation of infected B cells. It is estimated that in these healthy individuals, approximately 1 in 10^5–10^6 circulating B cells is infected. In immunosuppressed individuals, on the other hand, in the absence of a robust cytotoxic T lymphocyte response, the virus is free to revert to the growth program. The resultant unchecked proliferation of B cells provides ideal conditions for the development of lymphoproliferative disorders. In such individuals, EBV is associated with a number of such proliferative diseases, including Burkitt's lymphoma, B-cell lymphoproliferative disease, infectious mononucleosis, nasopharyngeal carcinoma, natural killer cell lymphoma, primary effusion lymphoma, Hodgkin's disease, X-linked lymphoproliferative disease and post-transplant lymphoproliferative disease. The immunosuppressed state leading to the development of the mentioned disorders in infected individuals could be either acquired, as is the case in AIDS, or it could be induced, such as in transplant patients. Accordingly, Burkitt's lymphoma, B lymphoproliferative disease, primary effusion lymphoma, and natural killer cell lymphoma have been shown to be exacerbated by dint of the patient being HIV positive (Table 1). EBV infection also poses a significant problem in transplant patients who are generally immunosuppressed in order to prevent organ rejection. Interestingly, patients that are EBV seronegative at the time of transplant are at greater risk of developing post-transplant lymphoprolif-erative disease than previously infected patients.[114] Other factors that are believed to contribute to the development of post-transplant lymphoma are the type of organ transplanted and the level and type of immunosuppression.[114] Therefore, children in general are at a higher risk of developing PTLD than adults, as they are more likely to be EBV seronegative before transplant. Early-onset PTLD is usually of a polyclonal origin, expresses the latency III repertoire of EBV anti-gens, and is often successfully treated by reduction of immunosuppression and/or administration of antiviral drugs. Late-onset PTLD is, on the other hand, usually monoclonal and has a down-regulated EBNA expression of latency type I or II. Patients with late-onset PTLD or recurring PTLD have a poor prognosis, and the disease is often fatal.[114] These diseases have been grouped into three categories based on the type of latency and the corresponding range of EBV antigens expressed by the infected cell (Table 1). In the subsequent sections, each of the viral antigens expressed in one or more disease associated with AIDS is discussed in detail.

EBV LATENT GENES EXPRESSED IN AIDS-ASSOCIATED MALIGNANCIES

All or some of the following genes are expressed in at least one of the AIDS-associated malignancies (Table 2).

EPSTEIN-BARR VIRUS NUCLEAR ANTIGEN 1

It is now well established that EBV is capable of maintaining its genome extra-chromosomally in dividing mammalian cells,[128,131] with only infrequent incidences of integration reported.[66] Replication of the viral genome using the host cell repli-cation machinery and segregation into daughter cells upon mitosis is the established form of replication in all types of EBV-associated latencies.[23,104,130] Two compo-nents are thought to be necessary to accomplish this, the viral origin of plasmid replication (oriP),[129] and the EBV nuclear antigen 1 (EBNA1)[70] (Fig. 2). Proof for this is provided by the fact that inclusion of both these components on conven-tional plasmids lacking the eukaryotic origin of replication is sufficient to allow them to replicate and be retained in dividing cells.[130,131]

EBNA1 is believed to maintain the viral genome by binding to both the episome and the host chromatin, thus acting as a protein anchor[78] (Table 3). EBNA1 is there-fore deemed the most important of all EBV nuclear antigens and is accordingly

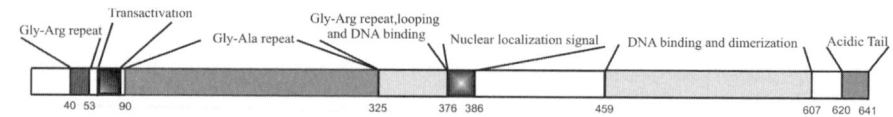

EBV Nuclear Antigen 1 (EBNA1)

Figure 2. EBV nuclear antigen 1. The EBV nuclear antigen 1 (EBNA1) is one of six nuclear antigens encoded by EBV. It is expressed in all EBV-infected cells and is the only nuclear antigen expressed in cells in type I latency.[93] Its most important function is maintenance of the viral episome by binding both the viral genome and the host cell chromosome.[126] Accordingly, it has two identified DNA-binding domains and a nuclear localization signal.[78] The glycine-alanine repeat domain of EBNA1 is responsible for its resistance to proteosomal degradation and relative nonimmunogenicity.[9,72] The transactivation domain plays a role in EBNA1 mediated regulation of transcription.[126]

Table 3. EBV latency proteins. The major function and other associated roles of the EBV latency proteins

EBV protein	Latency	Major function	Other functions	Required for Immortalization	Size
EBNA1	I, II, III	Episomal maintenance	Tethers viral genome to host cell chromosome, viral DNA replication, transcriptional activation, immune evasion	Yes	641 aa (64–69 kDa)
EBNA 2	III	Transcriptional modulation	Interacts with RBP-Jκ and modulates notch signaling Targets: cMyc, LMP1 and 2, CD21, CD23	Yes	487 aa (85–87 kDa)
EBNA 3A	III	Transcriptional modulation	EBNA2 antagonist and coactivator	Yes	944 aa (~190 kDa)
EBNA 3B	III	Transcriptional modulation	Interacts with RBP-Jκ	No	938 aa (~195 kDa)
EBNA 3C	III	Transcriptional modulation	EBNA2 antagonist and coactivator, recruits SCF^{SKP2} ubiquitin ligase to degrade pRb, p27 (cell cycle deregulation)	Yes	992 aa (~192 kDa)
EBNA-LP	III	Transcriptional modulation	EBNA2 coactivation	Yes/no	506 aa
LMP1	II, III	Transcriptional modulation	Induces constitutive NF-κB activation, cytotoxic T lymphocyte modulation	Yes	386 aa (63 kDa)
LMP2A	II, III	Maintenance of latency	Disrupts B-cell receptor signaling	No	497 aa (54 kDa)
BARTs	I, II, III	Function unclear	Modulation of notch signaling	No	–
EBERs	I, II, III	Antiapoptotic	Disrupt α-Interferon-mediated apoptosis	No	–

expressed in all EBV-infected cells.[100] Indeed, in latency type I, which has the lowest viral antigen expression profile of all known latency types, EBNA1 is the only nuclear antigen expressed.[100] Until recently, it was believed that EBV persists in this EBNA1-only latency program without detection by the immune system.

EBNA1 is generated from the *Bam* HI-Q/U/K fragments of EBV genome, and is approximately 64–69 kDa (641 aa), depending on number of the internal repeats.[56] A number of domains fulfilling specific functions have been identified within EBNA1 (Fig. 2). It has two separate glycine-arginine repeats, a transactivation domain, a glycine-alanine repeat, two separate DNA-binding domains, a nuclear localization signal, and a dimerization domain.[87] The glycine-alanine repeat of EBNA1 is thought to confer resistance of the viral protein to proteasomal degradation and therefore prevent antigen presentation.[9,72] This has been demonstrated by fusing the glycine-alanine repeat with labile cellular proteins, which then become resistant to degradation.[72] This apparent lack of antigen presentation is believed to be the primary reason for cytotoxic T lymphocyte escape by EBV in latency type I and is thought to be crucial for maintenance of the viral reservoir in healthy infected individuals.

Several cellular proteins that interact with EBNA1 have been identified. These include the cellular origin recognition complex[17] and other components of the prereplication complex, replication protein A, and the telomere repeat binding factor 2 (TRF2).[21] EBNA1 also binds specifically to two clusters within the EBV viral origin of plasmid replication (oriP). This is a 1.7 kb *cis*-acting region in the EBV genome, and is composed of 21 families of repeat sequences (FR) and four dyad symmetry elements (DS). Binding of EBNA1 to oriP and its recruitment of the host cell replication machinery elements demonstrate its importance in mediating EBV genome replication.[30]

Additionally, EBNA-1 has also been shown to play a role in transcriptional regulation of the three EBNA promoters, Wp, Cp, and its own latent promoter Qp.[100] EBNA1 has also been shown to bind to the metastasis suppressor protein Nm23-H1[87] and to specific viral RNA sequences, EBER1.[110] It has a demonstrated tumorigenic potential in vivo.[124] These experimental data suggest that EBNA1 has a critical role in EBV-associated human cancers.

EPSTEIN-BARR VIRUS NUCLEAR ANTIGEN 2

EBNA2 is one of the first viral proteins expressed following EBV infection.[4] It is encoded by the *Bam* HI W/Y/H region of the EBV genome (BHRF1) and its size depends on the length of the polyproline domain, which is, 82–87 kDa in type I: B95-8 and 75 kDa in type II: AGS and P3HR-1 (Fig. 3). EBNA2 is essential for transformation of human B lymphocytes and is a transactivator of the EBV genome as well as many host cell encoded genes.[18,35] In a positive feedback loop, it is known to transactivate its own promoter.[125] Other genes known to be upregulated by EBNA2 include EBV proteins EBNA1, EBNA3A, 3B, 3C, LMP1, and LMP2A,[1,27,94,123,139] and cellular genes c-myc, CD21, c-fgr, and CD23.[65,112,121,122] Consistent with this observation is the fact that EBNA 2, 3A, 3B, and 3C are either

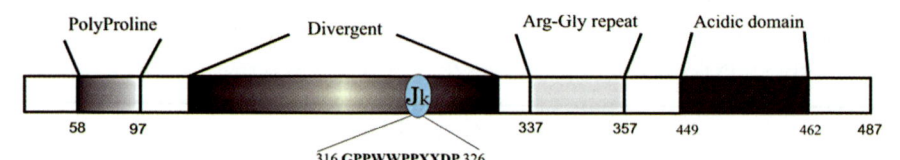

EBV nuclear antigen 2 (EBNA2)

Figure 3. EBV nuclear antigen 2. EBV nuclear antigen 2 (EBNA2) along with EBNA-LP are the first two genes that are expressed following EBV infection.[4] EBNA2 is a potent transcriptional regulator and controls the transcription of various viral as well as cellular genes.[1,27,68,139] It does not bind directly to DNA and exerts its effect by binding to and regulating the activity of cellular transcription factors.[68] In infected cells EBNA2 mimics a constitutively active intracellular notch, interacts with and releases transcriptional suppression mediated by the corepressor RBP-Jκ {Huen, 1995 #110}. It also binds other cellular DNA-binding proteins CBP1 and PU.1.[68] It has been shown to transactivate the HIV LTR via its NF-κB sites.[103]

coexpressed, as in latency-III, or not at all, as in latency programs-I and II. Despite its functions as a major transcriptional modulator, EBNA2 does not directly bind DNA.[138] Its transcriptional modulation is a result of its association with various cellular DNA-binding proteins such as CBF1, RBP-Jκ, and PU.1.[37,68,74]

Although a wide array of genes is regulated by EBNA2 via its association with cellular transcription factors, its predominant mechanism of activation is its ability to mimic the activities of activated notch-I.[127] RBP-Jκ/CBF-1/CSL is a ubiquitous DNA-binding protein which recognizes the sequence CGTGGGAA.[81] The intracellular region of notch (ICN) has been shown to possess transactivation ability when overexpressed in various cell lines. ICN directly binds the nuclear protein RBP-Jκ at two regions: the RAM domain located immediately C terminal to the transmembrane region[118] and the CDC 10/ankyrin repeats. EBNA2 has been shown to bind to this same part of RBP-Jκ and can functionally replace the intracellular region of notch.[54,99] Therefore, EBNA2 acts as a constitutively activated form of notch-I and results in activation of the notch-I signaling pathway and its downstream genes.[54] The ability of EBNA2 for B-cell immortalization correlates directly with its ability to ameliorate transcriptional repression by RBP-Jκ.[127] Also, the ability of EBNA2 to functionally mimic ICN is further demonstrated by the fact that ICN can modulate expression of EBNA2 target genes.[39] Interestingly, EBNA2 transactivates the human immunodeficiency virus (HIV) LTR, and this transactivation is dependent on the NF-κB sites in the HIV LTR.[103] The exact repercussions of this transactivation are, however, not well understood.

EPSTEIN-BARR VIRUS NUCLEAR ANTIGEN LEADER PROTEIN

The Epstein-Barr virus nuclear antigen leader protein (EBNA-LP) contains multiple copies of a 66 aa repeat domain encoded by two exons in the internal repeat 1 (IR1), repeats W1 (22 amino acids), and W2 (44 amino acids) followed by a unique

45 aa domain encoded by the Y1 and Y2 exons located within the *Bam*Y fragment just downstream of the IR1 repeats.[101] EBNA-LP and EBNA2 are the first viral proteins expressed after EBV infection.[4,97] EBNA-LP is phosphorylated on many serine residues, and this phosphorylation occurs during late G2 phase of the cell cycle.[58,59,90] The kinases implicated in mediating the phosphorylation of EBNA-LP are casein kinase II (CKII) and cyclin dependent p34/cdc2 kinase.[58,59] The main function of EBNA-LP seems to be co-stimulation of EBNA-2-mediated activation of cellular and viral gene expression, which has been shown to be critical for immortalization of B cells by EBV.[45,108] Therefore, EBNA-LP is potentially an important contributor in EBV-induced immortalization, but thus far the data do not support the claim that it is essential.

EBNA-LP is expressed as one of two species and is localized in nuclear structures called nuclear domain 10 (ND10) bodies[47] or promyelocytic leukemia-associated protein oncogenic domains.[90] EBNA-LP is localized to both the cytoplasm and nucleus, and a major part of the nuclear localized EBNA-LP is associated with the nuclear matrix.[90] It is known to form complexes with cellular proteins HA95 and protein kinase A (PKA), and has been shown to bind to tumor suppressor proteins (p53) and retinoblastoma protein (pRb).[36] However, coexpression of EBNA-LP and pRb or p53 did not result in any functional effect on pRb- or p53-dependent transcription from reporter plasmids.[45] EBNA-LP also interacts with viral and cellular bcl-2 proteins through HS1-associated protein X-1 (HAX-1).[49] Thus, EBNA-LP has a potential function in the regulation of EBV-infected B-cell signal transduction and apoptosis. Recently, microarray expression profiling identified thymus- and activation-regulated chemokine (TARC), a member of the CC chemokine family, as a novel cellular gene induced by EBNA-LP.[48] TARC is a functional ligand for CC chemokine receptor 4, which is selectively expressed on Th2 cells and has been shown to induce chemotaxis of Th2-type CD4$^+$ T lymphocytes.[44] Thus expression of EBNA-LP may have an important role in the immune evasion of EBV-infected B cells.

EPSTEIN-BARR VIRUS NUCLEAR ANTIGEN 3A, 3B, AND 3C

Although the three EBNA3 genes, EBNA3A, 3B, and 3C, have dissimilar amino acid sequences, they share a similar gene structure (Fig. 4).[53] They lie in tandem on the EBV genome, indicating that they may have evolved from a common ancestral gene to perform similar and in some cases divergent functions.[95] An indication of similarity in function comes from the fact that the most highly conserved domain in the N-terminus of all three proteins interacts with RBP-Jκ.[95] It is via this interaction that the EBNA3 genes differentially modulate EBNA2-mediated release of transcriptional repression by RBP-Jκ.[94] In some cases they seem to act as coactivators of EBNA2-activated genes and at the same time they block its activation of EBV LMP1 and LMP2 genes.[94] An indication of the divergence in function of the three EBNA3 genes comes from the fact that only EBNA3A and 3C have been found to be essential for B-cell transformation.[28]

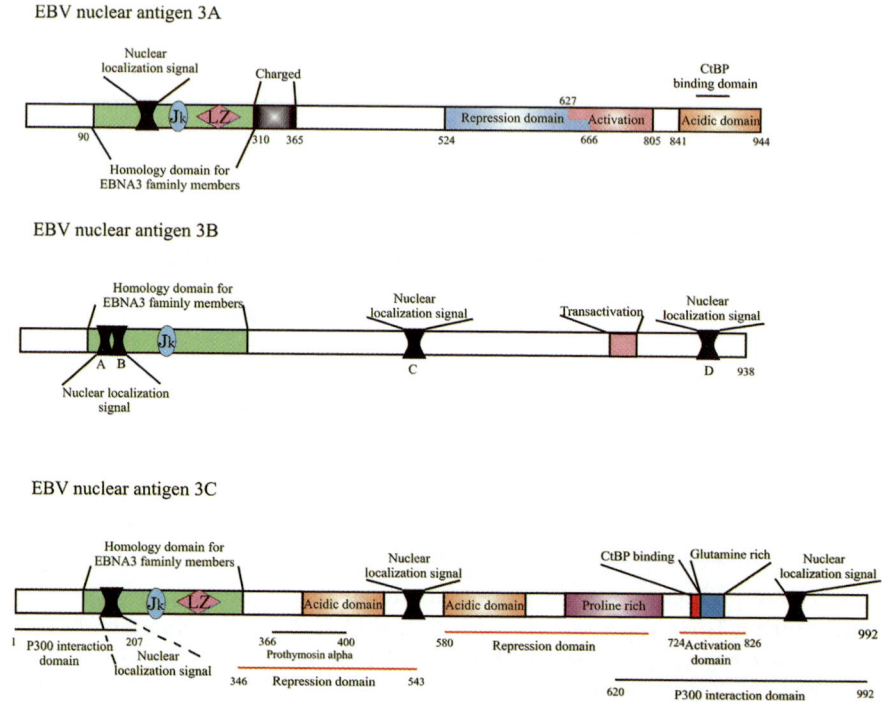

Figure 4. EBV nuclear antigens 3A, 3B, and 3C. The three EBNA3 genes encoded by EBV are expressed from adjacent loci in the EBV genome.[95] It is believed that these genes have evolved from a common ancestral gene to mediate slightly divergent functions. The amino termini of all three genes have a conserved domain that binds to the transcriptional corepressor RBP-Jκ and enables these proteins to differentially regulate EBNA2-mediated transcription.[95] EBNA3A and 3C are critical for EBV-mediated in vitro transformation of B cells and both proteins bind Ct-BP, although via different sequences. EBNA3C associates with various transcription factors and regulates transcription of various cellular genes[60,113] It also recruits the E3 ubiquitin ligase complex SCF[SKP2] and mediates the degradation of critical cell cycle regulators contributing the increased proliferation of EBV-infected cells.[61–64]

EBNA3C is the most extensively studied out of the three EBNA3 genes. Its C-terminus has been shown to recruit histone deacetylase (p300) and interact with Prothymosin alpha (ProTalpha), both of which put together suggest a possible role for EBNA3C in modulation of histone acetylation and in chromatin remodeling.[60,113] EBNA3C has a potential basic leucine zipper (bZIP) domain characteristic of transcriptional activators,[91] and it has been shown that EBNA3C associates with the transcription factor PU1/Spi-1.[137]

A series of recent studies challenge the notion that EBNA3C is merely a large transcription factor. Here, EBNA3C was shown to physically interact with the CyclinA/cdk2 complex and possibly play a direct role in regulating cell cycle.[61,62] Further proof of this was obtained, when EBNA3C was shown to recruit the ubiquitin E3 ligase SCF[SKP2] complex,[64] and mediate the ubiquitylation of its known targets retinoblastoma (Rb) and p27 leading to their degradation.[63,64] Both these proteins are critical regulators of the cell cycle and reduction in their cellular levels is likely to play a critical role in unregulated cell growth associated with EBV infections. Interestingly, HPV E7 recruits the same ubiquitin ligase complex and causes the degradation of Rb,[10] indicating a degree of conservation in pathways adopted by seemingly unrelated viruses to circumvent cell cycle check points.

LATENT MEMBRANE PROTEIN 1

LMP1 is encoded by the *Bam* HI N region of the EBV genome (BNLF1),[41] and the size of full length LMP1 is 63 kDa (386 aa) (Fig. 5). Another gene transcript starts at methionine-129 of the full-length LMP1, and encodes for a truncated

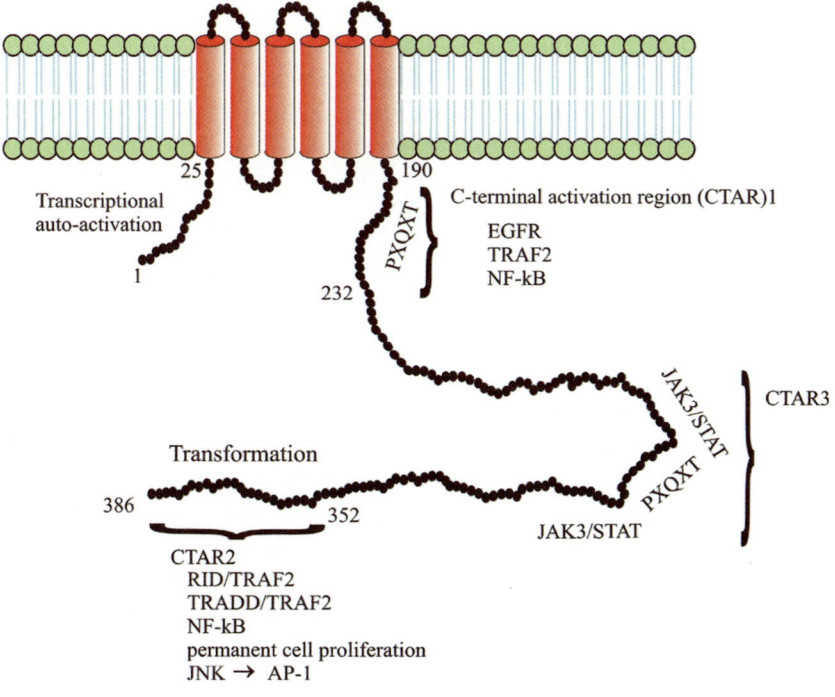

LMP1

Transcriptional auto-activation

25

190

PXQXT

C-terminal activation region (CTAR)1

EGFR
TRAF2
NF-kB

1

232

JAK3/STAT

CTAR3

PXQXT

Transformation

386

352

JAK3/STAT

CTAR2
RID/TRAF2
TRADD/TRAF2
NF-kB
permanent cell proliferation
JNK → AP-1

Figure 5. Latent membrane protein 1. The latent membrane protein 1 is a transmembrane protein with six transmembrane domains, a short intracellular N-terminus and a relatively long intracellular C-terminus.[42,52] The C-terminus of the protein is functionally homologous to a constitutively active CD40 receptor.[42] It binds to members of the TRAF family, interacts with JAK3 and members of the STAT family, and activates the AP-1 transcription complex.[22,25,34,55,83] All these interactions contribute to uncontrolled proliferation of LMP1 expressing cells. Its degradation by the proteosome is believed to result in an immunosuppressive peptide homologous to the retrovirus-encoded transmembrane protein p15E, which strongly tempers cytotoxic T lymphocyte response.[6]

LMP1 (258 aa).[3] The truncated LMP1 is made up of the fifth and sixth transmembrane domains and cytoplasmic carboxy terminus of full length LMP1.[3] LMP1 is one of the essential proteins for B-cell transformation, and has been identified as a viral oncogene due to experimental evidence both in vitro and in vivo.[46,50–52] It is an integral membrane protein composed of a 24 aa long cytoplasmic amino terminus domain, six hydrophobic transmembrane spanning domains separated by short turns, and a 200 aa long cytoplasmic carboxy terminus domain.[73] LMP1 is functionally homologous to CD40, a member of the tumour necrosis factor receptor (TNFR) family, and substitutes for CD40 providing growth and differentiation responses in B cells.[25] Thus, LMP1 has the ability to activate the NF-κB signaling pathway[42] and, similar to CD40, has been shown to interact with various members of the TNFR-associated factor (TRAF) family,[22,25,83] as well as with Janus-activated kinase 3 (JAK3) to activate signal transducers and activators of transcription (STATs).[34] Moreover, LMP-1 has been known to activate the activator protein 1 (AP-1) transcription complex through the c-Jun N-terminal kinase (JNK) signaling pathway.[55] Activation and interaction of these various pathways suggest that LMP1 is functionally similar to CD40 and leads to the same B-cell fate. Due to the lack in any distinct sequence homology, LMP1 and CD40 do not interact with exactly the same sets of molecules.[85] Unlike CD40 however LMP1 fails to target TRAF-2 and -3 for proteasomal degradation, resulting in their constitutive activation.[11] Thus, LMP1 acts as a constitutively active CD40 molecule. This ability to drive B-cell proliferation and prevent apoptosis supports the view that LMP-1 is essential for B-cell transformation by EBV.

LMP-1 is a relatively short-lived protein, with a reported half-life of 2 h.[84] Its ubiquitylation has been shown to occur at its N-terminus leading to rapid degradation by ubiquitin-dependent proteolysis.[6] Paradoxically, several components of the antigen presentation pathway have also been shown to be upregulated in LMP-1-expressing cells. Coupling of rapid degradation of a viral antigen and enhanced antigen presentation would normally result in robust cytotoxic T-cell activation and rapid clearance of antigen expressing cells. However, LMP-1 has been found to contain two peptides with strong homology to an immunosuppressive peptide found in the retrovirus-encoded transmembrane protein p15E.[24] Recombinant peptides corresponding to these sequences have been shown to strongly inhibit cytotoxic T lymphocyte and natural killer cell responses in vitro.[24] Therefore, proteasomal processing of LMP-1 probably results in the generation of these suppressive peptides and the corresponding inhibition of immune response.

LATENT MEMBRANE PROTEIN 2A AND 2B

LMP2A and 2B are transcribed from the same strand as the EBNA genes and opposite to the strand from which LMP1 is transcribed.[13] Thus, LMP2A transcript is antisense to the LMP1 transcript and vice versa. The first exon of LMP2A and 2B are the only unique exons, with all the other exons being shared by two molecules.[13]

The first exon of LMP2A encodes a 119 aa hydrophilic amino terminal cytoplasmic domain.[76] The translation of LMP2B initiates at a methionine in the common second exon, almost at the start of the first transmembrane sequence.[69,102] The remaining exons encode 12 hydrophobic integral membrane sequences separated by short turns and a 27 aa cytoplasmic carboxy terminal domain (Fig. 6). The sizes of LMP2A and 2B protein are 54 kDa and 40 kDa, respectively. The first exon of LMP2A bears considerable homology to the B-cell receptor and contains the same immunoreceptor tyrosine-based activation motifs (ITAMs).[31,92] The B-cell receptor has been shown to associate with Lyn, a member of the Src family of tyrosine kinases.[120] Phosphorylation of tyrosine residues within the ITAM by Lyn leads to recruitment of the Syk tyrosine kinase, a 70 kDa phosphoprotein identified in LMP2A immunoprecipitates, and the activation of nonreceptor tyrosine kinases.[120] These downstream effectors of the B-cell receptor can lead to B-cell proliferation and/or survival.[120] Despite the homology in sequence and interaction with the same effectors, LMP2 fails to activate either of these signals. In fact, LMP2 was found to effectively block B-cell receptor signaling.[120] This opposite effect of LMP2 on B-cell receptor-mediated signaling has been attributed to its recruitment of HECT-domain E3 ubiquitin ligases and the consequent degradation of the effector nonreceptor tyrosine kinases.[43] Triggering of the B-cell receptor has been associated with the activation of lytic viral replication.[67] Thus, LMP2A prevents effective B-cell receptor activation and thus plays a critical role in maintenance of viral latency by blocking normal receptor signaling.

EBV ENCODED SMALL RNA'S 1 AND 2

EBER1 and EBER2 (Fig. 7) are EBV encoded small nonpolyadenylated RNA's that are transcribed by the RNA polymerase III.[117] They do not encode for any protein, but an estimated 10^7 copies of EBER's are present per infected cell.[40,93] EBER1 and 2 are 167 bp and 173 bp long, respectively.[117] Most of the molecules localize to the infected cell's nucleus where they form complexes with the cellular La protein,[71] the L22 ribosomal protein and a novel nuclear protein EAP.[119] Thus they are used as markers for detection of EBV infection. EBERs, the adenovirus VA, and the U6 cell RNA have similar primary sequences, secondary structures and associate with the La protein to accomplish similar functions. In adenovirus infection, VA1 RNA acts to directly inhibit activation of an interferon-induced protein kinase.[105] Recent studies have shown EBERs contribute to the induction of human interleukin-10,[57] and resistance to alpha interferon induced apoptosis by inhibition of the PKR pathway[88] in B-cell lines. In epithelial cells, they have been found to bind PKR and mediate resistance to Fas-mediated apoptosis.[89]

BamHI-A RIGHTWARD TRANSCRIPTS

The EBV BamHI-A rightward transcripts (BARTs) were first detected in nasopharyngeal carcinomas, where they are the most abundant viral transcripts.[109] It was soon discovered, however, that BARTs are expressed in all EBV latently infected

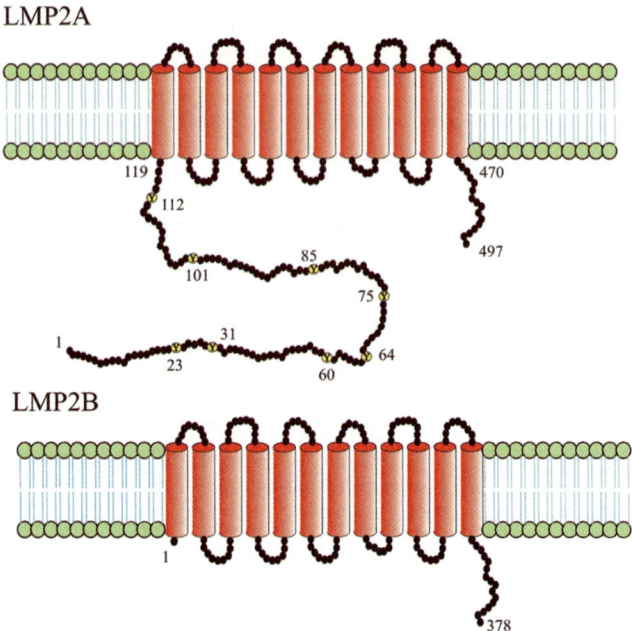

Figure 6. Latent membrane proteins 2A and 2B. Latent membrane protein 2A (LMP2A) is a transmembrane protein with 12 transmembrane domains, a short intracellular C-terminal domain and a relatively long intracellular N-terminal domain.[69,102] LMP2B is an amino terminal truncated form of LMP2A, with transcription beginning just before the first transmembrane domain.[69,102] The N-terminus of LMP2A is homologous to the B-cell receptor cytoplasmic domain and contains the same ITAM motifs.[31,92] LMP2A inhibits B-cell receptor signaling by sequestering nonreceptor tyrosine kinases, which are downstream effectors of the B-cell receptor.[31] LMP2A also mediates the degradation of these same effectors by recruiting a HECT-domain E3 ubiquitin ligase.[43] Signaling via the B-cell receptor leads to lytic reactivation of the virus,[67] and LMP2A by preventing B-cell receptor signaling helps preserve latent replication of the virus.

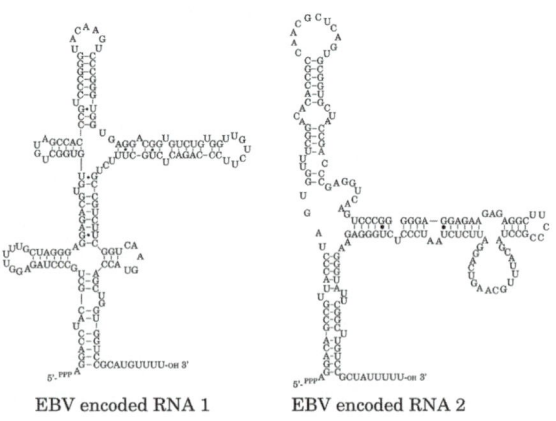

Figure 7. EBV encoded RNAs 1 and 2. EBV encoded RNAs 1 and 2 (EBERs) do not encode for any known protein product.[93] The sheer numbers (10^7 per infected cell) of EBERs present in an infected cell points to a potentially important role for these RNAs. Preliminary evidence suggests a role in induction of IL-10 and prevention of IFNα mediated apoptosis.[57,89] Investigators are currently looking at a potential shRNA like silencing of target cellular genes by EBERs.

cells.[135] The region containing BARTs can be deleted from the EBV genome, without any effect on the transformation potential of the resulting virus.[96] BARTs are therefore not essential for EBV-mediated B-cell transformation.

Three alternatively spliced ORFs have been identified in the region of the genome containing BARTs.[33] There is an ongoing debate about whether the genes encoded by these ORFs are actually expressed. The products of two of the ORFs, RPMS, and BARF0, when overexpressed have been shown to interact with the notch pathway.[109] RPMS appears to be an EBNA2 antagonist,[136] and prevents it from displacing the transcription corepressor RBP-Jκ and BARF0 interacts with notch4 and potentiates its function.[109]

LYTIC REACTIVATION OF EBV

A small fraction of cells in Latency type III malignancies undergo lytic replication at any given time. It is thus important from a disease standpoint to have some understanding of the events leading to and following lytic reactivation of the virus. Clues to cellular, viral, and environmental triggers of lytic reactivation in vivo are provided by experiments showing how this reactivation can be achieved in cell lines derived from EBV-associated malignancies in vitro. Cells in latency type I (e.g., Akata cells from Burkitt's lymphoma) can be reactivated by crosslinking of the B-cell receptor by anti-IgG antibodies.[106,116] This may mimic physiological events in B-cell differentiation and consequent lytic reactivation in vivo. As briefly mentioned previously, viral gene expression upon lytic reactivation follows a sequential order. The first genes to be expressed are the immediate early genes, which are mostly transcriptional inducers of early gene expression. In the specific case of EBV, two genes that fall in this category, BZLF1 and BRLF1, have been shown to contribute to its reactivation.[8] BZLF1, also known as Zta, is expressed by direct signaling from the B-cell receptor, independent of any other gene expression. Expression of Zta alone is sufficient to trigger the entire lytic cascade.[19,98] Accordingly, transfection with plasmids encoding for Zta is another way to induce lytic reactivation in latently infected cells in vitro.[79] Zta is a transcription factor that has been shown to activate several viral early gene promoters, cellular gene promoters and bind to the origins of viral genome replication.[111] In addition Zta has also been shown to activate its own promoter, resulting in a positive feedback loop that ensures full fledged lytic reactivation upon trigger.[75] In contrast, BRLF1 by itself is not sufficient to trigger the entire lytic cascade.[20] Zta and BRLF1 together are sufficient to activate most of the viral early genes.[20] Early genes are involved in facilitating host cell independent virus DNA replication and late genes facilitate production of viral structural genes, virus assembly and egress.[5] Other means to trigger lytic reactivation include treatment with chemicals like TPA and sodium butyrate.[15] These chemical induction methods, while useful, are unlikely to have a physiological equivalent in EBV reactivation in vivo. More importantly, in vitro lytic reactivation provides a means for production of infectious virus from infected cell lines and is therefore an invaluable tool in EBV-related research.

IN VITRO MODELS FOR EBV INFECTION

Peripheral blood lymphocytes from an infected individual give rise to EBV-transformed cell lines in culture in the presence of cytotoxic T-cell suppression agents. Such cell lines can also be generated by infecting primary B cells in vitro, with EBV obtained from lytic reactivation of infected cells. In addition, tumor cell lines have been established from EBV positive B-cell lymphomas and epithelial cell carcinomas. A number of such cell lines have been generated and are available to EBV researchers (Table 4). The type of latency and therefore the consequent viral antigen expression profile has been well characterized for these cell lines. While, most EBV tumor cell lines exhibit a strict control over latency, some cell lines do contain a small fraction of cells undergoing lytic replication.

Cell lines identified above are invaluable tools for EBV researchers. Comparison with EBV negative Burkitt's lymphoma cells and/or resting primary B cells can answer questions about differential regulation of cellular genes/processes in these cells. Isolation of the viral genome and manipulation by genetic recombination can arm researchers with mutant viruses that can be used to address questions about functions mediated by various viral genes.

FUTURE RESEARCH DIRECTION

The availability of many excellent in vitro model systems makes EBV a prime candidate for helping us understanding many of the more fundamental questions about our own biology. Viruses have coevolved along with humans since the very beginning of humanoid life-forms and are therefore quite adept at utilizing cellular and molecular processes in a more energetically efficient manner than us humans. Viral research has already provided important insights into cancer biology and has lead to the discovery of many important proto-oncogenes. The evolution of EBV to coexist in the face of host immunity is undoubtedly a perfect mirror, in which we can see how host immunity works. This will not only have repercussions for EBV pathogenesis, but also give us insights into autoimmune diseases and immunosuppressive diseases.

Table 4. In vitro models of EBV infection

Cell line	Type of cells	Type of latency	Reference
Akata	Burkitt's lymphoma	Type I	107
MUTU-1	Burkitt's lymphoma	Type I	14
C666-1	Nasopharyngeal carcinoma	Type II	14
C15	Nasopharyngeal carcinoma	Type II	14
Raji	Burkitt's lymphoma	Type III	16
Namalwa	Burkitt's lymphoma	Type III	115
GT-38	Gastric carcinoma	Type III	86
GT-39	Gastric carcinoma	Type III	86
B95.8	Marmoset LCL	Type III	80
Human LCLs	In vitro transformed B cells	Type III	132

SUMMARY

Epstein–Barr virus (EBV) is a gammaherpesvirus of the *Lymphocryptovirus* genus, which infects greater than 90% of the world's population. Infection is nonsymptomatic in healthy individuals, but has been associated with a number of lymphoproliferative disorders when accompanied by immunosuppression. Like all herpesviruses, EBV has both latent and lytic replication programs, which allows it to evade immune clearance and persist for the lifetime of the host. Latent infection is characterized by replication of the viral genome as an integral part of the host cell chromosomes, and the absence of production of infectious virus. A further layer of complexity is added in that EBV can establish three distinct latency programs, in each of which a specific set of viral antigens is expressed. In most malignant disorders associated with EBV, the virus replicates using one of these three latency programs. In the most aggressive latency program, only 11 of the hitherto 85 identified open reading frames in the EBV genome are expressed. The other two latency programs express even smaller subsets of this repertoire of latent genes. The onset of the AIDS pandemic and the corresponding increase in individuals with acquired immunodeficiency resulted in a sharp increase in EBV-mediated AIDS-associated malignancies. This has sparked a renewed interest in EBV biology and pathogenesis.

REFERENCES

1. Abbot, S. D., M. Rowe, K. Cadwallader, A. Ricksten, J. Gordon, et al. 1990. Epstein-Barr virus nuclear antigen 2 induces expression of the virus-encoded latent membrane protein. J Virol **64**:2126–34.
2. Adams, A. 1987. Replication of latent Epstein-Barr virus genomes in Raji cells. J Virol **61**:1743–6.
3. Ahsan, N., T. Kanda, K. Nagashima, and K. Takada. 2005. Epstein-Barr virus transforming protein LMP1 plays a critical role in virus production. J Virol **79**:4415–24.
4. Allday, M. J., D. H. Crawford, and B. E. Griffin. 1989. Epstein-Barr virus latent gene expression during the initiation of B cell immortalization. J Gen Virol **70**(Pt 7):1755–64.
5. Amon, W., and P. J. Farrell. 2005. Reactivation of Epstein-Barr virus from latency. Rev Med Virol **15**:149–56.
6. Aviel, S., G. Winberg, M. Massucci, and A. Ciechanover. 2000. Degradation of the Epstein-Barr virus latent membrane protein 1 (LMP1) by the ubiquitin-proteasome pathway. Targeting via ubiquitination of the N-terminal residue. J Biol Chem **275**:23491–9.
7. Baer, R., A. T. Bankier, M. D. Biggin, P. L. Deininger, P. J. Farrell, et al. 1984. DNA sequence and expression of the B95–8 Epstein-Barr virus genome. Nature **310**:207–11.
8. Biggin, M., M. Bodescot, M. Perricaudet, and P. Farrell. 1987. Epstein-Barr virus gene expression in P3HR1-superinfected Raji cells. J Virol **61**:3120–32.
9. Blake, N., S. Lee, I. Redchenko, W. Thomas, N. Steven, et al. 1997. Human CD8+ T cell responses to EBV EBNA1: HLA class I presentation of the (Gly-Ala)-containing protein requires exogenous processing. Immunity **7**:791–802.
10. Boyer, S. N., D. E. Wazer, and V. Band. 1996. E7 protein of human papilloma virus-16 induces degradation of retinoblastoma protein through the ubiquitin-proteasome pathway. Cancer Res **56**:4620–4.
11. Brown, K. D., B. S. Hostager, and G. A. Bishop. 2001. Differential signaling and tumor necrosis factor receptor-associated factor (TRAF) degradation mediated by CD40 and the Epstein-Barr virus oncoprotein latent membrane protein 1 (LMP1). J Exp Med **193**:943–54.
12. Burkitt, D. 1958. A sarcoma involving the jaws in African children. Br J Surg **46**:218–23.
13. Busson, P., R. McCoy, R. Sadler, K. Gilligan, T. Tursz, and N. Raab-Traub. 1992. Consistent transcription of the Epstein-Barr virus LMP2 gene in nasopharyngeal carcinoma. J Virol **66**:3257–62.
14. Cai, X., A. Schafer, S. Lu, J. P. Bilello, R. C. Desrosiers, et al. 2006. Epstein-Barr virus microRNAs are evolutionarily conserved and differentially expressed. PLoS Pathog **2**:e23.
15. Cen, H., and J. L. McKnight. 1994. EBV-immortalized isogenic human B-cell clones exhibit differences in DNA-protein complex formation on the BZLF1 and BRLF1 promoter regions among latent, lytic and TPA-activated cell lines. Virus Res **31**:89–107.

16. Cerimele, F., T. Battle, R. Lynch, D. A. Frank, E. Murad, et al. 2005. Reactive oxygen signaling and MAPK activation distinguish Epstein-Barr Virus (EBV)-positive versus EBV-negative Burkitt's lymphoma. Proc Natl Acad Sci USA **102**:175–9.

17. Chaudhuri, B., H. Xu, I. Todorov, A. Dutta, and J. L. Yates. 2001. Human DNA replication initiation factors, ORC and MCM, associate with oriP of Epstein-Barr virus. Proc Natl Acad Sci USA **98**:10085–9.

18. Cohen, J. I., F. Wang, J. Mannick, and E. Kieff. 1989. Epstein-Barr virus nuclear protein 2 is a key determinant of lymphocyte transformation. Proc Natl Acad Sci USA **86**:9558–62.

19. Countryman, J., and G. Miller. 1985. Activation of expression of latent Epstein-Barr herpesvirus after gene transfer with a small cloned subfragment of heterogeneous viral DNA. Proc Natl Acad Sci USA **82**:4085–9.

20. Cox, M. A., J. Leahy, and J. M. Hardwick. 1990. An enhancer within the divergent promoter of Epstein-Barr virus responds synergistically to the R and Z transactivators. J Virol **64**:313–21.

21. Deng, Z., L. Lezina, C. J. Chen, S. Shtivelband, W. So, and P. M. Lieberman. 2002. Telomeric proteins regulate episomal maintenance of Epstein-Barr virus origin of plasmid replication. Mol Cell **9**:493–503.

22. Devergne, O., E. Hatzivassiliou, K. M. Izumi, K. M. Kaye, M. F. Kleijnen, et al. 1996. Association of TRAF1, TRAF2, and TRAF3 with an Epstein-Barr virus LMP1 domain important for B-lymphocyte transformation: Role in NF-kappaB activation. Mol Cell Biol **16**:7098–108.

23. Dhar, S. K., K. Yoshida, Y. Machida, P. Khaira, B. Chaudhuri, et al. 2001. Replication from oriP of Epstein-Barr virus requires human ORC and is inhibited by geminin. Cell **106**:287–96.

24. Dukers, D. F., P. Meij, M. B. Vervoort, W. Vos, R. J. Scheper, et al. 2000. Direct immunosuppressive effects of EBV-encoded latent membrane protein 1. J Immunol **165**:663–70.

25. Eliopoulos, A. G., C. W. Dawson, G. Mosialos, J. E. Floettmann, M. Rowe, et al. 1996. CD40-induced growth inhibition in epithelial cells is mimicked by Epstein-Barr Virus-encoded LMP1: Involvement of TRAF3 as a common mediator. Oncogene **13**:2243–54.

26. Epstein, M. A., B. G. Achong, and Y. M. Barr. 1964. Virus particles in cultured lymphoblasts from Burkitt's lymphoma. Lancet **15**:702–3.

27. Fahraeus, R., A. Jansson, A. Ricksten, A. Sjoblom, and L. Rymo. 1990. Epstein-Barr virus-encoded nuclear antigen 2 activates the viral latent membrane protein promoter by modulating the activity of a negative regulatory element. Proc Natl Acad Sci USA **87**:7390–4.

28. Farrell, P. J., I. Cludts, and A. Stuhler. 1997. Epstein-Barr virus genes and cancer cells. Biomed Pharmacother **51**:258–67.

29. Faulkner, G. C., A. S. Krajewski, and D. H. Crawford. 2000. The ins and outs of EBV infection. Trends Microbiol **8**:185–9.

30. Frappier, L., and M. O'Donnell. 1992. EBNA1 distorts oriP, the Epstein-Barr virus latent replication origin. J Virol **66**:1786–90.

31. Fruehling, S., and R. Longnecker. 1997. The immunoreceptor tyrosine-based activation motif of Epstein-Barr virus LMP2A is essential for blocking BCR-mediated signal transduction. Virology **235**:241–51.

32. Gerber, P., S. Lucas, M. Nonoyama, E. Perlin, and L. I. Goldstein. 1972. Oral excretion of Epstein-Barr virus by healthy subjects and patients with infectious mononucleosis. Lancet **2**:988–9.

33. Gilligan, K., H. Sato, P. Rajadurai, P. Busson, L. Young, et al. 1990. Novel transcription from the Epstein-Barr virus terminal EcoRI fragment, DIJhet, in a nasopharyngeal carcinoma. J Virol **64**:4948–56.

34. Gires, O., F. Kohlhuber, E. Kilger, M. Baumann, A. Kieser, et al. 1999. Latent membrane protein 1 of Epstein-Barr virus interacts with JAK3 and activates STAT proteins. Embo J **18**:3064–73.

35. Hammerschmidt, W., and B. Sugden. 1989. Genetic analysis of immortalizing functions of Epstein-Barr virus in human B lymphocytes. Nature **340**:393–7.

36. Han, I., Y. Xue, S. Harada, S. Orstavik, B. Skalhegg, and E. Kieff. 2002. Protein kinase A associates with HA95 and affects transcriptional coactivation by Epstein-Barr virus nuclear proteins. Mol Cell Biol **22**:2136–46.

37. Henkel, T., P. D. Ling, S. D. Hayward, and M. G. Peterson. 1994. Mediation of Epstein-Barr virus EBNA2 transactivation by recombination signal-binding protein J kappa. Science **265**:92–5.

38. Henle, W., V. Diehl, G. Kohn, H. Zur Hausen, and G. Henle. 1967. Herpes-type virus and chromosome marker in normal leukocytes after growth with irradiated Burkitt cells. Science **157**:1064–5.

39. Hofelmayr, H., L. J. Strobl, C. Stein, G. Laux, G. Marschall, et al. 1999. Activated mouse Notch1 transactivates Epstein-Barr virus nuclear antigen 2-regulated viral promoters. J Virol **73**:2770–80.

40. Howe, J. G., and J. A. Steitz. 1986. Localization of Epstein-Barr virus-encoded small RNAs by in situ hybridization. Proc Natl Acad Sci USA **83**:9006–10.

41. Hudson, G. S., P. J. Farrell, and B. G. Barrell. 1985. Two related but differentially expressed potential membrane proteins encoded by the EcoRI Dhet region of Epstein-Barr virus B95–8. J Virol **53**:528–35.

42. Huen, D. S., S. A. Henderson, D. Croom-Carter, and M. Rowe. 1995. The Epstein-Barr virus latent membrane protein-1 (LMP1) mediates activation of NF-kappa B and cell surface phenotype via two effector regions in its carboxy-terminal cytoplasmic domain. Oncogene **10**:549–60.

43. Ikeda, M., A. Ikeda, L. C. Longan, and R. Longnecker. 2000. The Epstein-Barr virus latent membrane protein 2A PY motif recruits WW domain-containing ubiquitin-protein ligases. Virology **268**:178–91.
44. Imai, T., M. Nagira, S. Takagi, M. Kakizaki, M. Nishimura, et al. 1999. Selective recruitment of CCR4-bearing Th2 cells toward antigen-presenting cells by the CC chemokines thymus and activation-regulated chemokine and macrophage-derived chemokine. Int Immunol **11**:81–8.
45. Inman, G. J., and P. J. Farrell. 1995. Epstein-Barr virus EBNA-LP and transcription regulation properties of pRB, p107 and p53 in transfection assays. J Gen Virol **76**(Pt 9):2141–9.
46. Izumi, K. M., K. M. Kaye, and E. D. Kieff. 1994. Epstein-Barr virus recombinant molecular genetic analysis of the LMP1 amino-terminal cytoplasmic domain reveals a probable structural role, with no component essential for primary B-lymphocyte growth transformation. J Virol **68**:4369–76.
47. Jiang, W. Q., L. Szekely, V. Wendel-Hansen, N. Ringertz, G. Klein, and A. Rosen. 1991. Co-localization of the retinoblastoma protein and the Epstein-Barr virus-encoded nuclear antigen EBNA-5. Exp Cell Res **197**:314–8.
48. Kanamori, M., S. Watanabe, R. Honma, M. Kuroda, S. Imai, et al. 2004. Epstein-Barr virus nuclear antigen leader protein induces expression of thymus- and activation-regulated chemokine in B cells. J Virol **78**:3984–93.
49. Kawaguchi, Y., K. Nakajima, M. Igarashi, T. Morita, M. Tanaka, et al. 2000. Interaction of Epstein-Barr virus nuclear antigen leader protein (EBNA-LP) with HS1-associated protein X-1: Implication of cytoplasmic function of EBNA-LP. J Virol **74**:10104–11.
50. Kaye, K. M., K. M. Izumi, and E. Kieff. 1993. Epstein-Barr virus latent membrane protein 1 is essential for B-lymphocyte growth transformation. Proc Natl Acad Sci USA **90**:9150–4.
51. Kaye, K. M., K. M. Izumi, H. Li, E. Johannsen, D. Davidson, et al. 1999. An Epstein-Barr virus that expresses only the first 231 LMP1 amino acids efficiently initiates primary B-lymphocyte growth transformation. J Virol **73**:10525–30.
52. Kaye, K. M., K. M. Izumi, G. Mosialos, and E. Kieff. 1995. The Epstein-Barr virus LMP1 cytoplasmic carboxy terminus is essential for B-lymphocyte transformation; fibroblast cocultivation complements a critical function within the terminal 155 residues. J Virol **69**:675–83.
53. Kelly, G. L., A. E. Milner, R. J. Tierney, D. S. Croom-Carter, M. Altmann, et al. 2005. Epstein-Barr virus nuclear antigen 2 (EBNA2) gene deletion is consistently linked with EBNA3A, -3B, and -3C expression in Burkitt's lymphoma cells and with increased resistance to apoptosis. J Virol **79**:10709–17.
54. Kempkes, B., L. J. Strobl, G. W. Bornkamm, and U. Zimber-Strobl. 2005. EBNA2 and notch signaling. Norfolk: Caister Academic Press, pp. 463–99.
55. Kieser, A., C. Kaiser, and W. Hammerschmidt. 1999. LMP1 signal transduction differs substantially from TNF receptor 1 signaling in the molecular functions of TRADD and TRAF2. Embo J **18**:2511–21.
56. Kirchmaier, A. L., and B. Sugden. 1995. Plasmid maintenance of derivatives of oriP of Epstein-Barr virus. J Virol **69**:1280–3.
57. Kitagawa, N., M. Goto, K. Kurozumi, S. Maruo, M. Fukayama, et al. 2000. Epstein-Barr virus-encoded poly (A)(−) RNA supports Burkitt's lymphoma growth through interleukin-10 induction. Embo J **19**:6742–50.
58. Kitay, M. K., and D. T. Rowe. 1996. Cell cycle stage-specific phosphorylation of the Epstein-Barr virus immortalization protein EBNA-LP. J Virol **70**:7885–93.
59. Kitay, M. K., and D. T. Rowe. 1996. Protein–protein interactions between Epstein-Barr virus nuclear antigen-LP and cellular gene products: Binding of 70-kilodalton heat shock proteins. Virology **220**:91–9.
60. Knight, J. S., K. Lan, C. Subramanian, and E. S. Robertson. 2003. Epstein-Barr virus nuclear antigen 3C recruits histone deacetylase activity and associates with the corepressors mSin3A and NCoR in human B-cell lines. J Virol **77**:4261–72.
61. Knight, J. S., and E. S. Robertson. 2004. Epstein-Barr virus nuclear antigen 3C regulates cyclin A/p27 complexes and enhances cyclin A-dependent kinase activity. J Virol **78**:1981–91.
62. Knight, J. S., N. Sharma, D. E. Kalman, and E. S. Robertson. 2004. A cyclin-binding motif within the amino-terminal homology domain of EBNA3C binds cyclin A and modulates cyclin A-dependent kinase activity in Epstein-Barr virus-infected cells. J Virol **78**:12857–67.
63. Knight, J. S., N. Sharma, and E. S. Robertson. 2005. Epstein-Barr virus latent antigen 3C can mediate the degradation of the retinoblastoma protein through an SCF cellular ubiquitin ligase. Proc Natl Acad Sci USA **102**:18562–6.
64. Knight, J. S., N. Sharma, and E. S. Robertson. 2005. SCFSkp2 complex targeted by Epstein-Barr virus essential nuclear antigen. Mol Cell Biol **25**:1749–63.
65. Knutson, J. C. 1990. The level of c-fgr RNA is increased by EBNA-2, an Epstein-Barr virus gene required for B-cell immortalization. J Virol **64**:2530–6.
66. Koliais, S. I. 1979. Mode of integration of Epstein-Barr virus genome into host DNA in Burkitt lymphoma cells. J Gen Virol **44**:573–6.

67. Laichalk, L. L., and D. A. Thorley-Lawson. 2005. Terminal differentiation into plasma cells initiates the replicative cycle of Epstein-Barr virus in vivo. J Virol **79**:1296–307.
68. Laux, G., B. Adam, L. J. Strobl, and F. Moreau-Gachelin. 1994. The Spi-1/PU.1 and Spi-B ets family transcription factors and the recombination signal binding protein RBP-J kappa interact with an Epstein-Barr virus nuclear antigen 2 responsive cis-element. Embo J **13**:5624–32.
69. Laux, G., M. Perricaudet, and P. J. Farrell. 1988. A spliced Epstein-Barr virus gene expressed in immortalized lymphocytes is created by circularization of the linear viral genome. Embo J **7**:769–74.
70. Lee, M. A., M. E. Diamond, and J. L. Yates. 1999. Genetic evidence that EBNA-1 is needed for efficient, stable latent infection by Epstein-Barr virus. J Virol **73**:2974–82.
71. Lerner, M. R., N. C. Andrews, G. Miller, and J. A. Steitz. 1981. Two small RNAs encoded by Epstein-Barr virus and complexed with protein are precipitated by antibodies from patients with systemic lupus erythematosus. Proc Natl Acad Sci USA **78**:805–9.
72. Levitskaya, J., A. Sharipo, A. Leonchiks, A. Ciechanover, and M. G. Masucci. 1997. Inhibition of ubiquitin/proteasome-dependent protein degradation by the Gly-Ala repeat domain of the Epstein-Barr virus nuclear antigen 1. Proc Natl Acad Sci USA **94**:12616–21.
73. Liebowitz, D., D. Wang, and E. Kieff. 1986. Orientation and patching of the latent infection membrane protein encoded by Epstein-Barr virus. J Virol **58**:233–7.
74. Ling, P. D., D. R. Rawlins, and S. D. Hayward. 1993. The Epstein-Barr virus immortalizing protein EBNA-2 is targeted to DNA by a cellular enhancer-binding protein. Proc Natl Acad Sci USA **90**:9237–41.
75. Liu, P., and S. H. Speck. 2003. Synergistic autoactivation of the Epstein-Barr virus immediate-early BRLF1 promoter by Rta and Zta. Virology **310**:199–206.
76. Longnecker, R., and E. Kieff. 1990. A second Epstein-Barr virus membrane protein (LMP2) is expressed in latent infection and colocalizes with LMP1. J Virol **64**:2319–26.
77. Lu, C. C., Y. Y. Jeng, C. H. Tsai, M. Y. Liu, S. W. Yeh, et al. 2006. Genome-wide transcription program and expression of the Rta responsive gene of Epstein-Barr virus. Virology **345**:358–72.
78. Mackey, D., and B. Sugden. 1997. Studies on the mechanism of DNA linking by Epstein-Barr virus nuclear antigen 1. J Biol Chem **272**:29873–9.
79. Marschall, M., P. Alliger, F. Schwarzmann, C. Bogedain, M. Brand, et al. 1993. The lytic transition of Epstein-Barr virus is imitated by recombinant B-cells. Arch Virol **129**:23–33.
80. Masy, E., E. Adriaenssens, C. Montpellier, P. Crepieux, A. Mougel, et al. 2002. Human monocytic cell lines transformed in vitro by Epstein-Barr virus display a type II latency and LMP-1-dependent proliferation. J Virol **76**:6460–72.
81. Matsunami, N., Y. Hamaguchi, Y. Yamamoto, K. Kuze, K. Kangawa, et al. 1989. A protein binding to the J kappa recombination sequence of immunoglobulin genes contains a sequence related to the integrase motif. Nature **342**:934–7.
82. Miller, G. 1989. The switch between EBV latency and replication. Yale J Biol Med **62**:205–13.
83. Miller, W. E., G. Mosialos, E. Kieff, and N. Raab-Traub. 1997. Epstein-Barr virus LMP1 induction of the epidermal growth factor receptor is mediated through a TRAF signaling pathway distinct from NF-kappaB activation. J Virol **71**:586–94.
84. Moorthy, R., and D. A. Thorley-Lawson. 1990. Processing of the Epstein-Barr virus-encoded latent membrane protein p63/LMP. J Virol **64**:829–37.
85. Mosialos, G., M. Birkenbach, R. Yalamanchili, T. VanArsdale, C. Ware, and E. Kieff. 1995. The Epstein-Barr virus transforming protein LMP1 engages signaling proteins for the tumor necrosis factor receptor family. Cell **80**:389–99.
86. Murakami, M., Y. Hoshikawa, Y. Satoh, H. Ito, M. Tajima, et al. 2000. Tumorigenesis of Epstein-Barr virus-positive epithelial cell lines derived from gastric tissues in the SCID mouse. Virology **277**:20–6.
87. Murakami, M., K. Lan, C. Subramanian, and E. S. Robertson. 2005. Epstein-Barr virus nuclear antigen 1 interacts with Nm23-H1 in lymphoblastoid cell lines and inhibits its ability to suppress cell migration. J Virol **79**:1559–68.
88. Nanbo, A., K. Inoue, K. Adachi-Takasawa, and K. Takada. 2002. Epstein-Barr virus RNA confers resistance to interferon-alpha-induced apoptosis in Burkitt's lymphoma. Embo J **21**:954–65.
89. Nanbo, A., H. Yoshiyama, and K. Takada. 2005. Epstein-Barr virus-encoded poly(A)- RNA confers resistance to apoptosis mediated through Fas by blocking the PKR pathway in human epithelial intestine 407 cells. J Virol **79**:12280–5.
90. Petti, L., C. Sample, and E. Kieff. 1990. Subnuclear localization and phosphorylation of Epstein-Barr virus latent infection nuclear proteins. Virology **176**:563–74.
91. Radkov, S. A., M. Bain, P. J. Farrell, M. West, M. Rowe, and M. J. Allday. 1997. Epstein-Barr virus EBNA3C represses Cp, the major promoter for EBNA expression, but has no effect on the promoter of the cell gene CD21. J Virol **71**:8552–62.

92. Reth, M. 1989. Antigen receptor tail clue. Nature **338**:383–4.
93. Rickinson, A., and E. Kieff. 2001. Fields virology. Philadelphia: Lippincott Williams & Wilkins Publishers, pp. 2575–627.
94. Robertson, E. S., S. Grossman, E. Johannsen, C. Miller, J. Lin, et al. 1995. Epstein-Barr virus nuclear protein 3C modulates transcription through interaction with the sequence-specific DNA-binding protein J kappa. J Virol **69**:3108–16.
95. Robertson, E. S., J. Lin, and E. Kieff. 1996. The amino-terminal domains of Epstein-Barr virus nuclear proteins 3A, 3B, and 3C interact with RBPJ(kappa). J Virol **70**:3068–74.
96. Robertson, E. S., B. Tomkinson, and E. Kieff. 1994. An Epstein-Barr virus with a 58-kilobase-pair deletion that includes BARF0 transforms B lymphocytes in vitro. J Virol **68**:1449–58.
97. Rooney, C., J. G. Howe, S. H. Speck, and G. Miller. 1989. Influence of Burkitt's lymphoma and primary B cells on latent gene expression by the nonimmortalizing P3J-HR-1 strain of Epstein-Barr virus. J Virol **63**:1531–9.
98. Rooney, C. M., D. T. Rowe, T. Ragot, and P. J. Farrell. 1989. The spliced BZLF1 gene of Epstein-Barr virus (EBV) transactivates an early EBV promoter and induces the virus productive cycle. J Virol **63**:3109–16.
99. Sakai, T., Y. Taniguchi, K. Tamura, S. Minoguchi, T. Fukuhara, et al. 1998. Functional replacement of the intracellular region of the Notch1 receptor by Epstein-Barr virus nuclear antigen 2. J Virol **72**:6034–9.
100. Sample, J., E. B. Henson, and C. Sample. 1992. The Epstein-Barr virus nuclear protein 1 promoter active in type I latency is autoregulated. J Virol **66**:4654–61.
101. Sample, J., M. Hummel, D. Braun, M. Birkenbach, and E. Kieff. 1986. Nucleotide sequences of mRNAs encoding Epstein-Barr virus nuclear proteins: A probable transcriptional initiation site. Proc Natl Acad Sci USA **83**:5096–100.
102. Sample, J., D. Liebowitz, and E. Kieff. 1989. Two related Epstein-Barr virus membrane proteins are encoded by separate genes. J Virol **63**:933–7.
103. Scala, G., I. Quinto, M. R. Ruocco, M. Mallardo, C. Ambrosino, et al. 1993. Epstein-Barr virus nuclear antigen 2 transactivates the long terminal repeat of human immunodeficiency virus type 1. J Virol **67**:2853–61.
104. Schepers, A., M. Ritzi, K. Bousset, E. Kremmer, J. L. Yates, et al. 2001. Human origin recognition complex binds to the region of the latent origin of DNA replication of Epstein-Barr virus. Embo J **20**:4588–602.
105. Schneider, R. J., B. Safer, S. M. Munemitsu, C. E. Samuel, and T. Shenk. 1985. Adenovirus VAI RNA prevents phosphorylation of the eukaryotic initiation factor 2 alpha subunit subsequent to infection. Proc Natl Acad Sci USA **82**:4321–5.
106. Shimizu, N., and K. Takada. 1993. Analysis of the BZLF1 promoter of Epstein-Barr virus: Identification of an anti-immunoglobulin response sequence. J Virol **67**:3240–5.
107. Shimizu, N., A. Tanabe-Tochikura, Y. Kuroiwa, and K. Takada. 1994. Isolation of Epstein-Barr virus (EBV)-negative cell clones from the EBV-positive Burkitt's lymphoma (BL) line Akata: Malignant phenotypes of BL cells are dependent on EBV. J Virol **68**:6069–73.
108. Sinclair, A. J., I. Palmero, G. Peters, and P. J. Farrell. 1994. EBNA-2 and EBNA-LP cooperate to cause G0 to G1 transition during immortalization of resting human B lymphocytes by Epstein-Barr virus. Embo J **13**:3321–8.
109. Smith, P. 2001. Epstein-Barr virus complementary strand transcripts (CSTs/BARTs) and cancer. Semin Cancer Biol **11**:469–76.
110. Snudden, D. K., J. Hearing, P. R. Smith, F. A. Grasser, and B. E. Griffin. 1994. EBNA-1, the major nuclear antigen of Epstein-Barr virus, resembles 'RGG' RNA binding proteins. Embo J **13**:4840–7.
111. Speck, S. H., T. Chatila, and E. Flemington. 1997. Reactivation of Epstein-Barr virus: Regulation and function of the BZLF1 gene. Trends Microbiol **5**:399–405.
112. Strobl, L. J., H. Hofelmayr, G. Marschall, M. Brielmeier, G. W. Bornkamm, and U. Zimber-Strobl. 2000. Activated Notch1 modulates gene expression in B cells similarly to Epstein-Barr viral nuclear antigen 2. J Virol **74**:1727–35.
113. Subramanian, C., S. Hasan, M. Rowe, M. Hottiger, R. Orre, and E. S. Robertson. 2002. Epstein-Barr virus nuclear antigen 3C and prothymosin alpha interact with the p300 transcriptional coactivator at the CH1 and CH3/HAT domains and cooperate in regulation of transcription and histone acetylation. J Virol **76**:4699–708.
114. Swinnen, L. J. 2000. Transplantation-related lymphoproliferative disorder: A model for human immunodeficiency virus-related lymphomas. Semin Oncol **27**:402–8.
115. Szeles, A., K. I. Falk, S. Imreh, and G. Klein. 1999. Visualization of alternative Epstein-Barr virus expression programs by fluorescent in situ hybridization at the cell level. J Virol **73**:5064–9.
116. Takada, K. 1984. Cross-linking of cell surface immunoglobulins induces Epstein-Barr virus in Burkitt lymphoma lines. Int J Cancer **33**:27–32.

117. Takada, K., and A. Nanbo. 2001. The role of EBERs in oncogenesis. Semin Cancer Biol **11**:461–7.
118. Tamura, K., Y. Taniguchi, S. Minoguchi, T. Sakai, T. Tun, et al. 1995. Physical interaction between a novel domain of the receptor notch and the transcription factor RBP-J kappa/Su(H). Curr Biol **5**:1416–23.
119. Toczyski, D. P., A. G. Matera, D. C. Ward, and J. A. Steitz. 1994. The Epstein-Barr virus (EBV) small RNA EBER1 binds and relocalizes ribosomal protein L22 in EBV-infected human B lymphocytes. Proc Natl Acad Sci USA **91**:3463–7.
120. Tsubata, T., and J. Wienands. 2001. B cell signaling. Introduction. Int Rev Immunol **20**:675–8.
121. Wang, F., C. Gregory, C. Sample, M. Rowe, D. Liebowitz, et al. 1990. Epstein-Barr virus latent membrane protein (LMP1) and nuclear proteins 2 and 3C are effectors of phenotypic changes in B lymphocytes: EBNA-2 and LMP1 cooperatively induce CD23. J Virol **64**:2309–18.
122. Wang, F., C. D. Gregory, M. Rowe, A. B. Rickinson, D. Wang, et al. 1987. Epstein-Barr virus nuclear antigen 2 specifically induces expression of the B-cell activation antigen CD23. Proc Natl Acad Sci USA **84**:3452–6.
123. Wang, F., S. F. Tsang, M. G. Kurilla, J. I. Cohen, and E. Kieff. 1990. Epstein-Barr virus nuclear antigen 2 transactivates latent membrane protein LMP1. J Virol **64**:3407–16.
124. Wilson, J. B., J. L. Bell, and A. J. Levine. 1996. Expression of Epstein-Barr virus nuclear antigen-1 induces B cell neoplasia in transgenic mice. Embo J **15**:3117–26.
125. Woisetschlaeger, M., C. N. Yandava, L. A. Furmanski, J. L. Strominger, and S. H. Speck. 1990. Promoter switching in Epstein-Barr virus during the initial stages of infection of B lymphocytes. Proc Natl Acad Sci USA **87**:1725–9.
126. Wu, H., P. Kapoor, and L. Frappier. 2002. Separation of the DNA replication, segregation, and transcriptional activation functions of Epstein-Barr nuclear antigen 1. J Virol **76**:2480–90.
127. Yalamanchili, R., X. Tong, S. Grossman, E. Johannsen, G. Mosialos, and E. Kieff. 1994. Genetic and biochemical evidence that EBNA 2 interaction with a 63-kDa cellular GTG-binding protein is essential for B lymphocyte growth transformation by EBV. Virology **204**:634–41.
128. Yates, J., N. Warren, D. Reisman, and B. Sugden. 1984. A cis-acting element from the Epstein-Barr viral genome that permits stable replication of recombinant plasmids in latently infected cells. Proc Natl Acad Sci USA **81**:3806–10.
129. Yates, J. L., S. M. Camiolo, and J. M. Bashaw. 2000. The minimal replicator of Epstein-Barr virus oriP. J Virol **74**:4512–22.
130. Yates, J. L., and N. Guan. 1991. Epstein-Barr virus-derived plasmids replicate only once per cell cycle and are not amplified after entry into cells. J Virol **65**:483–8.
131. Yates, J. L., N. Warren, and B. Sugden. 1985. Stable replication of plasmids derived from Epstein-Barr virus in various mammalian cells. Nature **313**:812–5.
132. Young, L. S., and P. G. Murray. 2003. Epstein-Barr virus and oncogenesis: From latent genes to tumours. Oncogene **22**:5108–21.
133. Young, L. S., and A. B. Rickinson. 2004. Epstein-Barr virus: 40 years on. Nat Rev Cancer **4**:757–68.
134. Yuan, J., E. Cahir-McFarland, B. Zhao, and E. Kieff. 2006. Virus and cell RNAs expressed during Epstein-Barr virus replication. J Virol **80**:2548–65.
135. Zhang, C. X., P. Lowrey, S. Finerty, and A. J. Morgan. 1993. Analysis of Epstein-Barr virus gene transcription in lymphoma induced by the virus in the cottontop tamarin by construction of a cDNA library with RNA extracted from a tumour biopsy. J Gen Virol **74**(Pt 3):509–14.
136. Zhang, J., H. Chen, G. Weinmaster, and S. D. Hayward. 2001. Epstein-Barr virus BamHi-a rightward transcript-encoded RPMS protein interacts with the CBF1-associated corepressor CIR to negatively regulate the activity of EBNA2 and NotchIC. J Virol **75**:2946–56.
137. Zhao, B., and C. E. Sample. 2000. Epstein-Barr virus nuclear antigen 3C activates the latent membrane protein 1 promoter in the presence of Epstein-Barr virus nuclear antigen 2 through sequences encompassing an spi-1/Spi-B binding site. J Virol **74**:5151–60.
138. Zimber-Strobl, U., E. Kremmer, F. Grasser, G. Marschall, G. Laux, and G. W. Bornkamm. 1993. The Epstein-Barr virus nuclear antigen 2 interacts with an EBNA2 responsive cis-element of the terminal protein 1 gene promoter. Embo J **12**:167–75.
139. Zimber-Strobl, U., K. O. Suentzenich, G. Laux, D. Eick, M. Cordier, et al. 1991. Epstein-Barr virus nuclear antigen 2 activates transcription of the terminal protein gene. J Virol **65**:415–23.

6. EBV-ASSOCIATED DISEASES IN THE AIDS PATIENT

SCOTT M. LONG AND CLARE E. SAMPLE

Department of Biochemistry, St. Jude Children's Research Hospital, Memphis, TN

INTRODUCTION

Epstein–Barr virus (EBV) is a ubiquitous gammaherpesvirus, infecting 98% of people worldwide and, like all herpesviruses, establishing a life–long latent infection (for review see 78). These properties clearly increase its potential to be involved with a large number of diseases. Indeed, it is estimated that infectious agents may play a causal role in as many as 20% of human cancers, and EBV is one of a limited number of such agents.[121,175] Normally, however, the virus has a commensal relationship with its human host and is controlled by the immune system. Once the balance afforded by the immune system is removed, EBV is associated with a number of diseases, including as many as 50% of all AIDS-associated lymphomas. The exact roles of EBV and HIV in lymphomagenesis have not yet been fully elucidated. While HIV-related T-cell lymphomas have been described,[7] studies of immunoglobulin heavy–chain rearrangement have demonstrated that the majority of AIDS-related lymphomas are of a B-cell origin.[90] A variety of AIDS-associated lymphomas occur including B-cell immunoblastic lymphoma, small noncleaved lymphoma, i.e., Burkitt's lymphoma (BL) and primary effusion lymphoma (PEL),[43,79] and EBV likely plays differing roles in the development of each of these. In addition to these cancers, EBV has a strong association with leiomyosarcomas that occur most frequently in HIV-positive children. Although Hodgkin's lymphoma is not an AIDS-defining illness, there is a greater association with EBV and a more serious progression of disease in AIDS patients. Because HIV does not directly infect B cells or muscle cells and has not been detected in these lymphomas, HIV likely does not play a direct role in tumor development, although HIV does increase the levels of cytokines that affect B–cell proliferation.[22,143,144] A major contribution of HIV, however, is to promote a disrupted immunologic milieu with diminished regulatory T-cell functions that ultimately allows for the establishment of hyperplasia. Because the EBV genome can be detected in varying degrees in these malignancies and molecular analysis reveals that EBV infections precede the onset of clonal proliferation, it is predicted that EBV acts, at least minimally, as a cofactor in the establishment of some of these tumors.[158]

EBV LATENCY

There are two phases in the viral life cycle: the lytic cycle, where new virus is produced, and the latency program where no virus is produced. In the lytic cycle, the BZLF1 protein is a viral transcription factor that initiates a cascade of viral gene expression culminating in the formation of new viral particles (for review see 78). When these viral particles infect primary B lymphocytes in vitro, the virus establishes a latent infection where the full complement of latency-associated proteins is expressed and induces cellular proliferation and immortalization. Hence, this type of latency is known as the growth program or latency III. This is the least restrictive of the latency programs and is associated with the expression of all of the latency-associated genes: a group of nuclear proteins, EBNAs-1, -2, -3A, -3B, -3C, and -LP; three membrane proteins, LMPs-1, -2A, and -2B; and several virally encoded RNAs: the EBV-encoded RNAs (EBERs) and the *Bam*HI A rightward transcripts (BARTs). This pattern of expression is also believed to occur in newly infected naive B lymphocytes in vivo. However, the growth-promoting viral proteins are also key targets of the immune system, particularly the EBNA-3 proteins, which are predominant targets of cytotoxic T lymphocytes.[112] Thus, continued expression of these proteins would likely lead to an eradication of virally infected cells. The virus is adept at escaping detection from the immune system with one approach being the downregulation of expression of these immunogenic proteins to a more restricted pattern of latency-associated gene expression. In memory B cells in peripheral blood, cells that do not require the virus to maintain survival, latency (or latency I) is associated with expression of only EBNA-1 and none of the proteins that promote cellular proliferation, though the EBERs and BARTs are expressed, allowing the virus to escape detection by the immune system and persist indefinitely.[10] In nonproliferating memory B cells, it is believed that not even EBNA-1 is expressed. A slightly less restricted form of latency, where LMP-1 and LMP-2A are expressed, has also been detected in tonsillar germinal center cells and memory B cells, a pattern known as latency II.[9] These different latency programs are also found in different malignancies.

The major function of EBNA-1 in latency is to maintain the viral genome (described in more detail elsewhere in this volume; for review, see 77). EBNA-1 functions by binding to the origin of viral replication, oriP, and also to mitotic chromosomes.[118,132] Thus, EBNA-1 is expressed in all forms of latency in proliferating cells where the viral genome would otherwise be lost. EBNA-2 functions as a viral transactivator, targeting promoters through interactions with the cellular transcription factors PU.1 and CSL (alternatively known as CBF-1 or RBP-Jκ), to activate the promoters of all of the latency proteins as well as cellular genes.[55,60,72,85,93,174] The EBNA-3 proteins also bind to CSL but prevent its association with DNA.[134,172] In reporter gene assays, the EBNA-3 proteins counter regulation by EBNA-2 but this function may not occur in EBV-infected cells where CSL is present in excess.[71,73] EBNA-2's transactivating capacity is increased on some promoters by EBNA-LP and EBNA-3C.[58,115,173] One of these targets is LMP-1, often referred to as the viral oncoprotein. Although LMP-1 is not sufficient for transformation of B lymphocytes, its expression does result in transformation of rodent fibroblasts and tumor formation in nude mice.[166] Thus, it is often referred to as the viral oncoprotein. LMP-1 functions similarly to a constitutively activated CD40 molecule and activates NFκB.[110,161] In vitro, LMP-2A is not essential for EBV-mediated immortalization of B lymphocytes but it prevents activation of the lytic cycle.[94,95,107] However, expression of LMP-2A in mouse primary B cells results in inappropriate survival signals, suggesting that it may facilitate the persistence of EBV in vivo.[31] The EBERs are small nonpolyadenylated RNAs that are expressed in the cell nucleus during all forms of EBV latency.[65] Although the EBER transcripts bind to and inhibit activation of the double-stranded RNA-dependent protein kinase PKR in cell-free systems, it is not clear that this function operates in latently infected cells.[137,138] Thus, the function of the EBERs is largely unknown. Although the EBERs are dispensable for EBV-mediated immortalization of primary B lymphocytes, they augment the tumorigenicity of BL cells in xenograft assays.[81,138,152] A second group of alternatively spliced RNAs known as the BARTs are also expressed during all types of latency and within different EBV-associated malignancies.[26,38,39] These RNAs contain open reading frames but whether proteins are actually produced has been debated. Nevertheless, the functions of these putative protein products have been studied demonstrating that RPMS-1 and RK-BARF0 associate with CSL and NOTCH, respectively, whereas A73 binds to RACK1.[83,171]

IMMUNE CONTROL OF EBV

In normal individuals, EBV is readily controlled by the immune system. Early in infection, a robust T-cell response is directed against both latent and lytic antigens. In the memory response, the most dominant epitopes are within the EBNA-3 proteins.[30] EBNA-1 contains a large glycine-alanine repeat that inhibits proteosomal processing, and therefore, EBNA-1 does not efficiently induce a response by CD8 cytotoxic T cells, though EBNA-1-reactive cells can be detected.[88,92,157] It has long been known that immunodeficient populations are unable to control EBV-mediated proliferation. In X-linked lymphoproliferative syndrome, a rapid EBV-related lymphoproliferative disease frequently follows primary EBV infection, and lymphoblasts expressing a latency III program can be readily detected. Similar lymphoprolifera-tions are observed post-transplant (PTLD). These EBV-infected cells express the full complement of viral latency genes that are normally targeted by the immune system. The importance of immune control is underscored by the fact that reduc-tion of immunosuppression often results in regression. Furthermore, treatment of these patients with donor CTLS or EBV-specific CTLs rapidly decreases tumor growth.[64,142] It is not surprising that similar lymphoproliferations occur in AIDS patients whose T-cell counts are very low, suggesting that these tumors are a result of the loss of immune regulation normally controlling growth of the EBV-infected cells. Indeed, AIDS patients often exhibit an increased EBV load. As might be expected, the incidence of these tumors has decreased among patients receiving HAART. By contrast, the incidence of BL tumors, which only express EBNA-1 and not the more immunogenic viral proteins, has not appreciably changed since the introduction of HAART. Perhaps this finding is not surprising because BL often occurs before significant immunosuppression has occurred.

HOW IS THE PRESENCE OF EBV DETERMINED?

In order to conclude that EBV is associated with a particular disease, it is essential to demonstrate that EBV is present in the majority of cells in the tumor. The EBERs are present at very high copy numbers in all forms of latency are not homologous to any cellular RNAs, and, thus, represent a very sensitive measure of latent EBV infection.[6,89] In situ hybridization with EBER-specific probes has, therefore, become the method of choice for detection of EBV in tissue samples.[65,76] It is also important to determine whether any viral proteins are expressed, and anti-bodies for EBNA-2 and LMP-1, in addition to some lytic cycle proteins, are par-ticularly suitable for these studies.[136,170] Because EBV is a ubiquitous virus, whether EBV plays a specific role in disease development or is merely a carrier is an essen-tial consideration. One way that this has been addressed is to determine whether viral infection has preceded malignant transformation. To address this possibility, an assay that examines viral termini has been used.[128] This assay takes advantage of the fact that linear or replicating viral DNA contains variable numbers of terminal repeats. Once circularized, each distinct linear fragment can give rise to an episome with a distinct number of repeats. The number of repeats contained within the viral genome in a given tissue sample can be examined by Southern-blot analysis of DNA that is restricted near the terminal repeats. Replicating DNA will give rise to smaller fragments than episomal or latent DNA, with the size of the episomal (latent) DNA fragment dependent on the number of terminal repeats. A single size fragment is indicative of expansion of a single infected cell, suggesting that EBV may have played an initiating role in tumor development. Thus, this assay readily identifies a monoclonal or oligoclonal tumor with respect to the viral genome. Recently, it has been suggested that low numbers of terminal repeats may facilitate expression of LMP-2A.[108] While LMP-2A is dispensable for B-cell immortalization, it may have transforming effects in epithelial cells.[3,109,123,145] Thus, in some cells, it is possible that clonality may represent an event that occurs after infection of cells where tumorigenesis has already been initiated, suggesting that EBV may play a greater role in tumor progression rather than in initiation. Nevertheless, this scenario still argues for an important role of EBV in tumor development.

BURKITT'S LYMPHOMA

In 1958, Burkitt described a "sarcoma" affecting the jaw of African children which could also present as visceral tumors.[27,28] In 1961, these tumors were more fully characterized by O'Conor as malignant lymphomas, and the disease became known as Burkitt's lymphoma.[116,117] These lymphomas appeared endemic within equatorial Africa and Burkitt proposed that an infectious agent might be responsible. Electron microscopic analysis of these lymphomas revealed the presence of herpesviral particles, making EBV the first identified human tumor virus. As the first discovered EBV-associated disease, BL is one of the best studied. Nearly four decades later, it is becoming apparent that BL does not represent a single pathological entity but instead BL should be considered a family of closely related tumors.[140] Histologically, BL can be partly defined as a small noncleaved tumor of B-cell origin, i.e., tumor cells stain positive for the B-cell specific markers including CD19 and CD20. BL cells are monomorphic and contain well-rounded nuclei with multiple nucleoli and exhibit an extremely high rate of proliferation,[15,33] shown by staining for Ki-67 expression, a proliferation marker expressed in all active stages of the cell cycle but absent in resting cells.[146] Furthermore, a high rate of apoptotic death accompanies the overly high proliferative capacity of BL cells. Invading macrophages which phagocytose the apoptotic remnants can be viewed among the proliferating lymphoma cells, imparting a distinctive "starry sky" appearance to histological preparations.[15,33]

Overall, the clinical aspects of BL differ in different regions of the world. While classification of BL should take into consideration these geographical/clinical differences, the presence of EBV within tumor cells and the HIV status of the patient should also be weighed. Consequently, the World Health Organization now formally recognizes three subtypes of BL: (1) endemic or African (eBL), (2) sporadic (sBL), and (3) HIV or AIDS associated,[15,97] although some overlap does occur, e.g., both the sporadic and AIDS-associated forms of BL can occur in endemic areas.[86] Taken together, the subtypes are among the most aggressive lymphomas and account for 3–5% of all lymphomas worldwide.[14]

Virtually all cases of eBL (98–100%) are associated with clonal EBV.[11] The World Health Organization reports an annual eBL incidence of 6–7 cases per 100,000 with a peak incidence at 6–7 years of age. The highest prevalence of eBL, accounting for 50–70% of all pediatric malignancies,[2] occurs within the equatorial belt of Africa and Papua New Guinea in a region also known for a very high incidence of malaria.[14] This long-time association of malaria with the development of EBV-positive BL has led to the speculation that immune suppression, brought about by parasitic infection, leads to weakened anti-EBV surveillance, and ultimately to an increase in BL precursors. However, despite strong epidemiologic correlations, molecular connections that malaria predisposes one to BL through an interaction with components of EBV have not yet been made (reviewed in 140). Furthermore, the EBV growth-promoting latency genes are typically not expressed in BL.[155] Consequently, the role that EBV plays in the growth and development of these tumors has not yet been elucidated. Although abdominal masses do occur, the most common anatomical site of the primary tumor in eBL involves the jaw and other facial bones.[29]

Sporadic BL occurs predominately in the United States and Europe with an annual incidence of 2–3 cases per million.[122] Although sporadic BL has been reported in the head and neck of some pediatric patients, involvement of the jaw is quite rare, in contrast to eBL. Instead, a predilection for the abdomen, especially the terminal ileum, ascending colon, and mesentery has been reported.[91] Unlike eBL, where bone marrow and CNS involvement is rare, as many as two thirds of sporadic BL cases exhibit marrow involvement. Typically, the sporadic form affects children around the age of 12 years.[91] In the United States and Europe, only 20–30% of sporadic BL cases exhibit elevated EBV titers.[14] Interestingly, the sporadic BL typically seen in Brazil is associated with a much higher incidence of EBV infection (70–90%).[5,11,141]

Six out of every 1,000 AIDS patients develop AIDS-associated BL accounting for 30% of all HIV/AIDS-associated lymphomas. In the US, only 20–30% of these cases are believed to be EBV associated whereas a greater percentage are EBV associated in Africa. Like the sporadic form, AIDS-associated BL rarely presents as a jaw tumor; abdominal masses and marrow involvement are most frequently reported.[14] In contrast to eBL and sporadic BL, where the anatomical boundaries are more well defined (i.e., jaw or abdomen) uncharacteristic sites have been reported in AIDS-associated BL patients. These sites include heart, gingiva, anus, bile duct, and muscles.[150] Clinically, AIDS-associated BL is more aggressive and less responsive to chemotherapy than either the endemic or sporadic form and presently only 40% (compared to 90% in both the endemic and sporadic subtype) of patients diagnosed with an AIDS-associated lymphoma survive.[14] Treatment of AIDS-associated BL remains much less successful than treatment of the other BL subtypes further underscoring the need for novel therapeutics. Despite success with other lymphomas within the HAART era, there has been no significant decline in the incidence of AIDS-related NHL.[44,102,129,159]

Despite the differences in BL subtypes, all share a diagnostic chromosomal translocation that places the *MYC* allele under control of a B-cell-specific immunoglobulin enhancer (reviewed in 97). The consequence is constitutive, deregulated expression of MYC resulting in uncontrolled cell proliferation. Deregulated MYC expression is likely the underlying reason that BL cells share a common morphology regardless of their subtype. Finally, BL cells share the cellular phenotype of a germinal-center cell.[15,33] In this regard, tumor cells stain BCL-6-positive/MUM1-negative/syndecan-1-negative closely reflecting the phenotype of

B cells within germinal centers.[33] Consequently, the prevailing theory is that BL arises from precursors within germinal centers.[15]

As previously mentioned, the diagnostic determinant for BL regardless of its subtype is the translocation which joins the *MYC* allele from chromosome 8 with an Ig heavy chain locus (IgH) from chromosome 14, or with the κ or λ Ig light chain locus (IgL) from chromosomes 2 and 22, respectively.[97] The t(8;14) occurs in 80% of the translocations and results in the placement of the *MYC* allele downstream of the IgH locus. The t(2;8) and t(8;22) translocations comprise the remaining 15% and 5%, respectively, and result in the transfer of an IgL locus downstream of *MYC* on chromosome 8.[59] The result of all of these translocations is deregulated, constitutive expression of the *MYC* oncoprotein.

MYC is the human homolog of v-Myc, a viral oncoprotein encoded by the avian MC29 myelocytomatosis transforming virus.[148] Altered expression of MYC in greater than 70% of human cancers[114] underscores the importance of this oncogene, especially in BL where 100% of the cases[33] have aberrant MYC expression. The *MYC* gene comprises three exons. Generally, translation begins at an AUG start site within the second exon producing a 439 amino acid (64 kDa) protein although alternative initiation sites can result in the production of both longer (p67 MYC) and shorter (MYCS) isoforms.[17] Interestingly, the chromosomal breakpoints which allow for Ig/MYC juxtaposition differ for the different BL subtypes. For example, the breakpoint described for eBL can occur as far as 100 kb upstream of the first *MYC* exon whereas the breakpoints for both the sporadic and the AIDS-associated subtypes occur between the first and second *MYC* exon (reviewed in 140).

MYC is a member of a family of transcription factors known as the basic helix–loop–helix leucine zipper (bHLH-Zip) family. The HLH-Zip domain which defines this family of proteins resides in the C-terminal region of MYC and facilitates homotypic and heterotypic dimerization with other bHLH-Zip family members. Heterodimerization of MYC with MAX, another member of the bHLH-Zip family, allows for the binding of MYC:MAX to a hexanucleotide DNA sequence (CAYGTG) known as an E-box.[19–21,126] MYC:MAX heterodimers likely target transcriptional coactivators and histone acetyltransferases to the promoters of genes containing E-boxes.[47,105,106,119,120] In addition to activating transcription, MYC has also been reported to repress transcription of a number of gene products.[114] To date, MYC has been shown to affect the transcription, either directly or indirectly, of over 1,400 genes (http://www.myc-cancer-gene.org).

As an oncogene, MYC has often been referred to as the "master regulator" of the cell cycle and is required for sustained proliferation of tumor cells.[51,68] In this regard, MYC promotes cell proliferation via its effects on genes that directly regulate the cell cycle. This is accomplished through the upregulation of D cyclins, cyclin-dependent kinases (CDKs) 4 and 6, and cdc25A. As previously mentioned, MYC also effects the downregulation of a number of gene products. Central to its role in cell cycle regulation, MYC promotes the downregulation of the CDK inhibitors p21^{Cip1} and p27^{Kip1}.[52,59,62,100,111]

While MYC deregulation is without a doubt a founding event in the establishment of BL, aberrant MYC expression alone is not sufficient to cause BL. Despite the positive influences MYC imparts on cell cycle progression, MYC overexpression also results in profound programmed cell death or apoptosis.[50] Consequently, tumorigenesis necessitates the ability to disable MYC-mediated apoptotic pathways. MYC-induced apoptosis is mediated by the ARF-HDm2-p53 tumor suppressor pathway.[50] p53 is a transcription factor activated in response to aberrant oncogene expression (e.g., MYC) and DNA damage. Central to its role as a tumor suppressor, p53 can either induce cell cycle arrest or provide time for the cell to repair itself or (in the case of overly severe damage) initiate apoptosis. MYC has been shown to rapidly induce p19[ARF] which, in turn, effectively stabilizes p53 by binding HDm2 (the human equivalent of Mdm2) and neutralizing its functions, i.e., the ubiquitinization and shuttling of p53 from the nucleus to the cytoplasm for proteosomal degradation (reviewed in 50). An understanding of the role that MYC plays in p53-mediated apoptosis has been determined through studies involving Eμ-myc transgenic mice. In this model, the myc transgene is under control of the Eμ IgH enhancer.[1] Analogous to BL involving the t(8;14) translocation, following a protracted subclinical course, these transgenic mice develop clonal pre-B and B-cell lymphomas.[1,50] In this setting, early hyperproliferation of B cells in the peripheral blood and bone marrow are countered by an equally high apoptotic index.[67] Crosses of Eμ-myc mice with p53[−/+66] or Eμ-bcl-2[151] transgenic mice significantly enhanced lymphomagenesis, suggesting that disabling proapoptotic mechanisms is a requirement for progression to malignancy.

A majority of the tumors arising in Eμ-myc transgenic mice have demonstrated a disruption in the ARF-Mdm2-p53 pathway. For example, analysis of Eμ-myc-derived lymphomas revealed a loss of p53 function in 28% of the tumors examined. Of these, the majority (86%) harbored mutations in p53; rarely, biallelic deletion of p53 was recorded.[50] In this regard, the Eμ-myc mouse model system recapitulates BL where 30–40% of BL tumors have been shown to harbor mutations in p53.[127] The Mdm2 antagonist, ARF, was targeted for biallelic deletion in 24% of the lymphomas examined. In tumors that retain wildtype p53, Mdm2 (a negative regulator of p53) was frequently upregulated.[50] Clearly, the results obtained from the studies using Eμ-myc transgenics have demonstrated the requirement for inactivation of apoptotic pathways for the establishment of MYC-mediated lymphomagenesis. Consequently, the data gleaned from these studies is applicable to the development of BL. In this context, however, the contributions that EBV makes to the development of BL have not been addressed.

As previously mentioned, AIDS-associated BL most often does not involve EBV infection.[14] Consequently, the role that the virus plays in the development of this malignancy is indeterminate. However, because many of the lymphoproliferative disorders arising in immunocompromised individuals represent polyclonal proliferations of EBV-infected B cells,[57,98] the role that this transforming virus might play cannot be discounted. It is most probable that the inability of the immunosuppressed to mount an effective immune response against EBV-infected cells results in their uncontrolled, polyclonal proliferation. Moreover, it has been reported that some polyclonal lymphoproliferative processes evolve into monoclonal lymphomas[56] and that many of these bear the characteristic Ig:MYC translocations of BL.[98,169]

Treatment of BL often involves high-dose chemotherapy and radiation.[12] However, because AIDS patients do not tolerate high-dose chemotherapy well, these individuals are generally treated with lower-dose chemotherapeutic regimens.[45] Newer approaches for the treatment of BL advocate the combination of standard chemotherapy in conjunction with monoclonal antibodies that strictly target B cells. One such antibody currently in clinical trials for the treatment of AIDs-related BL is Rituximab, a monoclonal antibody that binds specifically to CD20 (human B-lymphocyte-restricted differentiation antigen). CD20 is expressed on the surface of pre-B and mature B lymphocytes including >90% of NHL,[4,49,163] but not on hematopoietic stem cells or plasma cells.[156] Antibody binding to CD20 positive B cells likely promotes cytolysis through antibody-dependent cell-mediated cytotoxicity and/or activation of the complement pathway.[131]

HODGKIN'S LYMPHOMA

Hodgkin's disease or lymphoma is characterized by the presence of Hodgkin's or Reed-Sternberg (HRS) cells, very large tumor cells that comprise only about 1% of the cells in the tumor. HRS cells are large multinucleate cells whose origin has been debated for many years. In many lymphomas, these cells bear few, if any, B cell markers and may in fact express markers of other cell lineages. Though HRS cells are the malignant cell type, the tumor comprises a minority of these cells, as low as 1%, in the presence of a large number of infiltrating lymphocytes. This fact made it difficult to analyze HRS cells in any detail until the advent of microdissection. This technique has allowed the analysis of immunoglobulin genes, which has revealed rearrangement of both heavy and light chain genes.[74,82] Because of this analysis, the disease has been renamed as Hodgkin's lymphoma. Further analysis of immunoglobulin V genes revealed extensive mutation.[25] Because hypermutation typically occurs in germinal centers, this finding suggested that HRS cells arise from a postgerminal center B cell. Curiously, many of these mutations have rendered the immunoglobulin gene product nonfunctional, an occurrence that would normally be associated with apoptosis within the germinal center.

In the absence of infection with HIV, approximately 50% of Hodgkin's lymphoma are associated with EBV although the role of the virus has not been definitively delineated. The incidence of HL increases between 10–20-fold in the context of HIV and almost all of the lymphomas are associated with EBV. This finding suggests that immunosuppression facilitates the development of EBV-positive HL. Interestingly, while HL is only 50% EBV associated in developed countries, the association is much higher in other countries: 57% in China, and the majority of cases in Peru and Kenya.[103,121] One study has detected rearranged genomes in a significant number of EBER-negative cases of HL raising the possibility that an even greater number of cases may be associated with EBV.[53]

What is the evidence that EBV is involved in this disease? Some of the first evidence was epidemiological data indicating that people who have had infectious mononucleosis, associated with a primary EBV infection, had an enhanced risk for the development of HL. In biopsies from EBV-positive tumors, EBV EBER RNAs, a diagnostic marker for EBV infection, can be detected within HRS cells. Furthermore, EBV genome analysis suggests that infection is monoclonal. HRS cells express the viral proteins LMP-1 and LMP-2A as detected by immunohistochemistry, and the BART RNAs have also been detected. Although LMP-1 is required for EBV-mediated transformation of B lymphocytes and its expression results in transformation of rodent fibroblasts, the contribution of LMP-1 to tumor development is not clear. On a molecular level, LMP-1 functions similar to a constitutively active CD40 molecule and increases NFκB activity.[110] Most HL, regardless of their EBV status, exhibit activated NFκB, which can occur as a result of mutations in IκB molecules, overexpression of c-Rel or by affecting signaling pathways that activate NFκB.[13] Thus, one contribution of EBV is likely to be the constitutive activation of NFκB by LMP-1. A second potential role stems from the observation that HRS cells contain crippling mutations in immunoglobulin genes that should result in apoptosis, suggesting the possibility that a viral gene product could protect the cells from apoptosis. Of considerable interest is the report that the majority of the HRS cells containing crippling mutations that would prevent BCR expression are EBV positive, suggesting that EBV plays a role in the survival of these cells. In support of this possibility, three groups have recently infected either tonsillar germinal center or BCR-negative B cells with EBV and isolated lymphoblastoid cell lines unable to express a functional B-cell receptor, suggesting that EBV is indeed capable of rescuing BCR-deficient GC B cells from apoptosis.[14,36,99] A good candidate for a gene product that may be involved in increasing cell survival is LMP-2A. In transgenic mice, LMP-2A provides inappropriate survival signals to primary B lymphocytes, allowing B cells that fail to express a function BCR to survive.[31] One of these signals could be activation of the Akt pathway, which is constitutively activated in the majority of HL and is also activated by LMP-2A.[48,145,153] Although it has been known for some time that HRS cells generally lack B-cell antigens, more recent gene profiling of HRS cells has demonstrated that there is a global downregulation of B-cell-specific genes.[147] The mechanism by which this change in gene expression occurs is not totally understood but recent reports have provided some insight. The downregulation of a large number of genes suggests that transcription factors that play important roles in B-cell-specific transcription might be downregulated. In fact, Oct-2, OBF-1, and PU.1 levels are very low or absent in HRS cells whereas Pax-5 is expressed although perhaps at decreased levels.[63] In addition, protein levels of E47/E12 was modestly downregulated but its DNA-binding ability, as measured by electrophoretic shift assay, was greatly decreased.[63] Two cellular proteins that decrease expression of B-cell-specific genes when transfected into B-cell lines are overexpressed in HRS cells. The inhibitor of differentiation and DNA-binding protein, Id2, is a helix–loop–helix transcription factor that can dimerize with other transcription factors but which itself lacks a DNA-binding domain.[101,133] ABF-1 is another helix–loop–helix protein that contains a potent transcriptional repression domain and is expressed in HRS cells and in EBV-transformed cells.[101] Perhaps as a result of transcriptional silencing, B-cell-specific genes are methylated in HRS cells.[162] Interestingly, expression of LMP-2A in B-cell lines results in downregulation of B-cell-specific genes similar to that found in HL, suggesting that in the EBV-positive tumors, EBV may play a role in gene silencing.[125] Although there are numerous changes in cellular genes, HL is unlike BL in that there are generally no mutations in p53.

PRIMARY CENTRAL NERVOUS SYSTEM LYMPHOMA

AIDS-related primary central nervous system lymphoma (PCNSL) occurs with profound immunosuppression and is typically fatal within a short time frame. As indicated by its name, PCNSL is limited to the central nervous system. Histologically, these lymphomas are classified as large cell lymphomas or immunoblastic lymphomas of B-cell origin. By contrast to other AIDS-related non-Hodgkin's lymphomas, PCNSL is universally associated with EBV infection. EBERs are detected in all biopsies from AIDS-related PCNSL, but are absent in many PCNSL occurring after renal transplant.[96] Both LMP-1 and EBNA-2 are expressed in almost all cases of AIDS-related PCNSL whereas LMP-1 may be occasionally detected in the absence of EBNA-2 in cases arising in other patients.[8,96] High levels of EBV DNA in the CSF is predictive of PCNSL and analysis of EBV DNA in CSF has become useful as a minimally invasive diagnostic tool.[42] The presence of very high levels of DNA in the CSF, together with the detection of some lytic cycle proteins within tumors, suggested that lytic replication may be occurring either in the CSF or in a subset of malignant cells. Because treatment options for PCNSL are few, novel therapeutics is desperately needed. Two patients were treated with low-dose hydroxyurea, a treatment suggested by in vitro studies that found that treatment of cell lines with low doses of hydroxyurea led to a loss of EBV episomes.[40,149] Both patients had a reduction in tumor size following treatment, with no further progression of the tumor up to 24 months posttreatment. A second study has corroborated these findings.[37] Agents that reduce EBV lytic replication have been reported to function in EBV-associated lymphomas, and a study in patients with AIDS-related PCNSL has found that ganciclovir dramatically reduced the amount of EBV in CSF of patients with PCNSL.[24] Control patients exhibited stable levels of EBV DNA in the CSF whereas levels increased in two patients removed from ganciclovir. Although the treated patients still succumbed to disease, they had a longer survival time, suggesting that antiviral therapies may be useful.

PRIMARY EFFUSION LYMPHOMAS

A minority of AIDS-associated lymphomas are those growing mainly in the pleural, pericardial, or abdominal cavities as a lymphomatous effusion. These lymphomas, known as PELs or body cavity-based lymphomas (BCBLs), represent a distinct entity occurring in AIDS patients and are categorized as primary effusions that are almost always associated with the Kaposi's sarcoma-associated herpesvirus (KSHV or HHV-8) (reviewed in more detail elsewhere in this volume).[34,75] Although the lymphomas lack many B lineage restricted antigens, the analysis of immunoglobulin genes indicates the consistent presence of immunoglobulin gene rearrangements, indicating a B-cell derivation.[34,75,80,113] Surprisingly, most of the AIDS-related BCBLs also contain EBV. Analysis of the EBV genomes reveals a monoclonal infection suggesting that EBV infection occurred at an early stage.[34] Analysis of EBV expression within these tumors, or in cell lines derived from the tumors, has revealed the expression of the EBERs, EBNA-1, and LMP-2A with variable expression of LMP-1.[32,34,84,154] While the role of EBV in the development of these tumors is currently unknown, the presence of EBV renders BCBL cell lines more tumorigenic in nude mice.[160] BCBLs are resistant to cytotoxic chemotherapy but it is possible that further study of the roles of the two viruses may allow use of specific treatments that target one of these viruses or their gene products.

ORAL HAIRY LEUKOPLAKIA

Oral hairy leukoplakia (HLP) is a benign lesion of oral epithelium, generally observed as white patches along the sides of the tongue. HLP occurs most commonly in HIV-infected persons. These lesions contain rapidly replicating EBV with virions readily detectable by ultrastructural analysis, and inhibition of EBV replication by acyclovir results in clinical remission.[16,61,130] Both EBV strains type 1 and type 2 have been found within a single lesion as well as multiple strains of each type.[164] Recombination between different strains can be detected, suggesting coinfection of a single cell by multiple viruses.[165] Curiously, despite the degree of lytic replication, latency-associated genes continue to be expressed. EBNA-2, EBNA-3B, and EBNA-3C have been demonstrated to be transcribed mainly from the W promoter (latency specific), whereas EBNA-1 is transcribed from either the W or F (lytic) promoters.[167,168] These lesions are unusual in that the EBERs, which are normally highly expressed in EBV-associated malignancies and are generally used as a diagnostic marker for EBV infection, are not expressed, which has suggested that these RNAs are a marker for latent infection.[54] LMP-1, LMP-2, and the BART transcripts are also expressed.[168] Moreover, LMP-1, EBNA-2, and EBNA-LP proteins can be detected.[167] Several pieces of data suggest that LMP-1 is functionally active within these cells. Normally, LMP-1 binds to cellular proteins such as the TNF receptor-associated proteins (TRAFs), an interaction that activates NFκB and results in altered cellular gene expression. TRAF-1, TRAF-2, and TRAF-3 are also expressed, and activated NFκB can be detected within nuclei in HLP biopsies.[167] Furthermore, cellular genes known to be induced by LMP-1 can be detected. One of these genes is A20 which blocks apoptosis induced by p53.[167] Levels of antiapoptotic proteins such as bcl-2 and bcl-xl do not appear to be altered but BHRF1, a viral homologue of bcl-2, is highly expressed in HLP.[41,167]

LEIOMYOSARCOMA

In the normal population, leiomyosarcomas are a rare malignant tumor of smooth muscle with a frequency of less than two cases per ten million children. These tumors occur more commonly in immunocompromised patients and are most commonly seen in HIV-1 infected children where they can occur in a variety of anatomic locations.[35,139] In fact, leiomyosarcomas represent the second most common cancer in HIV-1 infected children.[18,124] This tumor is less commonly seen in HIV-1 infected adults and the reason for this difference is unknown.[103] Despite the high prevalence of this tumor in HIV-infected populations, HIV-1 has not been detected in leiomyosarcomas, suggesting that the immunosuppression may be contributing to tumor development as is thought to be the case with HIV-associated lymphomas. Using in situ hybridization with EBER-specific probes, EBV can be detected in virtually all cells of tumors from immunocompromised patients but not in adjacent normal tissue or, in AIDS patients, in KS lesions located nearby.[23,104] By contrast, EBV cannot be detected in smooth muscle tumors from immunocompetent patients.[104] Using quantitative PCR, the viral load was estimated in one study to be approximately three or four EBV genomes per cell,[69] but a much larger study has shown that copy number can range from 1 to 100.[46] These data confirm that EBV is present within the tumor and not detected in infiltrating cells. To determine the contribution of EBV to the development of the tumors, the clonality of several tumors has been examined, and tumors are either monoclonal or contain a couple of distinct viruses.[46,69,87] Interestingly, immunosuppressed patients often develop multiple leiomyosarcomas at distinct sites that can contain genomes with different numbers of terminal repeats, suggesting that each has a distinct origin.[46,104] As a first step in defining the role of the virus in the development of these tumors, attempts have been made to determine which viral proteins are expressed. The most extensive analysis has been performed in cells explanted from a leiomyosarcoma.[70] In this study, EBNA-1 could be detected as well as various lytic cyle antigens such as BZLF1 (an immediate early gene) and late antigens such as the viral capsid antigen p160 and the glycoprotein gp350. Virus could be isolated from these cells and used to generate EBV-immortalized LCLs. Thus, infectious EBV can be produced by smooth muscle tumors though it is not clear whether virus can also be produced in vivo. Extensive RT-PCR was performed on RNA prepared from a tumor that arose following a heart transplant, and EBNA-1, EBNA-2, LMP-1, and LMP-2A could be detected whereas BZLF1 could not.[135] Immunohistochemistry has been used to detect EBNA-2 expression in nuclei of two patient samples whereas LMP-1 was detected in only a few cells in one sample.[87] Both strains of EBV have been detected with type 2 EBV detected in 4/4 samples in one study. Although the receptor that EBV might use to gain access to smooth muscle cells is not known, the EBV receptor used in lymphocytes, CD21, has been reported on the surface of leiomyosarcomas of HIV-1 infected persons, whereas tumors that develop in immunocompetent patients show little or no staining.[69,104] These findings suggest that EBV can infect smooth muscle cells and that it plays a causal role in the development of leiomyosarcomas in the immunocompromised population.

SUMMARY

EBV-associated malignancies remain a considerable problem in HIV-infected individuals, even in the era of HAART. Although EBV is a common factor, each disease has a unique pathogenesis. Study of these diseases reveals the viral proteins expressed in the malignancies that might contribute to the development of the disease as well as the molecular basis for pathogenesis. It is likely that this knowledge will contribute to the development of novel therapeutics that will result in more favorable outcomes in the future.

REFERENCES

1. Adams, J. M., A. W. Harris, C. A. Pinkert, L. M. Corcoran, W. S. Alexander, S. Cory, R. D. Palmiter, and R. L. Brinster. 1985. The c-myc oncogene driven by immunoglobulin enhancers induces lymphoid malignancy in transgenic mice. Nature **318**:533–38.
2. Adatia, A. K. 1968. Dental tissues and Burkitt's lymphoma: Histology, biology, clinical features, and treatment. Oral Surg **25**:221–34.
3. Allen, M. D., L. S. Young, and C. W. Dawson. 2005. The Epstein-Barr virus-encoded LMP2A and LMP2B proteins promote epithelial cell spreading and motility. J Virol **79**:1789–802.
4. Anderson, K. C., M. P. Bates, B. L. Slaughenhoupt, G. S. Pinkus, S. F. Schlossman, and L. M. Nadler. 1984. Expression of human B cell-associated antigens on leukemias and lymphomas: A model of human B cell differentiation. Blood **63**:1424–33.
5. Araujo, I., H. D. Foss, A. Bittencourt, M. Hummel, G. Demel, N. Mendonca, H. Herbst, and H. Stein. 1996. Expression of Epstein-Barr virus-gene products in Burkitt's lymphoma in Northeast Brazil. Blood **87**:5279–86.
6. Arrand, J. R., and L. Rymo. 1982. Characterization of the major Epstein-Barr virus-specific RNA in Burkitt lymphoma-derived cells. J Virol **41**:376–89.
7. Arzoo, K. K., X. Bu, B. M. Espina, L. Seneviratne, B. Nathwani, and A. M. Levine. 2004. T-cell lymphoma in HIV-infected patients. J Acquir Immune Defic Syndr **36**:1020–27.
8. Auperin, I., J. Mikolt, E. Oksenhendler, J. B. Thiebaut, M. Brunet, B. Dupont, and F. Morinet. 1994. Primary central nervous system malignant non-Hodgkin's lymphomas from HIV-infected and non-infected patients: Expression of cellular surface proteins and Epstein-Barr viral markers. Neuropathol Appl Neurobiol **20**:243–52.
9. Babcock, G. J., and D. A. Thorley-Lawson. 2000. Tonsillar memory B cells, latently infected with Epstein–Barr virus, express the restricted pattern of latent genes previously found only in Epstein-Barr virus-associated tumors. Proc Natl Acad Sci USA **97**:12250–55.
10. Babcock, J. G., D. Hochberg, and A. D. Thorley-Lawson. 2000. The expression pattern of Epstein-Barr virus latent genes in vivo is dependent upon the differentiation stage of the infected B cell. Immunity **13**:497–506.
11. Bacchi, M. M., C. E. Bacchi, M. Alvarenga, R. Miranda, Y. Y. Chen, and L. M. Weiss. 1996. Burkitt's lymphoma in Brazil: Strong association with Epstein-Barr virus. Mod Pathol **9**:63–7.
12. Banthia, V., A. Jen, and A. Kacker. 2003. Sporadic Burkitt's lymphoma of the head and neck in the pediatric population. Int J Pediatr Otorhinolaryngol **67**:59–65.
13. Bargou, R. C., F. Emmerich, D. Krappmann, K. Bommert, M. Y. Mapara, W. Arnold, H. D. Royer, E. Grinstein, A. Greiner, C. Scheidereit, and B. Dorken. 1997. Constitutive nuclear factor-kappaB-RelA activation is required for proliferation and survival of Hodgkin's disease tumor cells. J Clin Invest **100**:2961–69.
14. Bechtel, D., J. Kurth, C. Unkel, and R. Kuppers. 2005. Transformation of BCR-deficient germinal-center B cells by EBV supports a major role of the virus in the pathogenesis of Hodgkin and posttransplantation lymphomas. Blood **106**:4345–50.
15. Bellan, C., S. Lazzi, F. G. De, A. Nyongo, A. Giordano, and L. Leoncini. 2003. Burkitt's lymphoma: New insights into molecular pathogenesis. J Clin Pathol **56**:188–92.
16. Belton, C. M., and L. R. Eversole. 1986. Oral hairy leukoplakia: Ultrastructural features. J Oral Pathol **15**:493–99.
17. Bernard, O., S. Cory, S. Gerondakis, E. Webb, and J. M. Adams. 1983. Sequence of the murine and human cellular myc oncogenes and two modes of myc transcription resulting from chromosome translocation in B lymphoid tumours. EMBO J **2**:2375–83.

18. Biggar, R. J., M. Frisch, and J. J. Goedert. 2000. Risk of cancer in children with AIDS. AIDS-Cancer Match Registry Study Group. JAMA **284**:205–9.
19. Blackwell, T. K., J. Huang, A. Ma, L. Kretzner, F. W. Alt, R. N. Eisenman, and H. Weintraub. 1993. Binding of myc proteins to canonical and noncanonical DNA sequences. Mol Cell Biol **13**:5216–24.
20. Blackwell, T. K., L. Kretzner, E. M. Blackwood, R. N. Eisenman, and H. Weintraub. 1990. Sequence-specific DNA binding by the c-Myc protein. Science **250**:1149–51.
21. Blackwood, E. M., B. Luscher, and R. N. Eisenman. 1992. Myc and Max associate in vivo. Genes Dev **6**:71–80.
22. Blazevic, V., M. Heino, A. Lagerstedt, A. Ranki, and K. J. Krohn. 1996. Interleukin-10 gene expression induced by HIV-1 Tat and Rev in the cells of HIV-1 infected individuals. J Acquir Immune Defic Syndr Hum Retrovirol **13**:208–14.
23. Boman, F., H. Gultekin, and P. S. Dickman. 1997. Latent Epstein-Barr virus infection demonstrated in low-grade leiomyosarcomas of adults with acquired immunodeficiency syndrome, but not in adjacent Kaposi's lesion or smooth muscle tumors in immunocompetent patients. Arch Pathol Lab Med **121**:834–8.
24. Bossolasco, S., K. I. Falk, M. Ponzoni, N. Ceserani, F. Crippa, A. Lazzarin, A. Linde, and P. Cinque. 2006. Ganciclovir is associated with low or undetectable Epstein-Barr virus DNA load in cerebrospinal fluid of patients with HIV-related primary central nervous system lymphoma. Clin Infect Dis **42**:e21–e25.
25. Brauninger, A., R. Schmitz, D. Bechtel, C. Renne, M. L. Hansmann, and R. Kuppers. 2006. Molecular biology of Hodgkin's and Reed/Sternberg cells in Hodgkin's lymphoma. Int J Cancer **118**:1853–61.
26. Brooks, L. A., A. L. Lear, L. S. Young, and A. B. Rickinson. 1993. Transcripts from the Epstein-Barr virus BamHI A fragment are detectable in all three forms of virus latency. J Virol **67**:3182–90.
27. Burkitt, D. 1958. A sarcoma involving the jaws in African children. Br J Surg **46**:218–23.
28. Burkitt, D. 1962. A lymphoma syndrome in African children. Ann R Coll Surg Engl **30**:211–9.
29. Burkitt, D. P., and Wright D.H. 1970. Burkitt's lymphoma. Edinburgh: E&S Livingstone.
30. Burrows, S. R., T. B. Sculley, I. S. Misko, C. Schmidt, and D. J. Moss. 1990. An Epstein-Barr virus-specific cytotoxic T cell epitope in EBV nuclear antigen 3 (EBNA 3). J Exp Med **171**:345–9.
31. Caldwell, R. G., J. B. Wilson, S. J. Anderson, and R. Longnecker. 1998. Epstein-Barr virus LMP2A drives B cell development and survival in the absence of normal B cell receptor signals. Immunity **9**:405–11.
32. Callahan, J., S. Pai, M. Cotter, and E. S. Robertson. 1999. Distinct patterns of viral antigen expression in Epstein-Barr virus and Kaposi's sarcoma–associated herpesvirus coinfected body-cavity-based lymphoma cell lines: Potential switches in latent gene expression due to coinfection. Virology **262**:18–30.
33. Carbone, A. 2003. Emerging pathways in the development of AIDS-related lymphomas. Lancet Oncol **4**:22–9.
34. Cesarman, E., Y. Chang, P. S. Moore, J. W. Said, and D. M. Knowles. 1995. Kaposi's sarcoma–associated herpesvirus-like DNA sequences in AIDS-related body-cavity-based lymphomas. N Engl J Med **332**:1186–91.
35. Chadwick, E. G., E. J. Connor, I. C. Hanson, V. V. Joshi, H. bu-Farsakh, R. Yogev, G. McSherry, K. McClain, and S. B. Murphy. 1990. Tumors of smooth-muscle origin in HIV-infected children. JAMA **263**:3182–4.
36. Chaganti, S., A. I. Bell, N. B. Pastor, A. E. Milner, J. Drayson, J. Gordon, and A. B. Rickinson. 2005. Epstein-Barr virus infection in vitro can rescue germinal center B cells with inactivated immunoglobulin genes. Blood **106**:4249–52.
37. Chamberlain, M. C. 1994. Long survival in patients with acquired immune deficiency syndrome-related primary central nervous system lymphoma. Cancer **73**:1728–30.
38. Chen, H., P. Smith, R. F. Ambinder, and S. D. Hayward. 1999. Expression of Epstein-Barr virus BamHI-A rightward transcripts in latently infected B cells from peripheral blood. Blood **93**:3026–32.
39. Chen, H. L., M. M. Lung, J. S. Sham, D. T. Choy, B. E. Griffin, and M. H. Ng. 1992. Transcription of BamHI-A region of the EBV genome in NPC tissues and B cells. Virology **191**:193–201.
40. Chodosh, J., V. P. Holder, Y. J. Gan, A. Belgaumi, J. Sample, and J. W. Sixbey. 1998. Eradication of latent Epstein-Barr virus by hydroxyurea alters the growth-transformed cell phenotype. J Infect Dis **177**:1194–201.
41. Chrysomali, E., J. S. Greenspan, N. Dekker, D. Greenspan, and J. A. Regezi. 1996. Apoptosis-associated proteins in oral hairy leukoplakia. Oral Dis **2**:279–84.
42. Cinque, P., M. Brytting, L. Vago, A. Castagna, C. Parravicini, N. Zanchetta, M. A. d'Arminio, B. Wahren, A. Lazzarin, and A. Linde. 1993. Epstein-Barr virus DNA in cerebrospinal fluid from patients with AIDS-related primary lymphoma of the central nervous system. Lancet **342**:398–401.
43. Cote, T. R., R. J. Biggar, P. S. Rosenberg, S. S. Devesa, C. Percy, F. J. Yellin, G. Lemp, C. Hardy, J. J. Geodert, and W. A. Blattner. 1997. Non-Hodgkin's lymphoma among people with AIDS: Incidence, presentation and public health burden. AIDS/Cancer Study Group. Int J Cancer **73**:645–50.
44. Dal, M. L., D. Serraino, and S. Franceschi. 2001. Epidemiology of AIDS-related tumours in developed and developing countries. Eur J Cancer **37**:1188–201.

45. Dave, S. S., K. Fu, G. W. Wright, L. T. Lam, P. Kluin, E. J. Boerma, T. C. Greiner, D. D. Weisenburger, A. Rosenwald, G. Ott, H. K. Muller-Hermelink, R. D. Gascoyne, J. Delabie, L. M. Rimsza, R. M. Braziel, T. M. Grogan, E. Campo, E. S. Jaffe, B. J. Dave, W. Sanger, M. Bast, J. M. Vose, J. O. Armitage, J. M. Connors, E. B. Smeland, S. Kvaloy, H. Holte, R. I. Fisher, T. P. Miller, E. Montserrat, W. H. Wilson, M. Bahl, H. Zhao, L. Yang, J. Powell, R. Simon, W. C. Chan, and L. M. Staudt. 2006. Molecular diagnosis of Burkitt's lymphoma. N Engl J Med **354**:2431–42.

46. Deyrup, A. T., V. K. Lee, C. E. Hill, W. Cheuk, H. C. Toh, S. Kesavan, E. W. Chan, and S. W. Weiss. 2006. Epstein-Barr virus-associated smooth muscle tumors are distinctive mesenchymal tumors reflecting multiple infection events: A clinicopathologic and molecular analysis of 29 tumors from 19 patients. Am J Surg Pathol **30**:75–82.

47. Dugan, K. A., M. A. Wood, and M. D. Cole. 2002. TIP49, but not TRRAP, modulates c-Myc and E2F1 dependent apoptosis. Oncogene **21**:5835–43.

48. Dutton, A., G. M. Reynolds, C. W. Dawson, L. S. Young, and P. G. Murray. 2005. Constitutive activation of phosphatidyl-inositide 3 kinase contributes to the survival of Hodgkin's lymphoma cells through a mechanism involving Akt kinase and mTOR. J Pathol **205**:498–506.

49. Einfeld, D. A., J. P. Brown, M. A. Valentine, E. A. Clark, and J. A. Ledbetter. 1988. Molecular cloning of the human B cell CD20 receptor predicts a hydrophobic protein with multiple transmembrane domains. EMBO J **7**:711–7.

50. Eischen, C. M., J. D. Weber, M. F. Roussel, C. J. Sherr, and J. L. Cleveland. 1999. Disruption of the ARF-Mdm2-p53 tumor suppressor pathway in Myc-induced lymphomagenesis. Genes Dev **13**:2658–69.

51. Felsher, D. W., and J. M. Bishop. 1999. Reversible tumorigenesis by MYC in hematopoietic lineages. Mol Cell **4**:199–207.

52. Galaktionov, K., X. Chen, and D. Beach. 1996. Cdc25 cell-cycle phosphatase as a target of c-myc. Nature **382**:511–7.

53. Gan, Y. J., B. I. Razzouk, T. Su, and J. W. Sixbey. 2002. A defective, rearranged Epstein-Barr virus genome in EBER-negative and EBER-positive Hodgkin's disease. Am J Pathol **160**:781–6.

54. Gilligan, K., P. Rajadurai, L. Resnick, and N. Raab-Traub. 1990. Epstein-Barr virus small nuclear RNAs are not expressed in permissively infected cells in AIDS-associated leukoplakia. Proc Natl Acad Sci USA **87**:8790–4.

55. Grossman, S. R., E. Johannsen, X. Tong, R. Yalamanchili, and E. Kieff. 1994. The Epstein-Barr virus nuclear antigen 2 transactivator is directed to response elements by the Jk recombination signal binding protein. Proc Natl Acad Sci USA **91**:7568–72.

56. Hanto, D. W., G. Frizzera, K. J. Gajl-Peczalska, K. Sakamoto, D. T. Purtilo, H. H. Balfour, Jr., R. L. Simmons, and J. S. Najarian. 1982. Epstein-Barr virus-induced B-cell lymphoma after renal transplantation: Acyclovir therapy and transition from polyclonal to monoclonal B-cell proliferation. N Engl J Med **306**:913–8.

57. Hanto, D. W., K. Sakamoto, D. T. Purtilo, R. L. Simmons, and J. S. Najarian. 1981. The Epstein-Barr virus in the pathogenesis of posttransplant lymphoproliferative disorders. Clinical, pathologic, and virologic correlation. Surgery **90**:204–13.

58. Harada, S., and E. Kieff. 1997. Epstein-Barr virus nuclear protein LP stimulates EBNA-2 acidic domain-mediated transcriptional activation. J Virol **71**:6611–8.

59. Hecht, J. L., and J. C. Aster. 2000. Molecular biology of Burkitt's lymphoma. J Clin Oncol **18**:3707–21.

60. Henkel, T., P. D. Ling, S. D. Hayward, and M. G. Peterson. 1994. Mediation of Epstein-Barr virus EBNA2 transactivation by recombination signal-binding protein J kappa. Science **265**:92–5.

61. Herbst, J. S., J. Morgan, N. Raab-Traub, and L. Resnick. 1989. Comparison of the efficacy of surgery and acyclovir therapy in oral hairy leukoplakia. J Am Acad Dermatol **21**:753–6.

62. Hermeking, H., C. Rago, M. Schuhmacher, Q. Li, J. F. Barrett, A. J. Obaya, B. C. O'Connell, M. K. Mateyak, W. Tam, F. Kohlhuber, C. V. Dang, J. M. Sedivy, D. Eick, B. Vogelstein, and K. W. Kinzler. 2000. Identification of CDK4 as a target of c-MYC. Proc Natl Acad Sci USA **97**:2229–34.

63. Hertel, C. B., X. G. Zhou, S. J. Hamilton-Dutoit, and S. Junker. 2002. Loss of B cell identity correlates with loss of B cell-specific transcription factors in Hodgkin/Reed-Sternberg cells of classical Hodgkin lymphoma. Oncogene **21**:4908–20.

64. Heslop, H. E., M. Perez, E. Benaim, R. Rochester, M. K. Brenner, and C. M. Rooney. 1999. Transfer of EBV-specific CTL to prevent EBV lymphoma post bone marrow transplant. J Clin Apher **14**:154–6.

65. Howe, J. G., and J. A. Steitz. 1986. Localization of Epstein-Barr virus-encoded small RNAs by in situ hybridization. Proc Natl Acad Sci USA **83**:9006–10.

66. Hsu, B., M. C. Marin, A. K. el-Naggar, L. C. Stephens, S. Brisbay, and T. J. McDonnell. 1995. Evidence that c-myc mediated apoptosis does not require wild-type p53 during lymphomagenesis. Oncogene **11**:175–9.

67. Jacobsen, K. A.,V. S. Prasad, C. L. Sidman, and D. G. Osmond. 1994. Apoptosis and macrophage-mediated deletion of precursor B cells in the bone marrow of E mu-myc transgenic mice. Blood **84**:2784–94.
68. Jain, M., C. Arvanitis, K. Chu, W. Dewey, E. Leonhardt, M. Trinh, C. D. Sundberg, J. M. Bishop, and D. W. Felsher. 2002. Sustained loss of a neoplastic phenotype by brief inactivation of MYC. Science **297**:102–4.
69. Jenson, H. B., C. T. Leach, K. L. McClain,V.V. Joshi, B. H. Pollock, R. T. Parmley, E. G. Chadwick, and S. B. Murphy. 1997. Benign and malignant smooth muscle tumors containing Epstein-Barr virus in children with AIDS. Leuk Lymphoma **27**:303–14.
70. Jenson, H. B., E. A. Montalvo, K. L. McClain,Y. Ench, P. Heard, B. A. Christy, P. J. walt-Hagan, and M. P. Moyer. 1999. Characterization of natural Epstein-Barr virus infection and replication in smooth muscle cells from a leiomyosarcoma. J Med Virol **57**:36–46.
71. Jimenez-Ramirez, C., A. J. Brooks, L. P. Forshell, K. Yakimchuk, B. Zhao, T. Z. Fulgham, and C. E. Sample. 2006. Epstein-Barr virus EBNA-3C is targeted to and regulates expression from the bidirectional LMP-1/2B promoter. J Virol **80**:11200–8.
72. Johannsen, E., E. Koh, G. Mosialos, X. Tong, E. Kieff, and S. R. Grossman. 1995. Epstein-Barr virus nuclear protein 2 transactivation of the latent membrane protein 1 promoter is mediated by J kappa and PU.1. J Virol **69**:253–62.
73. Johannsen, E., C. L. Miller, S. R. Grossman, and E. Kieff. 1996. EBNA-2 and EBNA-3C extensively and mutually exclusively associate with RBPJkappa in Epstein-Barr virus-transformed B lymphocytes. J Virol **70**:4179–83.
74. Kanzler, H., R. Kuppers, M. L. Hansmann, and K. Rajewsky. 1996. Hodgkin and Reed-Sternberg cells in Hodgkin's disease represent the outgrowth of a dominant tumor clone derived from (crippled) germinal center B cells. J Exp Med **184**:1495–505.
75. Karcher, D. S., and S. Alkan. 1997. Human herpesvirus-8-associated body cavity-based lymphoma in human immunodeficiency virus-infected patients: A unique B-cell neoplasm. Hum Pathol **28**:801–8.
76. Khan, G., P. J. Coates, H. O. Kangro, and G. Slavin. 1992. Epstein Barr virus (EBV) encoded small RNAs: Targets for detection by in situ hybridisation with oligonucleotide probes. J Clin Pathol **45**:616–20.
77. Kieff, E. 1996. Epstein-Barr virus and its replication. In: Fields, B. N., D. M. Knipe, P. M. Howley, R. M. Chanock, J. L. Melnick, T. P. Monath, B. Roizman, and S. Straus (eds.),Virology. Philadelphia: Lippincott-Raven Publishers, pp. 2343–96.
78. Kieff, E., and A. B. Rickinson. 2001. Epstein-Barr virus and its replication. In: Knipe, D. M., and P. M. Howley (eds.),Virology. Philadelphia: Lippincott Williams & Williams, pp. 2511–74.
79. Knowles, D. M. 1997. Molecular pathology of acquired immunodeficiency syndrome-related non-Hodgkin's lymphoma. Semin Diagn Pathol **14**:67–82.
80. Knowles, D. M., G. Inghirami, A. Ubriaco, and R. la-Favera. 1989. Molecular genetic analysis of three AIDS-associated neoplasms of uncertain lineage demonstrates their B-cell derivation and the possible pathogenetic role of the Epstein-Barr virus. Blood **73**:792–9.
81. Komano, J., S. Maruo, K. Kurozumi, T. Oda, and K. Takada. 1999. Oncogenic role of Epstein-Barr virus-encoded RNAs in Burkitt's lymphoma cell line Akata. J Virol **73**:9827–31.
82. Kuppers, R., K. Rajewsky, M. Zhao, G. Simons, R. Laumann, R. Fischer, and M. L. Hansmann. 1994. Hodgkin disease: Hodgkin and Reed-Sternberg cells picked from histological sections show clonal immunoglobulin gene rearrangements and appear to be derived from B cells at various stages of development. Proc Natl Acad Sci USA **91**:10962–6.
83. Kusano, S., and N. Raab-Traub. 2001. An Epstein-Barr virus protein interacts with notch. J Virol **75**:384–95.
84. Lacoste,V., J. G. Judde, G. Bestett, J. Cadranel, M. Antoine, F. Valensi, E. Delabesse, E. Macintyre, and A. Gessain. 2000.Virological and molecular characterisation of a new B lymphoid cell line, established from an AIDS patient with primary effusion lymphoma, harbouring both KSHV/HHV8 and EBV viruses. Leuk Lymphoma **38**:401–9.
85. Laux, G., B. Adam, L. J. Strobl, and F. Moreau-Gachelin. 1994. The Spi-1/PU.1 and Spi-B ets family transcription factors and the recombination signal binding protein RBP-J kappa interact with an Epstein Barr virus nuclear antigen 2 responsive cis-element. EMBO J **13**:5624–32.
86. Lazzi, S., F. Ferrari, A. Nyongo, N. Palummo, M. A. de, M. Zazzi, L. Leoncini, P. Luzi, and P. Tosi. 1998. HIV-associated malignant lymphomas in Kenya (Equatorial Africa). Hum Pathol **29**:1285–9.
87. Lee, E. S., J. Locker, M. Nalesnik, J. Reyes, R. Jaffe, M. Alashari, B. Nour, A. Tzakis, and P. S. Dickman. 1995. The association of Epstein-Barr virus with smooth-muscle tumors occurring after organ transplantation. N Engl J Med **332**:19–25.
88. Lee, S. P., J. M. Brooks, H. Al-Jarrah, W. A. Thomas, T. A. Haigh, G. S. Taylor, S. Humme, A. Schepers, W. Hammerschmidt, J. L. Yates, A. B. Rickinson, and N. W. Blake. 2004. CD8 T cell recognition of endogenously expressed Epstein-Barr virus nuclear antigen 1. J Exp Med **199**:1409–20.

89. Lerner, M. R., N. C. Andrews, G. Miller, and J. A. Steitz. 1981. Two small RNAs encoded by Epstein-Barr virus and complexed with protein are precipitated by antibodies from patients with systemic lupus erythematosus. Proc Natl Acad Sci USA **78**:805–9.

90. Levine, A. M. 2006. AIDS-related lymphoma. Semin Oncol Nurs **22**:80–9.

91. Levine, P. H., L. S. Kamaraju, R. R. Connelly, C. W. Berard, R. F. Dorfman, I. Magrath, and J. M. Easton. 1982. The American Burkitt's Lymphoma Registry: Eight years' experience. Cancer **49**:1016–22.

92. Levitskaya, J., M. Coram, V. Levitsky, S. Imreh, P. M. Steigerwald-Mullen, G. Klein, M. G. Kurilla, and M. G. Masucci. 1995. Inhibition of antigen processing by the internal repeat region of the Epstein-Barr virus nuclear antigen-1. Nature **375**:685–8.

93. Ling, P. D., D. R. Rawlins, and S. D. Hayward. 1993. The Epstein-Barr virus immortalizing protein EBNA-2 is targeted to DNA by a cellular enhancer-binding protein. Proc Natl Acad Sci USA **90**:9237–41.

94. Longnecker, R., C. L. Miller, X. Q. Miao, A. Marchini, and E. Kieff. 1992. The only domain which distinguishes Epstein-Barr virus latent membrane protein 2A (LMP2A) from LMP2B is dispensable for lymphocyte infection and growth transformation in vitro; LMP2A is therefore nonessential. J Virol **66**:6461–9.

95. Longnecker, R., C. L. Miller, X. Q. Miao, B. Tomkinson, and E. Kieff. 1993. The last seven transmembrane and carboxy-terminal cytoplasmic domains of Epstein-Barr virus latent membrane protein 2 (LMP2) are dispensable for lymphocyte infection and growth transformation in vitro. J Virol **67**:2006–13.

96. MacMahon, E. M., J. D. Glass, S. D. Hayward, R. B. Mann, P. S. Becker, P. Charache, J. C. McArthur, and R. F. Ambinder. 1991. Epstein-Barr virus in AIDS-related primary central nervous system lymphoma. Lancet **338**:969–73.

97. Magrath, I. 1990. The pathogenesis of Burkitt's lymphoma. Adv Cancer Res **55**:133–270.

98. Magrath, I., J. Erikson, J. Whang-Peng, H. Sieverts, G. Armstrong, D. Benjamin, T. Triche, O. Alabaster, and C. M. Croce. 1983. Synthesis of kappa light chains by cell lines containing an 8;22 chromosomal translocation derived from a male homosexual with Burkitt's lymphoma. Science **222**:1094–8.

99. Mancao, C., M. Altmann, B. Jungnickel, and W. Hammerschmidt. 2005. Rescue of "crippled" germinal center B cells from apoptosis by Epstein-Barr virus. Blood **106**:4339–44.

100. Mateyak, M. K., A. J. Obaya, and J. M. Sedivy. 1999. c-Myc regulates cyclin D-Cdk4 and -Cdk6 activity but affects cell cycle progression at multiple independent points. Mol Cell Biol **19**:4672–83.

101. Mathas, S., M. Janz, F. Hummel, M. Hummel, B. Wollert-Wulf, S. Lusatis, I. Anagnostopoulos, A. Lietz, M. Sigvardsson, F. Jundt, K. Johrens, K. Bommert, H. Stein, and B. Dorken. 2006. Intrinsic inhibition of transcription factor E2A by HLH proteins ABF-1 and Id2 mediates reprogramming of neoplastic B cells in Hodgkin lymphoma. Nat Immunol **7**:207–15.

102. Matthews, G. V., M. Bower, S. Mandalia, T. Powles, M. R. Nelson, and B. G. Gazzard. 2000. Changes in acquired immunodeficiency syndrome-related lymphoma since the introduction of highly active anti-retroviral therapy. Blood **96**:2730–4.

103. Mbulaiteye, S. M., D. M. Parkin, and C. S. Rabkin. 2003. Epidemiology of AIDS-related malignancies an international perspective. Hematol Oncol Clin North Am **17**:673–96, v.

104. McClain, K. L., C. T. Leach, H. B. Jenson, V. V. Joshi, B. H. Pollock, R. T. Parmley, F. J. DiCarlo, E. G. Chadwick, and S. B. Murphy. 1995. Association of Epstein-Barr virus with leiomyosarcomas in children with AIDS. N Engl J Med **332**:12–8.

105. McMahon, S. B., H. A. Van Buskirk, K. A. Dugan, T. D. Copeland, and M. D. Cole. 1998. The novel ATM-related protein TRRAP is an essential cofactor for the c-Myc and E2F oncoproteins. Cell **94**:363–74.

106. McMahon, S. B., M. A. Wood, and M. D. Cole. 2000. The essential cofactor TRRAP recruits the histone acetyltransferase hGCN5 to c-Myc. Mol Cell Biol **20**:556–62.

107. Miller, C. L., R. Longnecker, and E. Kieff. 1993. Epstein-Barr virus latent membrane protein 2A blocks calcium mobilization in B lymphocytes. J Virol **67**:3087–94.

108. Moody, C. A., R. S. Scott, T. Su, and J. W. Sixbey. 2003. Length of Epstein-Barr virus termini as a determinant of epithelial cell clonal emergence. J Virol **77**:8555–61.

109. Morrison, J. A., and N. Raab-Traub. 2005. Roles of the ITAM and PY motifs of Epstein-Barr virus latent membrane protein 2A in the inhibition of epithelial cell differentiation and activation of {beta}-catenin signaling. J Virol **79**:2375–82.

110. Mosialos, G., M. Birkenbach, R. Yalamanchili, R. VanArsdale, C. Ware, and E. Kieff. 1995. The Epstein-Barr virus transforming protein LMP1 engages signaling proteins for the tumor necrosis factor receptor family. Cell **80**:389–99.

111. Muller, D., C. Bouchard, B. Rudolph, P. Steiner, I. Stuckmann, R. Saffrich, W. Ansorge, W. Huttner, and M. Eilers. 1997. Cdk2-dependent phosphorylation of p27 facilitates its Myc-induced release from cyclin E/cdk2 complexes. Oncogene **15**:2561–76.

112. Murray, R. J., M. G. Kurilla, J. M. Brooks, W. A. Thomas, M. Rowe, E. Kieff, and A. B. Rickinson. 1992. Identification of target antigens for the human cytotoxic T cell response to Epstein-Barr virus (EBV): Implications for the immune control of EBV-positive malignancies. J Exp Med **176**:157–68.

113. Nador, R. G., E. Cesarman, A. Chadburn, D. B. Dawson, M. Q. Ansari, J. Sald, and D. M. Knowles. 1996. Primary effusion lymphoma: A distinct clinicopathologic entity associated with the Kaposi's sarcoma-associated herpes virus. Blood **88**:645–56.

114. Nilsson, J. A., and J. L. Cleveland. 2003. Myc pathways provoking cell suicide and cancer. Oncogene **22**:9007–21.

115. Nitsche, F., A. Bell, and A. Rickinson. 1997. Epstein-Barr virus leader protein enhances EBNA-2-mediated transactivation of latent membrane protein 1 expression: A role for the W1W2 repeat domain. J Virol **71**:6619–28.

116. O'Conor, G. T. 1961. Malignant lymphoma in African children. II. A pathological entity. Cancer **14**:270–83.

117 O'Conor, G. T., and J. N. Davies. 1960. Malignant tumors in African children. With special reference to malignant lymphoma. J Pediatr **56**:526–35.

118. Ohno, S., J. Luka, T. Lindahl, and G. Klein. 1977. Identification of a purified complement-fixing antigen as the Epstein-Barr-virus determined nuclear antigen (EBNA) by its binding to metaphase chromosomes. Proc Natl Acad Sci USA **74**:1605–09.

119. Park, J., S. Kunjibettu, S. B. McMahon, and M. D. Cole. 2001. The ATM-related domain of TRRAP is required for histone acetyltransferase recruitment and Myc-dependent oncogenesis. Genes Dev **15**:1619–24.

120. Park, J., M. A. Wood, and M. D. Cole. 2002. BAF53 forms distinct nuclear complexes and functions as a critical c-Myc-interacting nuclear cofactor for oncogenic transformation. Mol Cell Biol **22**:1307–16.

121. Parkin, D. M. 2006. The global health burden of infection-associated cancers in the year 2002. Int J Cancer **118**:3030–44.

122. Parkins, D. M. 1998. International Incidence of Childhood Cancer, Volume II (IARC Scientific Publications No. 144). Lyon: International Agency for Research on Cancer.

123. Pegtel, D. M., A. Subramanian, T. S. Sheen, C. H. Tsai, T. R. Golub, and D. A. Thorley-Lawson. 2005. Epstein-Barr-virus-encoded LMP2A induces primary epithelial cell migration and invasion: Possible role in nasopharyngeal carcinoma metastasis. J Virol **79**:15430–42.

124. Pollock, B. H., H. B. Jenson, C. T. Leach, K. L. McClain, R. E. Hutchison, L. Garzarella, V. V. Joshi, R. T. Parmley, and S. B. Murphy. 2003. Risk factors for pediatric human immunodeficiency virus-related malignancy. JAMA **289**:2393–9.

125. Portis, T., P. Dyck, and R. Longnecker. 2003. Epstein-Barr Virus (EBV) LMP2A induces alterations in gene transcription similar to those observed in Reed–Sternberg cells of Hodgkin lymphoma. Blood **102**:4166–78.

126. Prendergast, G. C., and E. B. Ziff. 1991. Methylation-sensitive sequence-specific DNA binding by the c-Myc basic region. Science **251**:186–9.

127. Preudhomme, C., I. Dervite, E. Wattel, M. Vanrumbeke, M. Flactif, J. L. Lai, B. Hecquet, M. C. Coppin, B. Nelken, B. Gosselin, et al. 1995. Clinical significance of p53 mutations in newly diagnosed Burkitt's lymphoma and acute lymphoblastic leukemia: A report of 48 cases. J Clin Oncol **13**:812–20.

128. Raab-Traub, N., and K. Flynn. 1986. The structure of the termini of the Epstein-Barr virus as a marker of clonal cellular proliferation. Cell **47**:883–9.

129. Rabkin, C. S. 2001. AIDS and cancer in the era of highly active antiretroviral therapy (HAART). Eur J Cancer **37**:1316–9.

130. Reed, K. D., C. B. Fowler, and R. B. Brannon. 1988. Ultrastructural detection of herpes-type virions by negative staining in oral hairy leukoplakia. Am J Clin Pathol **90**:305–8.

131. Reff, M. E., K. Carner, K. S. Chambers, P. C. Chinn, J. E. Leonard, R. Raab, R. A. Newman, N. Hanna, and D. R. Anderson. 1994. Depletion of B cells in vivo by a chimeric mouse human monoclonal antibody to CD20. Blood **83**:435–45.

132. Reisman, D., J. Yates, and B. Sugden. 1985. A putative origin of replication of plasmids derived from Epstein-Barr virus is composed of two cis-acting components. Mol Cell Biol **5**:1822–32.

133. Renne, C., J. I. Martin-Subero, M. Eickernjager, M. L. Hansmann, R. Kuppers, R. Siebert, and A. Brauninger. 2006. Aberrant expression of ID2, a suppressor of B-cell-specific gene expression, in Hodgkin's lymphoma. Am J Pathol **169**:655–64.

134. Robertson, E. S., S. Grossman, E. Johannsen, C. Miller, J. Lin, B. Tomkinson, and E. Kieff. 1995. Epstein-Barr virus nuclear protein 3C modulates transcription through interaction with the sequence-specific DNA-binding protein J kappa. J Virol **69**:3108–16.

135. Rogatsch, H., H. Bonatti, A. Menet, C. Larcher, H. Feichtinger, and S. Dirnhofer. 2000. Epstein-Barr virus-associated multicentric leiomyosarcoma in an adult patient after heart transplantation: Case report and review of the literature. Am J Surg Pathol **24**:614–21.

136. Rowe, M., H. S. Evans, L. S. Young, K. Hennessy, E. Kieff, and A. B. Rickinson. 1987. Monoclonal antibodies to the latent membrane protein of Epstein-Barr virus reveal heterogeneity of the protein and inducible expression in virus-transformed cells. J Gen Virol **68**:1575–86.

137. Ruf, I. K., K. A. Lackey, S. Warudkar, and J. T. Sample. 2005. Protection from interferon-induced apoptosis by Epstein-Barr virus small RNAs is not mediated by inhibition of PKR. J Virol **79**:14562–9.

138. Ruf, I. K., P. W. Rhyne, C. Yang, J. L. Cleveland, and J. T. Sample. 2000. Epstein-Barr virus small RNAs potentiate tumorigenicity of Burkitt lymphoma cells independently of an effect on apoptosis. J Virol **74**:10223–8.

139. Sabatino, D., S. Martinez, R. Young, H. Balbi, P. Ciminera, and M. Frieri. 1991. Simultaneous pulmonary leiomyosarcoma and leiomyoma in pediatric HIV infection. Pediatr Hematol Oncol **8**:355–9.

140. Sample, J. T., and I. K. Ruff. 2006. Burkitt lymphoma. In: Tselis, A., and H. B. Jenson (eds.), Epstein-Barr Virus. New York: Taylor and Francis, pp. 187–222.

141. Sandlund, J. T., T. Fonseca, T. Leimig, L. Verissimo, R. Ribeiro, V. Lira, C. W. Berard, J. Sixbey, W. M. Crist, L. Mao, G. Chen, C. H. Pui, M. Heim, and F. Pedrosa. 1997. Predominance and characteristics of Burkitt lymphoma among children with non-Hodgkin lymphoma in northeastern Brazil. Leukemia **11**:743–6.

142. Savoldo, B., H. E. Heslop, and C. M. Rooney. 2000. The use of cytotoxic t cells for the prevention and treatment of Epstein-Barr virus induced lymphoma in transplant recipients. Leuk Lymphoma **39**:455–64.

143. Scala, G., I. Quinto, M. R. Ruocco, A. Arcucci, M. Mallardo, P. Caretto, G. Forni, and S. Venuta. 1990. Expression of an exogenous interleukin 6 gene in human Epstein Barr virus B cells confers growth advantage and in vivo tumorigenicity. J Exp Med **172**:61–8.

144. Scala, G., M. R. Ruocco, C. Ambrosino, M. Mallardo, V. Giordano, F. Baldassarre, E. Dragonetti, I. Quinto, and S. Venuta. 1994. The expression of the interleukin 6 gene is induced by the human immunodeficiency virus 1 TAT protein. J Exp Med **179**:961–71.

145. Scholle, F., K. M. Bendt, and N. Raab-Traub. 2000. Epstein-Barr virus LMP2A transforms epithelial cells, inhibits cell differentiation, and activates Akt. J Virol **74**:10681–9.

146. Scholzen, T., and J. Gerdes. 2000. The Ki-67 protein: From the known and the unknown. J Cell Physiol **182**:311–22.

147. Schwering, I., A. Brauninger, U. Klein, B. Jungnickel, M. Tinguely, V. Diehl, M. L. Hansmann, R. la-Favera, K. Rajewsky, and R. Kuppers. 2003. Loss of the B-lineage-specific gene expression program in Hodgkin and Reed-Sternberg cells of Hodgkin lymphoma. Blood **101**:1505–12.

148. Sheiness, D., and J. M. Bishop. 1979. DNA and RNA from uninfected vertebrate cells contain nucleotide sequences related to the putative transforming gene of avian myelocytomatosis virus. J Virol **31**:514–21.

149. Slobod, K. S., G. H. Taylor, J. T. Sandlund, P. Furth, K. J. Helton, and J. W. Sixbey. 2000. Epstein-Barr virus-targeted therapy for AIDS-related primary lymphoma of the central nervous system. Lancet **356**:1493–94.

150. Sparano, J. A. 2001. Clinical aspects and management of AIDS-related lymphoma. Eur J Cancer **37**:1296–1305.

151. Strasser, A., A. W. Harris, M. L. Bath, and S. Cory. 1990. Novel primitive lymphoid tumours induced in transgenic mice by cooperation between myc and bcl-2. Nature **348**:331–3.

152. Swaminathan, S., B. Tomkinson, and E. Kieff. 1991. Recombinant Epstein-Barr virus with small RNA (EBER) genes deleted transforms lymphocytes and replicates in vitro. Proc Natl Acad Sci USA **88**:1546–50.

153. Swart, R., I. K. Ruf, J. Sample, and R. Longnecker. 2000. Latent membrane protein 2A-mediated effects on the phosphatidylinositol 3-Kinase/Akt pathway. J Virol **74**:10838–45.

154. Szekely, L., F. Chen, N. Teramoto, B. Ehlin-Henriksson, K. Pokrovskaja, A. Szeles, A. Manneborg-Sandlund, M. Lowbeer, E. T. Lennette, and G. Klein. 1998. Restricted expression of Epstein-Barr virus (EBV)-encoded, growth transformation-associated antigens in an EBV- and human herpesvirus type 8-carrying body cavity lymphoma line. J Gen Virol **79**(Pt 6):1445–52.

155. Takada, K. 2001. Role of Epstein-Barr virus in Burkitt's lymphoma. Curr Top Microbiol Immunol **258**:141–51.

156. Tedder, T. F., A. W. Boyd, A. S. Freedman, L. M. Nadler, and S. F. Schlossman. 1985. The B cell surface molecule B1 is functionally linked with B cell activation and differentiation. J Immunol **135**:973–9.

157. Tellam, J., G. Connolly, K. J. Green, J. J. Miles, D. J. Moss, S. R. Burrows, and R. Khanna. 2004. Endogenous presentation of CD8+ T cell epitopes from Epstein-Barr virus-encoded nuclear antigen 1. J Exp Med **199**:1421–31.

158. Thorley-Lawson, D. A., and A. Gross. 2004. Persistence of the Epstein-Barr virus and the origins of associated lymphomas. N Engl J Med **350**:1328–37.

159. Tirelli, U., M. Spina, G. Gaidano, E. Vaccher, S. Franceschi, and A. Carbone. 2000. Epidemiological, biological and clinical features of HIV-related lymphomas in the era of highly active antiretroviral therapy. AIDS **14**:1675–88.

160. Trivedi, P., K. Takazawa, C. Zompetta, L. Cuomo, E. Anastasiadou, A. Carbone, S. Uccini, F. Belardelli, K. Takada, L. Frati, and A. Faggioni. 2004. Infection of HHV-8+ primary effusion lymphoma cells with a recombinant Epstein-Barr virus leads to restricted EBV latency, altered phenotype, and increased tumorigenicity without affecting TCL1 expression. Blood **103**:313–6.

161. Uchida, J., T. Yasui, Y. Takaoka-Shichijo, M. Muraoka, W. Kulwichit, N. Raab-Traub, and H. Kikutani. 1999. Mimicry of CD40 signals by Epstein-Barr virus LMP1 in B lymphocyte responses. Science **286**:300–3.

162. Ushmorov, A., F. Leithauser, O. Sakk, A. Weinhausel, S. W. Popov, P. Moller, and T. Wirth. 2006. Epigenetic processes play a major role in B-cell-specific gene silencing in classical Hodgkin lymphoma. Blood **107**:2493–500.

163. Valentine, M. A., K. E. Meier, S. Rossie, and E. A. Clark. 1989. Phosphorylation of the CD20 phosphoprotein in resting B lymphocytes. Regulation by protein kinase C. J Biol Chem **264**:11282–7.

164. Walling, D. M., S. N. Edmiston, J. W. Sixbey, M. bdel-Hamid, L. Resnick, and N. Raab-Traub. 1992. Coinfection with multiple strains of the Epstein-Barr virus in human immunodeficiency virus-associated hairy leukoplakia. Proc Natl Acad Sci USA **89**:6560–4.

165. Walling, D. M., A. G. Perkins, J. Webster-Cyriaque, L. Resnick, and N. Raab-Traub. 1994. The Epstein-Barr virus EBNA-2 gene in oral hairy leukoplakia: Strain variation, genetic recombination, and transcriptional expression. J Virol **68**:7918–26.

166. Wang, D., D. Liebowitz, and E. Kieff. 1985. An EBV membrane protein expressed in immortalized lymphocytes transforms established rodent cells. Cell **43**:831–40.

167. Webster-Cyriaque, J., J. Middeldorp, and N. Raab-Traub. 2000. Hairy leukoplakia: An unusual combination of transforming and permissive Epstein-Barr virus infections. J Virol **74**:7610–8.

168. Webster-Cyriaque, J., and N. Raab-Traub. 1998. Transcription of Epstein-Barr virus latent cycle genes in oral hairy leukoplakia. Virology **248**:53–65.

169. Whang-Peng, J., E. C. Lee, H. Sieverts, and I. T. Magrath. 1984. Burkitt's lymphoma in AIDS: Cytogenetic study. Blood **63**:818–22.

170. Young, L., C. Alfieri, K. Hennessy, H. Evans, C. O'Hara, K. C. Anderson, J. Ritz, R. S. Shapiro, A. Rickinson, E. Kieff, et al. 1989. Expression of Epstein-Barr virus transformation-associated genes in tissues of patients with EBV lymphoproliferative disease. N Engl J Med **321**:1080–5.

171. Zhang, J., H. Chen, G. Weinmaster, and S. D. Hayward. 2001. Epstein-Barr virus BamHi-a rightward transcript-encoded RPMS protein interacts with the CBF1-associated corepressor CIR to negatively regulate the activity of EBNA2 and NotchIC. J Virol **75**:2946–56.

172. Zhao, B., D. M. Marshall, and C. E. Sample. 1996. A conserved domain of the Epstein-Barr virus nuclear antigens 3A and 3C binds to a discrete domain of Jk. J Virol **70**:4228–36.

173. Zhao, B., and C. E. Sample. 2000. Epstein-Barr virus nuclear antigen 3C activates the latent membrane protein 1 promoter in the presence of Epstein-Barr virus nuclear antigen 2 through sequences encompassing an spi-1/Spi-B binding site. J Virol **74**:5151–60.

174. Zimber-Strobl, U., L. J. Strobl, C. Meitinger, R. Hinrichs, T. Sakai, T. Furukawa, T. Honjo, and G. W. Bornkamm. 1994. Epstein-Barr virus nuclear antigen 2 exerts its transactivating function through interaction with recombination signal binding protein RBP-J kappa, the homologue of Drosophila Suppressor of Hairless. EMBO J **13**:4973–82.

175. zur, Hausen, H. 1999. Viruses in human cancers. Eur J Cancer **35**:1878–85.

7. HUMAN PAPILLOMAVIRUS INFECTION AND DISEASE IN THE HIV+ INDIVIDUAL

JENNIFER E. CAMERON[*], MICHAEL E. HAGENSEE[†]

[*]Tulane Health Sciences Center, Tulane Medical School, New Orleans, LA
[†]Department of Medicine, Louisiana State University Health Sciences Center, New Orleans, LA

BASIC BIOLOGY AND EPIDEMIOLOGY OF HUMAN PAPILLOMAVIRUS AND CERVICAL CANCER

Infection with human papillomavirus (HPV) has been implicated and now is widely accepted as a prerequisite for the development of the majority of anogenital malignancies, accounting specifically for more than 90% of cervical carcinomas.[119] Although the incidence of cervical cancer has decreased in the United States over the past 40 years, it is the second most common malignancy of women worldwide with about 500,000 new cases reported each year.[84] The lower incidence of cervical cancer in the United States is presumably due chiefly to the increased screening of women via Papanicolaou (Pap) smears. Despite the drop in cervical cancer rates over the past 30–40 years, in recent years a disturbing rise in cancer rates has occurred from a low of 7.8 per 100,000 in 1994 to a level of 8.8 per 100,000 in 2003 (Centers for Disease Control and Prevention Web site, www.cdc.gov). The reasons for this rise are unclear but may be related to an increase in HPV-related disease in HIV-infected (HIV+) women. As described in detail below, men and women infected with HIV have higher rates of HPV infection, longer persistence of HPV, increased incidence of genital warts, cervical dysplasia, anal dysplasia, cervical cancer, and anal cancer as well as more recurrences after definitive therapy.[91,99] Recent data from Africa have demonstrated that cervical cancer is the most common AIDS-defining malignancy in women.[21] The impact of highly active antiretroviral therapy (HAART) has led to some improved resolution of abnormal Pap smears but still leaves the majority of HIV+ women at risk for cervical cancer[55] and has had little to no impact on HIV+ men with HPV-related disease.

HPV is a small double-stranded DNA virus, 8 kb in length, which is difficult to propagate in culture (Fig. 1). The genes E6 and E7 are responsible for the oncogenic potential of the high-risk HPV types with E6 protein binding and degrading the p53 tumor suppressor gene and the E7 protein inactivating the retinoblastoma (Rb) tumor suppressor gene.[62] These viral proteins interact and inhibit their cellular targets thus allowing for immortalization of the cervical epithelial cell. During the oncogenic process, the viral genome integrates into the host chromosome leading to unchecked expression of the E6 and E7 oncoproteins.[62] There are over 100 types of HPV, with approximately 40 of them infecting the anogenital tract. Types of HPV are defined by a greater than 10% difference in DNA sequence of L1, E6, and E7.[124] The genital types of HPV are subdivided according to their oncogenic potential. HPV types 16, 18, 31, and 45 are the most common cancer-causing types worldwide, accounting for more than 75% of the HPV types found in tumors (Fig. 2).[62] The low oncogenic risk viruses HPV 6 and 11 are found in approximately 80% of genital warts.

HPV is also the most common viral sexually transmitted infection (STI), with prevalence ranging from 10–50% in sexually active men and women. The vast majority

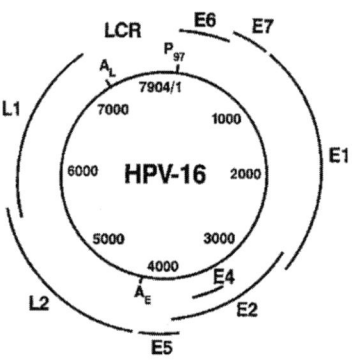

Figure 1. HPV genome.

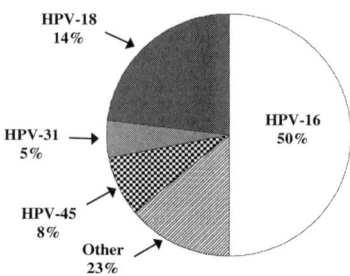

Figure 2. HPV types in cervical cancer.

of these infections clear without sequellae with only 7% of all women having an abnormal Pap smear each year.[62] The development of cervical abnormalities occurs gradually with precursor lesions, squamous intraepithelial lesions (SIL) on Pap smear or cervical intraepithelial neoplasms (CIN) on biopsy being detectable clinically and in the research laboratory. Detection of these precursor lesions is the goal of routine cervical cancer prevention efforts comprising repeat speculum examination, Pap smear analysis, and colposcopy with directed cervical biopsy.[12] The majority (75%) of low-grade SIL (LSIL) Pap smears have CIN I on biopsy, and the majority of these cases regress (60%). This is felt by many to be the natural history of cervical HPV infection (Fig. 3).[100] In contrast, 70% of high-grade SIL (HSIL) have CIN II or III on biopsy and the majority of these lesions progress to cancer (Fig. 4). Thus,

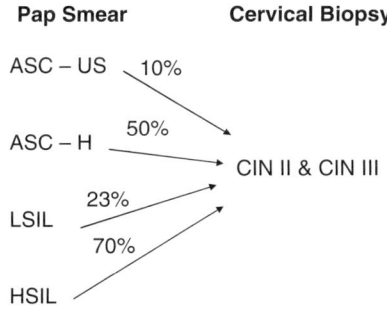

Figure 3. Pap smear vs biopsy.
ASC-US-atypical squamous cells of unclear significance
ASC-H-atypical squamous cells favoring high grade
LSIL-Low grade squamous intraepithelial lesion
HSIL-High grade squamous intraepithelial lesion
CIN-cervical intraepithelial neoplasm

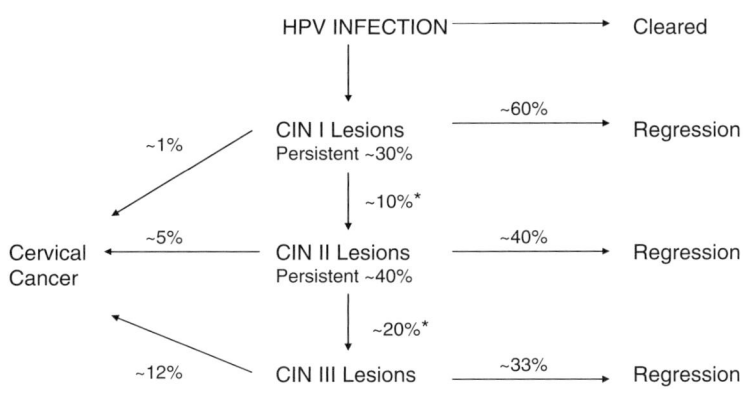

Figure 4. Regression of CIN.

therapeutic intervention is recommended for CIN II and III.[123] Atypical cells of unclear significance (ASCUS) refer to cervical cells that appear somewhat abnormal but not to the degree that would classify them as LSIL or HSIL. Studies have found that roughly 33% of ASCUS cases are due to HPV infection and these women are at higher risk for the development of cervical abnormalities.[107] ASC-H is atypical squamous cells favoring high-grade dysplasia. This is an uncommon Pap smear diagnosis (<5%) but is associated with significant pathology (CIN II and greater) on biopsy in 50% of cases. Men who have sex with men (MSM) have higher rates of HPV-related anal disease[29] and anal dysplasia is felt to progress in stages in a similar manner as HPV-related cervical disease.[88]

HIV+ WOMEN

Increased Detection of HPV Infection in HIV+ Women

There have been numerous studies that have shown an increase in HPV DNA detection from genital samples in the HIV+ woman. The prevalence rates of HPV in these women depend on the population studied, the method to collect the sample, and the assay used to detect HPV. The initial studies utilized relatively insensitive tests (hybrid capture 1, Viratype, and Southern blot) and demonstrated HPV infection rates in HIV+ women to be 37–52% as compared to 8–24% in HIV-negative women.[65,66,97,115] These initial studies examined relatively few women (HIV+, $n = 47$–124, HIV-negative, $n = 48$–126) and were cross sectional in design. These studies did show statistically significant increases in HPV detection in HIV+ women with odds ratios (OR) between 1.7 and 7.5. Subsequent studies that utilized more sensitive assays [polymerase chain reaction, (PCR) and hybrid capture II] to detect HPV showed higher rates of infection in both HIV+ (60–69%) and HIV-negative (27–49%) women.[51,109,113,122] Indeed, most studies from this time period (mid to late 1990s) showed that about two thirds of HIV+ women had detectable HPV from cervical samples.[15,51,109,122]

Additional small studies from diverse areas have also demonstrated an increased HPV infection rate in HIV+ women but with rates that differ from the 60 to 69% mentioned above. Levi et al. showed extremely high rate of HPV infection (98%), utilizing the SPF10 primer system and PCR in a cohort of women from Brazil.[69] These high rates either reflect an extremely at risk population, extreme sensitivity of this primer set or concern for contamination of clinical samples. Taiwanese HIV+ women were HPV infected at a rate of 48% as compared to 20% of HIV-negative women.[17] Thai HIV+ women were HPV+ 35% of the time as compared to 23% for HIV-negative women.[7]

The demonstration of increased HPV detection in HIV+ women as well as other important questions led to the establishment of large multicenter cohorts. The Women's Interagency Health Study, WIHS, is a prospective multicenter study from six cities in the United States (San Francisco, New York City – Bronx/Manhattan and Brooklyn, Chicago, Los Angeles, Washington, DC). The goal of the WIHS is to characterize the natural history of HIV infection in women. This has included extensive studies on HPV infections and HPV-related disease. From the initial report of this study, HPV infection ranged in HIV+ women from 53% (Los Angeles) to 73% (New York City) and HIV-negative women from 12% (Chicago) to 36% (Washington, DC).[91] Similarly the HIV Epidemiological Research Study (HERS) is a longitudinal study of HIV+ women from four US cities (Providence Rhode Island, Baltimore, New York City – The Bronx, Detroit) and has shown similar rates of HPV infection in HIV+ women (64% cumulative) as compared to HIV-negative women (28%).[26] In a more recent update of the HERS, similar rates of HPV initial infection were seen in HIV+ (64%) versus HIV-negative (27%) women.[58]

Risk Factors for HPV Infection

The risk factors for HPV infection in HIV+ women have been examined by the initial smaller studies and more extensively studied in the subsequent larger cohorts. For the most part, these studies have shown an increase in HPV DNA detection in women with lower CD4 T cell counts and higher peripheral HIV viral loads. Sun et al.[109] studied 344 HIV+ women and in multivariate analysis found that HIV+ status, low CD4 (<200), younger age (<30 years old) and being single were risk factors for cervical HPV infection. Hankins et al.,[51] studying the Canadian Women's HIV study group, showed that lower CD4 cell count (<200), non-white race, inconsistent condom use in past 6 months and lower age (<30 years old) were all independent predictors of HPV cervical status. Studies from the WIHS and HERS confirm these findings with ORs as high as 10.1 for CD4 cell counts less than 200[58,91] and show other risk factors being non-white race, younger age, and smoking. The more recent studies have shown similar associations for increased HIV viral load and increased HPV cervical prevalence.[7,91] In a nice study by Minkoff et al.,[82] examining a large number of the women in the WIHS, smoking was found to be positively associated with HPV infection rates with a specific increased risk for HPV-18 cervical infection. The most consistent risk factors for prevalent HPV cervical infection in HIV+ women are lower CD4 cell counts, higher peripheral HIV viral loads, non-white race, younger age, and smoking.

HPV Persistence

Longitudinal studies have been performed that have also examined the influence of HIV status on the persistence of HPV cervical infections.[8,27,58,108,115] The initial study by Vernon et al.[115] studied 124 HIV+ women and followed them at monthly intervals for 8 months. They showed an increased prevalence rate of HPV in HIV+ women (43%) versus HIV-negative women (13%) but no significant association with lower CD4 cell counts. Persistent HPV infection (defined as being HPV at >1 visit) was seen more often in HIV+ women, especially those with lower CD4 cell counts. Branca et al.[8] followed 89 HIV+ and 48 HIV-negative women over a 14-month period. HIV+ women had a higher incidence of HPV infection (27% vs. 3%), and less clearance of prevalent HPV infections (23% vs. 69%). The study by Strickler et al.[108] followed 1848 HIV+ and 514 HIV-negative women enrolled in the WIHS. They showed increased HPV prevalence, persistence, and decreased resolution of infection in HIV+ women as compared to HIV-negative women. They showed that the CD4 cell counts and HIV peripheral viral load are highly interactive. This study followed sufficient numbers of women so that both of these factors could be shown to contribute to the persistence and/or resolution of HPV infection. The greatest risk for HPV cervical infection was in those women with the lowest CD4 cell count and the highest systemic HIV viral load. However, even women with undetectable HIV peripheral viral load and the highest CD4 cell counts had higher rates of HPV infection than HIV-negative women. Greater increases in incident infection and larger decreases in resolution of HPV infection were seen in the women with the lowest CD4 cell counts and the highest HIV viral loads. The study also strongly suggested that a significant proportion of the incident HPV infections were in women with little to no sexual activity. This implies either nonsexual routes of transmission or reactivation of latent virus.

HPV Types in HIV+ Women

Increases in both high oncogenic risk HPV types and nononcogenic risk HPV types have been seen in HIV-infected women. However, the majority of studies either utilized less-specific HPV tests such as hybrid capture which detects a group of high oncogenic risk viruses or the specific type(s) of HPV infection were not analyzed separately. The study by Palefsky et al.[91] did report the individual prevalence of 29 HPV genotypes in HIV+ and HIV-negative women. In general, no specific type or types of HPV were found exclusively in HIV+ women. Instead, increases of most, if not all, HPV types were seen in the HIV-infected women. Similar findings have been found in the HERS[58]; e.g., HPV types 6 and 11 were both found to be more prevalent in HIV+ women than HIV-negative women.[104] A few reports have shown unique HPV types to be found only in HIV+ women.[49] On a genotype basis, most HPV infections were seen in increased numbers in women with lower CD4 cell counts.[91] In addition, many studies have shown an increase in multiple HPV infections (defined as simultaneous infection with more than one HPV genotype) in HIV+ women (23–45%) as compared to HIV-negative women (11–26%).[16,58,91,109] These multiple HPV infections appear to be acquired independently of each other without any clear evidence of synergy or antagonism between types.[16]

HPV Types Variants in HIV+ Women

An HPV type variant is defined as a certain strain of HPV whose DNA sequence varies less than 10% from the parental strain. Type variants for HPV 16 have been implicated in an increased risk of cervical dysplasia. It has been hypothesized that the type variant seen in HIV+ women could be unique and would dictate the increased persistence and pathogenicity of HPV types in these women. Indeed, the study by Chaturvedi et al. demonstrated an increase frequency of the HPV-16 non-European variants in HIV+ women in the New Orleans cohort.[14] However, the presence of these variants did not correlate with an abnormal Pap smear and merely may reflect the sexual networking of these women. Other studies by Schlect et al. for HPV 16 and 18 variants[101] and Gagnon for HPV-31[43] demonstrate associations of specific type variants with ethnicity and not HIV status. The study by Schlect et al. also demonstrated a relatively high frequency (6%) of multiple variants in the same HIV+ women.[101]

HPV Viral Load in HIV+ Women

It is not surprising that studies have shown an increase in viral load of a specific HPV type viral load in HIV+ women as compared to HIV-negative women.[39,67,110,120,122] Initial comparison utilized the strength of the hybrid capture signal as a crude measure of HPV viral load[122] or defined those positive by Southern blot and PCR being of higher viral load than those positive by PCR alone. Using more sophisticated real time PCR assays, HPV 16 viral load was not only increased in HIV+ women but also higher in those HIV+ women with low CD4 cell counts[67,110,120] (Brinkman and Hagensee unpublished observation).

Detection of an Abnormal Pap Smear in the HIV+ Woman

A large number of studies have focused on the detection and resolution of abnormal Pap smears in HIV+ and HIV-negative women.[65,66,109] Similar to the detection of HPV infection, the initial studies enrolled relatively few women in a cross-sectional design. These studies demonstrated an increase in abnormal Pap smears in HIV+ women as compared to HIV-negative women. Kreiss et al.[65] showed that 50% of HIV+ women had CIN as compared to 8% of HIV-negative women whereas Laga et al.[66] showed 27% of HIV+ women had an abnormal Pap smear as compared to 3% of HIV-negative women. Studies from the HERS showed an increase in ASCUS diagnosis in HIV+ women as compared to HIV-negative women[36] and these HIV+ women were more likely to progress to higher grade lesions. Initial studies from the WIHS showed increased rates of abnormal Pap smears in HIV+ women (38% as compared to 16% in similarly risked HIV-negative women).[78] This study also showed increases in each Pap category – ASCUS, LSIL, and HSIL for HIV+ women as compared to HIV-negative women. Independent risk factors for an abnormal Pap smear included high systemic HIV viral load, low CD4 cell counts, detection of cervical HPV DNA, prior history of an abnormal Pap smear and increased number of recent sex partners. Studies on African HIV+ cohorts showed similar findings with lower CD4 cell counts and higher HIV viral loads with higher rates of abnormal Pap smears particularly with increases in high-grade lesions.[54] Furthermore, this study showed higher rates of HSIL in women infected with HIV-2 as compared to HIV-1. This may be explained by women with HIV-2 having slower CD4 cell count declines and overall lower mortality so that these women survive to have the HSIL diagnosed.

From these cross-sectional studies, larger cohorts were examined longitudinally to describe not only the prevalence of abnormal Pap smears in HIV+ women but also the incidence of SIL and rates of resolution. Six et al. followed 271 HIV+ women and 71 HIV-negative women over a 1 year period and showed higher rates of abnormal Pap smears in HIV+ women (31% vs. 7%).[105] In addition, the incidence of SIL was elevated in HIV+ women (27% vs. 5%) and progression seen only in women with lower CD4 cell counts (<500 cells). Regression was seen more often in the HIV+ women with higher (>500 cells) CD4 cell counts (52% vs. 20%). Risk factors for SIL were the detection of high oncogenic risk HPV infection, increased number of lifetime sexual partners, younger age, past history of SIL and lack of cervical screening. Ellerbrock et al.[37] showed increased incidence of SIL in HIV+ women (8.3 vs. 1.8 cases per 100 person years) with the vast majority of these lesions being low grade (91%). Risks for SIL development were again HIV infection, high oncogenic risk HPV infection (transient or persistent), and younger age. Delmas et al.[31] described similar findings from the European cohort with a twofold increases in prevalence, incidence, and lack of resolution of SIL in HIV+ women as compared to HIV-negative women. Multivariate models of HIV+ women showed that the presence of HPV, a low CD4 cell count, younger age, history of untreated SIL, history of warts, and Nordic location were all independent risk factors for having an abnormal Pap smear. Branca et al.[8] also showed an increase in progression of an abnormal Pap smear (13% vs. 4%) and less resolution of an abnormal Pap smear in HIV+ women (8% vs. 13%). The HERS group again showed similar findings but with a larger population of 774 HIV+ women and 391 HIV-negative women with an average follow-up of 5.5 years.[102] The incidence of SIL was 35% (11.5 cases/100 person years) in HIV+ women versus 9% (2.6 cases/100 person years) for HIV-negative women. Women with detectable HPV and those with lower CD4 cell counts had higher rates of dysplasia. Finally, one study showed that life stress was a risk factor for the progression and persistence of an abnormal Pap smear.[92]

A series of articles examined the rates of abnormal Pap smears in women enrolled in the WIHS. HIV+ women without detectable HPV DNA had rates of abnormal Pap smears similar to HIV-negative women.[53] In addition, no cases of HSIL developed in these women. This implies that these HIV+ women can be followed in a similar manner to HIV-negative women. Women (HIV+-391, HIV-negative-103) with ASCUS or LSIL were followed for 4 years.[77] Progression was seen more often in HIV+ women (12%) as compared to HIV-negative women (4%). Detection of highly oncogenic HPV and African-American race. The rates of progression were similar for the first two years of the study but there were more late progressors in the HIV+ group. Interestingly, on average, it took longer for the HIV+ women to progress (2.7 years) as compared to the HIV-negative women (1.1 years). Finally, Massad et al. examined the natural history of CIN I in 202 HIV+ women and 21 HIV-negative women.[79] Progression occurred only in HIV+ women (8 out of 202). Regression occurred in 67% of the HIV-negative women and only 33% of the HIV+ women. Progression included hormonal contraception, cervical HPV infection and persistent LSIL. Again, these lesions appear to be relatively rare and clinical management of these women can be similar to HIV-negative women.

Finally, two recent studies focused on the development and resolution of HSIL in HIV+ populations. Moscicki et al.[83] followed 182 HIV+ and 84 HIV-negative adolescents for the development of HSIL. HSIL was seen more frequently in the HIV+ girls (21%) as compared to the HIV-negative girls (5%). Independent risk factors for the development of HSIL included use of hormonal birth control, HPV cervical infection, persistent LSIL, and a high concentration of cervical Interleukin IL-12. Hawes et al.[54] studied 627 women from Senegal for the development of HSIL. HSIL developed in 11% of these women; whose risk correlated with detection of oncogenic HPV either transient or persistent. Rates were higher for women with lower CD4 cell counts and higher HIV viral loads but these factors were not independent of the presence of HPV DNA. In summary, the HIV+ woman is at higher risk for cervical abnormalities – in terms of prevalence, incidence, and lack of regression. Fortunately, established guidelines for the care of both HIV+ and HIV-negative women appear to be able to prevent cervical cancer in both populations.

Cervical Cancer in HIV+ Women

Kaposi's sarcoma and non-Hodgkin's lymphoma are increased in HIV+ individuals. Similarly, HIV+ positive women are at greater risk for progression of HPV-related genital pathology. Chin-Hong et al.[19] demonstrated that HIV+ women had a rate of invasive cervical cancer of 10.4/1,000 women as compared to HIV-negative women's rate of 6.2/1,000 women. The rate of cancer was higher for younger women (aged 20–34), and women of African-American or Hispanic decent. However with the extremely high rates of cervical dysplasia seen in advanced HIV+ women, the overall increases in cervical cancer have been rather minimal.[38] Frisch et al.[41] reviewed the cases of HPV-related cancers in the HIV+ population. They demonstrated increases in cancer rates in HIV+ individuals for in situ tumors from the cervix, vulva, anus, and penis. They also noted increases in invasive cervical, vulvar, anal, penile, tonsillar, and conjunctival cancer which all may be caused, in part, by HPV. They saw no relationship to lower CD4 cell counts for these cancers. In a meta-analysis of 15 studies, HIV+ women (OR 8.8) were more likely to have HPV-related cancer than HIV-negative women (OR 5.0).[74] Finally, in a review by Chin-Hong and Palefsky,[19] HIV+ individuals are at increased risk for anogenital malignancy but there is no correlation between lower CD4 cell counts and cancer development. This is a very interesting conundrum which has not been well explained. Increased risk of oncogenic HPV cervical infection which increases with lower CD4 cell count. They also have increased risk of abnormal Pap smears and severity of the cervical lesions with lower CD4 cell counts. The lack of progression to cancer in these women with lower CD4 cell counts implies that either these women die of HIV/AIDS complications before invasive cervical cancer (ICC) can be detected or there are other, yet unknown, risks for the development of ICC. Additional studies focusing on HIV+ women with LSIL and HSIL are necessary to explain these findings.

Other Genital Sites of HPV Infection and Disease

HIV+ women have been shown by a few studies to have an increased prevalence of genital warts as compared to HIV-negative women. A rate of 8.2 cases/100 person years of genital warts was seen in New York City HIV+ population, which was significantly increased as compared to the 0.8/100 person years for HIV-negative individuals.[20] Others have also seen a 6–10-fold increase in the prevalence of genital warts in HIV+ women as compared to HIV-negative women.[18,22,104]

Vaginal neoplasia is overall quite rare but the premalignant changes seen in vulvar intraepithelial neoplasia (VIN) have been shown to be more prevalent in the HIV+ woman. VIN developed in 3–6% of HIV+ women as compared to 0–0.8% of HIV-negative women.[18,22] These lesions tend to be in the upper one third of the vagina and it is not clear if this arises from a separate HPV genital infection or as an extension of current or previous cervical disease. Indeed, the HIV+ woman needs a thorough examination of the vaginal and surrounding vulvar area for any suspicious premalignant lesions. Furthermore, since vulvar cancer is so rare even in the HIV+ population, there are not clear recommendations as to treatment modalities for these women. One study[22] showed that HAART had no impact on the prevalence of vulvar disease in the HIV+ women.

Anal dysplasia and cancer is more common in women than men and is also increased in the HIV+ woman. Hillemanns et al.[57] followed 102 HIV+ and 96 HIV-negative women and noted anal cytological abnormalities in 26% of the HIV+ women and only 7% of the HIV-negative women. HPV DNA was detected in 58% of the anal samples from HIV+ women as compared to 8% of the HIV-negative women. For both anal dysplasia and anal HPV infection, HIV and lower CD4 cell counts were the most significant risk factors. Palefsky et al.[90] followed 251 HIV+ and 68 HIV-negative women showing increased rates of anal HPV detection (76% vs. 42%) with significant risk factors being younger than 36 years old, and being Caucasian. The prevalence of anal HPV infection was greater than the rates of cervical HPV infection collected simultaneously. These rates of anal disease make routine examination and anal Pap smear evaluation recommended at least for those women with current cervical HPV disease.

The Role of Highly Active Antiretroviral Therapy in HPV Infection and Disease in the HIV+ Woman

A number of studies have examined the role of HAART on the prevalence of HPV infection and HPV-related disease in HIV infected women. Heard et al.[55] followed 49 women over a 5-month time period. Sixty-eight percent of the women had an abnormal Pap smear at the start of the study which was significantly reduced to 54% with HAART. Improvements were seen in the women with SIL but still the majority of women maintained an abnormal Pap smear and remained at risk for the development of cervical cancer. This study also showed no significant change in the rate of HPV cervical infection on or off HAART. Women with improvement in Pap smear status were found to have higher CD4 cell counts. A follow-up study from this same group[56] followed a larger cohort ($n = 168$) of women with

CIN and examined the rates of regression or progression after 18 months of follow-up. Regression was observed in 40% of the women at 12 months with twofold higher regression rates in those women on HAART. However, again the majority of the women (60%) either had a persistent abnormal Pap smear or progressed. Other studies have demonstrated conflicting results. An initial study from Italy[70] showed no impact of HAART on HPV infection rates or HPV-related SIL. Another study from Italy showed an increased rate of LSIL regression for those women on HAART but no impact on HSIL regression rates.[30]

The two large cohorts of women from the United States, the WIHS and HERS, also showed conflicting results. Minkoff et al.[81] found that women with persistent HPV infection were more likely to have their abnormal Pap smear progress whereas women on HAART were 40% more likely to regress and less likely to progress. On the other hand, Schuman et al.[102] found in the HERS that SIL was more likely in women with lower CD4 cell counts and higher HIV viral loads but there was no impact of HAART. Both of these studies focused on at risk inner city women and followed them for an extended period of time (5 years or more). It is not clear why the disparate results were obtained. Clearly, HAART is not a panacea when it comes to cervical dysplasia – it may have a role in improving abnormal Pap results but these women clearly need close follow-up and frequent Pap smear evaluations to prevent cervical cancer.

Identification and Treatment of Cervical Dysplasia in HIV+ Women

Since the HIV+ woman is more likely to be infected with HPV, this infection persists for a longer period of time and may have more virus being shed, it is possible that self-sampling techniques could be developed that would reliably reflect the HPV infection status of the cervix. Indeed, the study by Brinkman et al. showed comparable HPV infection frequencies between physician obtained cervical swabs and self-collected urine.[10] In addition, the study by Petignat et al. showed an increase prevalence of HPV and more types detected by a self-vaginal swab as compared to physician obtained vaginal swab.[93]

The treatment of the HIV+ women with dysplasia is essentially the same as for the HIV-negative women (Fig. 5). The initial management of the newly diagnosed HIV+ woman is to obtain Pap smears at 6-month intervals. If two sequential tests are normal then she can return to yearly testing. This approach has been shown to be cost-effective.[46] There is a lower threshold for referring the HIV+ women for colposcopy and biopsy for an ASC-US or LSIL diagnosis with most authors recommending that all of these women be referred. At this time, screening at longer intervals than yearly cannot be recommended as a relatively recent study[37] found that 20% of HIV+ women with no prior abnormal Pap history developed SIL within 3 years. This underscores the need for close follow-up of these women. A study by Belafsky et al.[4] noted 20% of the HIV+ women with LSIL would progress to high-grade lesions and this was not significantly improved by the application of isoretinoid.[98] The reliable HIV+ women with higher CD4 cell counts who has LSIL, it is appropriate to watch them closely for signs of progression rather

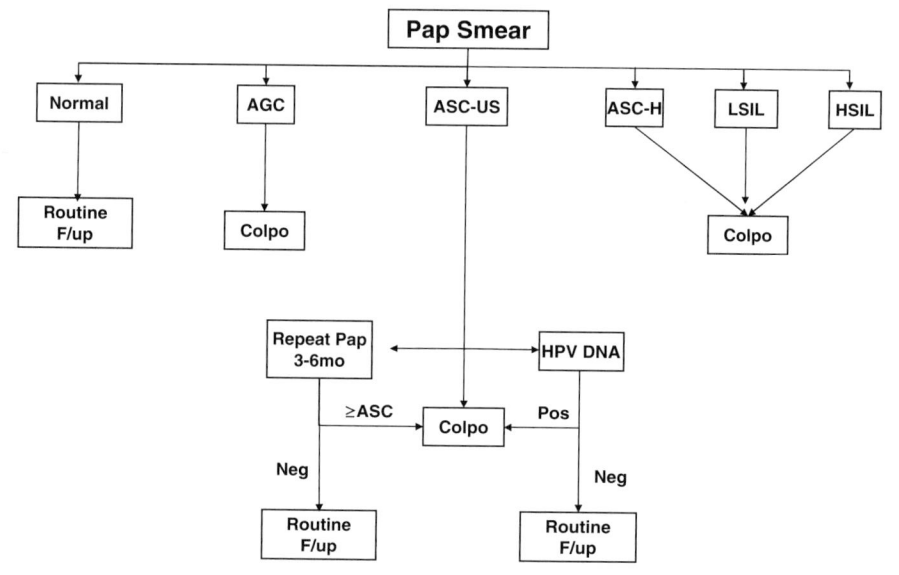

Figure 5. Management of an abnormal pap smear (AGC – atypical glandular cells).

than referring for colposcopy and biopsy. On the other hand, if the patient is less reliable or has any other complicating situation then referral to colposcopy and biopsy is indicated. Indeed, those younger women (<30 years old) and those with lower CD4 cell counts are more likely to progress and should be referred to colposcopy. Treatment of higher grade lesions, CIN II and III, leads to recurrence rates in the 50% range. Thus, attention to follow-up is critical in the management of these women. The study by Fruchter et al.[42] demonstrated that 62% of the 127 HIV+ women experienced a recurrence within 3 years of therapy. Subsequent recurrence rates were 42% for the second recurrence and 50% for the third. Risk factors for recurrence included HIV infection and lower CD4 cell count. Indeed 25% of these women progressed on Pap smear as compared to 2% of HIV-negative women. These findings are corroborated by the study by Robinson et al. which demonstrated a 57% rate of recurrence. These women may benefit from treatment with 5-flourouracil (5-FU) which has been shown to reduce the recurrence rate from 47 to 28%.[73]

HIV+ MEN

Detection of Anal HPV in HIV+ Men

Similar to the studies in women, rates of HPV from anal samples in men initially started with small cross-sectional prevalence studies which expanded larger studies and then a few longitudinal studies. The vast majority of these studies have come from the San Francisco area with Dr. Joel Palefsky being the leader in this field. The initial studies[3,9,13,60,80] estimated an anal HPV prevalence rate in HIV+ men to be between 15 and 61% with the lower rates reflecting use of less sensitive assays to detect HPV (Virapap, Southern blot) and the higher rates reflecting the use of PCR-based assays. This is compared to rates in the HIV-negative male of 12–29% using similar HPV detection assays. Critchlow et al.[24] examined 322 HIV+ and 287 HIV-negative men in a cohort from Seattle, Washington. Utilizing a PCR-based assay to detect HPV, they noted an extremely high rate of anal HPV infection in HIV+ (92%) as compared to HIV-negative gay men (66%). Additional high rates of HPV anal infection were also seen in HIV+ men from the Netherlands (65%), New York City (61%), and San Francisco (95%).[90,121] High rates of HPV infection (46%) were also seen in a substantial number of men who deny anal intercourse and acquired HIV by intravenous drug use. This may imply that HPV can either spread throughout the genital tract or can be transferred by shared objects. Finally, Moscicki et al.[83] studied a unique population of adolescent boys and girls with HIV infection. They noted a high rate of HPV from anal samples from HIV+ boys (48%) but this was not statistically different than HIV-negative boys (36%). These high rates of oncogenic HPV infection lend support to the routine screening of men for anal cytological abnormalities.

Risk Factors for Detection of Anal HPV

Initial studies did note an increase in anal HPV DNA detection in those HIV+ men with lower CD4 cell counts.[3,13] A larger cohort determined risk factors for anal HPV to be lower CD4 cell count and a history of anal warts for HIV+ men and lifetime number of sexual partners, recent anal sex, and anal warts for HIV-negative men.[24] Other studies also demonstrated these risk factors and also noted increased HPV anal DNA detection in those HIV+ men not on HAART.[121] The adolescent male was more likely to have anal HPV if there was a history of anal warts.[83]

HPV Types in HIV+ Men

There are fewer studies looking at the specific HPV types in the anal canal of HIV+ and HIV-negative men. Smaller studies did note an increase in HPV 16, 18, 31, 33, and 52 in anal samples from HIV+ men as compared to HIV-negative men.[9] Multiple infections were quite common in the anal canal of HIV+ men. Reported rates are between 47 and 88% with some men having six or more different HPV types detected in anal samples.[24,89,95] In contrast to the case of HIV+ women with cervical HPV-16 infection, there has only been a single study looking at the HPV 16 type variants in HIV+ men. This study[28] analyzed 628 anal specimens from 193 HIV+ and 59 HIV-negative men. The G131 variant was present in only 6% of samples but these men had an increased risk of HSIL with an OR of 3.4.

HPV Persistence in HIV+ Men

HPV detection from anal samples was also more persistent in HIV+ as compared to HIV-negative men. An initial study followed men only for 6 months but saw more persistence in those HIV+ men with a clinical diagnosis of AIDS as compared to those HIV+ without AIDS.[9] Palefsky et al.[87] followed 346 HIV+ and 262 HIV-negative men and noted increased persistence of anal HPV in HIV+ (63%) versus HIV-negative (23%) men over a 4-year period. Critchlow reported similar findings from Seattle,[24] with 38% of the HIV+ men having persistent anal HPV as compared to 23% of the HIV-negative men over a period of 28 months. This study also noted an increase in anal HPV DNA acquisition of 1.8-fold in HIV+ as compared to HIV-negative men and less clearance of anal HPV in HIV+ men (13%) versus HIV-negative men (45%). In summary, anal infection by HPV in HIV+ men is very common, is found more frequently than in HIV-negative men with more multiple infections and longer persistence. In general, these rates are much higher than the rates of cervical infection in HIV+ women.

Detection of Abnormal Anal Pap Smears in the HIV+ Man

The majority of the studies on HIV+ men have focused on the prevalence, incidence, and progression of anal squamous intraepithelial lesions (ASIL). The earliest study by Frazer et al.[40] noted anal dysplasia in 9 of 14 HIV+ men with associations with lower CD4 cell counts, anal warts and receptive anal intercourse. Other cross-sectional studies noted significant anal dysplasia in HIV+ men (16–24%) as compared to HIV-negative men (7%).[13,80] The subsequent larger studies not only confirmed these findings but those longitudinal in design were also able to measure the incidence rate of anal lesions and the rates of progression to higher grade lesions. Palefsky et al.[87] noted a 3.7 increased relative risk for HIV+ men in the development of HSIL with incidence rate of HSIL being 49% after 4 years of follow-up. This compares to an incidence rate of only 7% for HIV-negative men. Overall, HSIL incidence rates as high as 62% have been reported by this group.[87] Risk factors for anal dysplasia include lower CD4 cell counts, anal HPV DNA detection, and multiple HPV types detected from anal specimens.[80,87,89] In one recent study, HAART actually increased the risk of anal Pap smear abnormalities.[89] These findings support the need for active and longitudinal following of anal Pap smears and appropriate treatment for high-grade anal lesions.

Anal Cancer in HIV+ Men

The rates of anal cancer have been reported to be as high as 37 cases per 100,000 men who have sex with men (MSM). This is comparable to the estimated rate of cervical cancer in women (40–50 per 100,000) prior to the implementation of cervical Pap smear screening.[19] The relative risk for invasive anal cancer in HIV+ individuals is 37-fold in men and sevenfold in women as compared to HIV-negative individuals. The risks for in situ disease is similar at 70-fold in HIV+ men and tenfold in HIV+ women as compared to HIV-negative individuals.[76] Tumors in the HIV+ individuals tend to occur at a younger age and are larger at initial diagnosis.[44] Anal cancers are not increased in HIV+ men with lower CD4 cell counts which is similar to the study of cervical cancer in the HIV+ woman. One author[19] has proposed that other factors such as host chromosomal mutations may be responsible for determining who progresses to anal malignancy. Conversely, the critical CD4 cell count threshold for control of HPV may be higher such as over 500 cells/ml. Furthermore, there may be other, yet unidentified risk factors for the development of invasive anal cancer in the HIV+ man.

Other Genital Sites of Infection in HIV+ Men

There is less data on the prevalence of other genital sites infected with HPV in HIV+ men. One study noted rates on the penis to be 23% in HIV+ men versus 15% in HIV-negative men. A small prevalence study was performed in New Orleans on 50 HIV+ men. These men were tested for HPV infection by swabbing six different locations in the anogenital tract. The combination of data from all six locations indicated that HPV DNA could be detected in 50% of the men. By far the anal (44%) and perianal (36%) regions were the most commonly infected; however, the coronal sulcus (18%) and shaft (16%) of the penis were also commonly infected (Fig. 6) (Teitz and Hagensee et al., unpublished observation). These infections may be important for both future development of anal and penile cancers and the ability to transmit the virus to others. It is somewhat surprising that relatively few penile cancers have arisen in the HIV-infected males to date, given the large numbers of HIV+ men with oncogenic HPV penile infection. Nevertheless, the frequency has been reported to be two to three times greater than the rate of occurrence in HIV-negative men.[1,111]

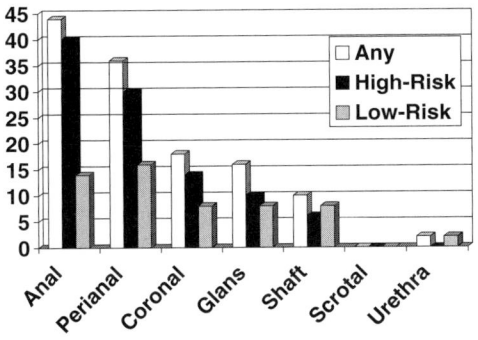

Figure 6. Site of HPV infection in men.

Treatment of HIV+ Men with HPV Infection and HPV-Related Disease

The role of HAART in the disease course of anal HPV infection and anal dysplasia has been investigated. For most of the studies, there has been no impact of HAART on either HPV anal infection or disease. Palefsky et al. noted that 75% of the men with anal disease did not improve despite being on HAART.[87] Piketty et al.[96] intensely studied 45 HIV+ men and noted no difference in the prevalence of HPV anal infection or disease in those with a significant increase in CD4 cell count after initiating HAART as compared to those with no CD4 cell count increase. Finally, there has been no decline in the rates of anal cancer in HIV+ individuals since the routine use of HAART in 1996.[96]

The management of the HIV+ male with a history of receptive anal intercourse is not clear. There have been numerous studies from the Palefsky group that indicate the need for routine and longitudinal screening for anal abnormalities in this population. A cost analysis study by Goldie et al. indicates that yearly anal Pap smears appears to be the most cost-effective.[45] Who to screen is also not clear; rates of anal HPV infection and dysplasia indicate that at least the HIV+ male who is a MSM should be screened. Furthermore, all HIV+ men and all MSM could be considered for routine screening. The treatment guidelines are similar to that of women with cervical disease in that AIN I can be watched carefully whereas AIN II and III should be followed by high resolution anoscopy and biopsy of abnormal areas. Treatment of these anal lesions is also not delineated. Dr. Palefsky is the expert in this field and he recommends that smaller lesions be treated by local treatment such as liquid nitrogen, purified podophyllin toxin, TCA, coagulation, or imiquimod.[89] Larger and more circumferential lesions are much more difficult to treat and may require surgical excision. However, the rates of side effects are greater than similar cervical treatment with the concern of rectal stenosis and pain. In addition, recurrent rates are high and these men need to be followed at 4–6-month intervals. Invasive cancer is usually treated with 5-FU, mitomycin D, or cisplatinin with radiation. All of these treatments have more significant side effects in the HIV+ individual. Alternatives include continuous infusion of 5-FU, and lower doses of mitomycin D.

HPV INFECTION AND DISEASE IN THE ORAL CAVITY OF THE HIV+ INDIVIDUAL

HIV-Associated Oral Warts: Pathology and Etiology

The widespread administration of HAART in the United States has had a profound impact on the incidence of HIV-associated oral-opportunistic infections. Since the advent of HAART, the incidences of oral thrush (oropharyngeal candidiasis, OPC) and oral hairy leukoplakia (OHL) have significantly decreased.[47] In stark contrast, Dr. Janet Leigh, Director of the HIV Dental Clinic at the Medical Center of Louisiana, New Orleans, reported an increase in incidence of oral warts since the advent of HAART.[68] Dr. Leigh's report was followed by a report in *Lancet* by

Greenspan et al. (San Francisco), which clearly demonstrated not only a rise in the incidence of oral warts in the 1990s but also a significant association between the rise in warts and HAART.[47] Increased incidence of oral warts concurrent with a drop in plasma HIV viral load was subsequently reported in the Atlanta HIV cohort by King et al.[59]

Unlike OHL, which occurs almost exclusively on the tongue, and OPC, which occurs on either the tongue or buccal mucosa, papillomas due to HPV infection can occur on virtually all oral mucosal surfaces.[48] While the majority of papillomas occur on the labial mucosa, they can also occur on the buccal mucosa, the tongue, the soft palate, and the gingiva. While the histopathology of oral warts almost invariably demonstrates poorly differentiated, large, vacuolated koilocytic cells, the gross appearance varies greatly. Often the clinical appearance of lesions is reflective of the specific HPV genotype causing the lesion. For instance, HPV genotypes 6 and 11, the most common causes of genital warts, tend to cause soft, sessile, cauli-flower-like lesions (condyloma accuminatum) in the oral cavity. HPV genotypes 1, 2, and 7, which are associated with cutaneous warts, cause firm, sessile, oral common warts (verruca vulgaris). HPV genotypes 13 and 32, which have been described exclusively in the oral cavity, are the cause of oral focal epithelial hyper-plasia, a dysplastic lesion characterized by multiple small, flat papules generally found on the lower lip. While there is some degree of HPV genotype-specific clinical presentation, unusual manifestations of oral HPV disease in the HIV+ patient frequently occur.[48] Examinations of oral–wart biopsies from HIV+ individ-uals prior to the routine use of HAART contain a range of HPV genotypes, includ-ing cutaneous type 2; genital types 6, 11, 16, and 18; and oral type 13.[48] However, the most common HPV genotypes identified in HIV-associated oral warts are the oral-specific HPV type 32 and the cutaneous HPV type 7.[48]

Genotype Prevalence and Risk Factors for Oral HPV Infection

The majority of oral HPV infections are asymptomatic, and the development of HPV pathology is a rare event that occurs only after prolonged infection. Therefore, epidemiological studies often examine the risk factors for asymptomatic detection of oral HPV. As for cervical and anal HPV infection, HIV+ individuals have higher rates of oral HPV carriage than do HIV-negative individuals.[23] Oral HPV infection is associated with markers of sexual risk, such as homosexuality, unprotected oral sex, and a history of previous sexually transmitted disease, indicating that HPV is likely transmitted to the oral cavity through sexual contact. Kreimer et al.[64] examined tonsillar brushes and oral rinses from 190 HIV+ and 396 HIV-negative individuals. They noted an increased prevalence of oral HPV infection in HIV+ individuals. In the HIV+ individual, independent risk factors for having oral HPV infection were low CD4 cell counts (<200), being seropositive for HSV-2, having an oral mucosal abnormality, and having more than one oral sex partner over the last year. In the HIV-negative population, increased age, male gender, and HSV-2 seropositivity were associated with oral HPV infection. The oral cavity may serve as a reservoir of HPV infection, allowing for transmission of HPV to the genital tract of a receptive partner.

A recent study by Cameron et al. demonstrated 36% of saliva samples from HIV+ individuals to be HPV positive as compared to 8% of HIV-negative individuals.[11] HPV types 16, 52, and 83 were the most common types detected. Risk factor analysis for oral HPV infection in HIV+ individuals from this study noted increases in Caucasians (57% vs. 24%) and males (68% vs. 25%). It is of interest that neither CD4 cell count nor HIV viral load correlated with the presence of oral HPV. Interestingly, in the Caucasian population, there was a sixfold increase in oral HPV detected in those on HAART (71% vs. 28%). An additional study examined the types of HPV involved in oral warts in the HIV+ individual in the setting of routine use of HAART. The vast majority of these samples were HPV-32 positive with other HPVs detected being 6, 7, 53, 73, and 84. (Fig. 7). Examination of local cytokine and chemokine production yielded no specific pattern found in the HIV+ individual with oral warts.[71]

A follow-up study examined the site-specific prevalence of various HPV types from various locations in the oral cavity (buccal mucosa, tonsil, labia, lingula, palate, and gingival, Fig. 8). Data from 180 HIV+ individuals noted 37% of these people had HPV detected on at least one oral swab and 20% of all samples collected were positive. Most common site of infection (~30% positive) were the labial and lingual tissue. The addition of specific screening tests for HPV-32 demonstrated that HPV-32 was, in fact, the most common type detected (Fig. 9).

The detection of genital HPV genotypes may be masking concurrent infections with oral-specific genotypes in the oral cavity. The role of genital HPV infections of the oral cavity in disease is uncertain, though most of these infections are high-oncogenic-risk HPV genotypes (16, 18, etc.), which have been identified in up to 50% of head and neck carcinomas.[103] There is a potential for an increase in HPV-related oral malignancies with long-term HAART leading to an increased lifespan of HIV+ individuals. Indeed, some recent studies have noted an increase in head

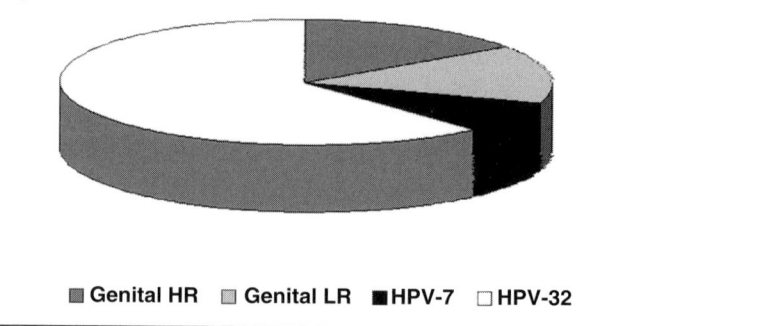

■ Genital HR   ▨ Genital LR   ■ HPV-7   □ HPV-32

Figure 7. HPV types in oral warts.

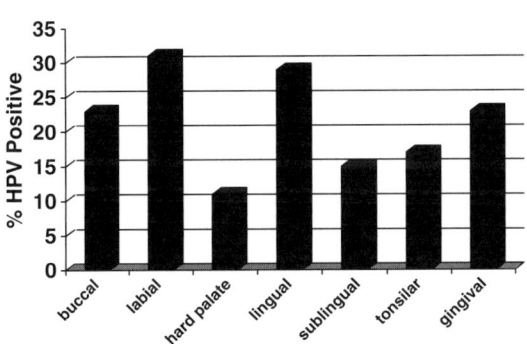

Figure 8. Site of oral HPV infection.

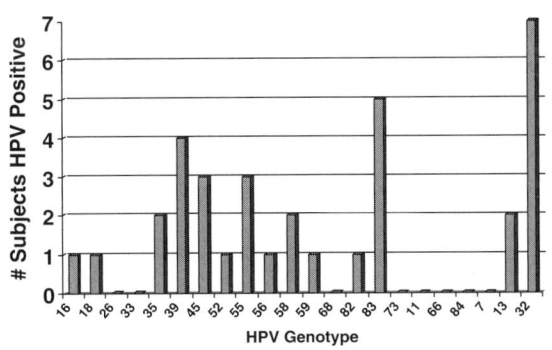

Figure 9. HPV genotypes found in the oral cavity.

and neck cancer in HIV+ individuals[6] and these cancers tend to occur at a younger age than in HIV-negative individuals.[32] It has been reported that oral warts in HIV+ individuals containing HPV 16 or 18 can undergo dysplastic changes. The role of HPV in these tumors is not yet known.

Treatment and Prognosis of HIV-Associated Oral Warts

The treatment of oral warts in the HIV patient is difficult due to both the wide distribution of lesions throughout the oral mucosa and the high recurrence rate. Although these lesions are generally painless, they can become traumatized and interfere with eating and talking and can also look unsightly when present externally on the lips. Treatments utilized include both medical and surgical modalities, depending upon the site of the wart, the characteristics of the wart, and the number of lesions. Surgical techniques include excision, electrosurgery, cryosurgery, and

CO_2 laser, whereas medical modalities include podophyllin resin and, more recently, interferon-α injections.[35] Despite recent advances in the treatment of cutaneous and genital warts with the introduction of imiquimod, this topical medicine is not approved for use in the oral cavity.

Surgical excision is difficult when multiple warts cover a large area. However, surgical or electrosurgical debulking of the wart is useful prior to use of topical agents or the CO_2 laser. The use of the CO_2 laser has proven problematic because the dispersal of HPV in the laser plume can lead to nasal warts in either operator or patient.[35] The use of surgical techniques does not often lead to postoperative scarring of the intraoral mucosa, but scarring of the lips can occur, leading to stricture and diminished opening. The use of podophyllin resin 25% as a topical agent has proven disappointing. Recent case reports showed promising results with the use of interferon-α as a topical and systemic combination with weekly intralesional injections in addition to twice-weekly subcutaneous injections. Follow-up ranged from 12 months to three years, with no recurrence of warts at the site of treatment.[72] A recent case report noted the utility of topical cidofovir gel 1% which successfully treated a case of recalcitrant oral warts in a HIV+ individual.[33]

Other Sites of HPV-Related Cancers

Conjunctival cancer has emerged as a possible cancer caused by HPV. Studies have noted an increase in conjunctiva tumors in HIV+ individuals from Africa and Waddell[118] had noted HPV-16 in 35% of conjunctival tumors. Interestingly, Tornesello et al.[112] found HPV in 25% of conjunctival lesions but this was more common in low-grade lesions than high-grade lesions. This implied that HPV may not be related to these tumors. This type of data needs to be carefully interpreted and additional larger studies need to be done to better define the role of HPV in these tumors.

MECHANISM OF HPV–HIV INTERACTION

There are two main theories regarding the mechanism by which HIV and HPV interact to lead to the increase in rates of HPV infection, persistence, and pathology. The first theory is that HIV causes immune dysfunction that leads to unchecked HPV infection leading to pathology and the second theory is that HIV enhances HPV-induced pathology by directly interacting with HPV infected cells. A direct interaction between these two viruses is an intriguing proposition. There is some epidemiological evidence that HPV persistence and HPV-related pathology is increased in individuals with higher HIV plasma viral loads as an independent risk factor.[108] In addition, some studies showed that women with higher CD4 cell counts still had an increased risk of HPV infection compared to similarly sexually active HIV-negative women, implying a potential direct effect of HIV. The HIV tat protein has been shown to directly increase gene expression by regulation of HPV-16 transcriptional control elements in vitro and in SiHa cervical cancer cell line.[116] Similar experiments demonstrated that tat could transactivate HeLa cells and HPV-18 E7 expression.[112] In addition, Dolei et al.[34] were able to show that expression of HIV tat protein could increase HPV-18 E1 and L1 protein expression. However, there is much evidence that does not support a direct interaction. First, HIV predominately infects lymphoid tissue and HPV infects epithelial cells so it is not likely that they infect the same cell. Since tat has been noted to be able to diffuse through cell membranes, a co-infection state is not a requirement for an association, but Unger et al. demonstrated that there was no colocalization of HIV and HPV in cervical tissue making it less likely for a direct interaction to occur.[114]

On the other hand, there is much evidence that immune suppression could lead to cervical disease. Numerous studies mentioned above for both HIV+ men and women have shown an increase in HPV infection and HPV-related pathology in individuals with lower CD4 cell counts. One certainly does not need to be HIV infected to be afflicted with HPV-related disease but immune suppression seems to potentially accelerate the progression from infection to disease. The exact mechanism of how lower CD4 cells leads to a lack of HPV-related immunity is not known. This is, in part, due to the lack of a simple cultivation system for HPV. This makes it rather difficult to generate large quantities of highly purified antigens that are needed for immune studies. HPV infection is felt to be controlled by CD8 activated T cells but the exact role of CD4 cells, macrophages, Langerhans cells (LC), and various cytokines and chemokines is still unknown. One study[5] showed an increase in lymphocytes in cervical biopsies from HIV+ women with CIN. These lymphocytes demonstrated an inverse CD4/CD8 ratio which has also been seen by other investigators.[86] On the other hand, a more recent study by Kobayashi et al.[61] showed increases in B and T cells in women with HPV-related CIN 2/3 but these numbers were decreased in HIV–HPV coinfected women. Why these two studies appear to have opposite results is not clear but this discrepancy points to the need for more research in this area.

There have been a number of studies that analyze the local cytokines milieu in order to get a better understanding of HPV-related immunity. HPV cervical infection and disease has been shown to cause no significant change in any local Th1 or Th2 cytokine measured,[106] with another study showing increases in IL-6, INF-gamma, TNF-alpha, increased macrophages, B lymphocytes, T lymphocytes, and NK cells.[61,85] In HIV+ women with cervical dysplasia, the increases in IL-6, INF-gamma, and TNF-alpha are less but levels of IL-4, 8, and 10 are higher.[85] In addition, CD8 T cells were increased and macrophages were seen in lower quantities.[61] Also higher levels of IL-10 were found in HIV, HPV coinfected women.[25,106] Furthermore IL-10 has been shown to be able to increase HPV-16 E7 expression in vitro utilizing a Stat3 pathway.[2]

Other studies have shown an increase in LC in SIL but HIV+ women with SIL have fewer LCs. As CIN increases in severity, one study showed that there were increases in p53, Ki-67, apoptosis parameters, and BCL-2 expression. HIV+ women have a rate of apoptosis that was fivefold higher than HIV-negative. All of these studies suffer from a limited sample size, which is unfortunately necessary due to the complexity and expense of these assays. It may indeed be possible that a combination of immune deficiencies and direct HIV–HPV interactions could explain the clinical findings.

The detection of serum antibodies directed against the L1 protein of HPV capsids has been measured as an attempt to better understand the immune response to HPV. Silverberg et al. have measured changes in antibody levels over time from women enrolled in the WIHS.[104] An increase in antibody levels was seen in HIV+ women and was associated with current and past HPV-16 infection and lower CD4 cell count; potentially implicating reactivation of a previously acquired HPV infection. On the other hand, antibody increases in HIV-negative women correlated with past HPV-16 infection, smoking, and the number of sexual partners and may be more closely linked to new exposure to HPV. Viscidi et al.[117] measured the antibody levels against various HPV types in 829 HIV+ and 413 HIV-negative women enrolled in the HERS. In general, there was no protection from future HPV infection if one had serum antibodies against HPV. There was a modest amount of protection from HPV-45 infection about 4.5 years after developing a HPV 45 serum antibody response. At a local level, cervical antibodies against HPV can be detected.[50,75,106] These have been found to be increased in HIV+ women[75] but the presence of these antibodies did not lead to increased clearance of this infection. The exact role of the local anti–HPV antibodies is not clear.

Impact of HPV Prophylactic Vaccines on the HIV+ Individuals

Both Merck and GlaxoSmithKline are in the final developmental stages of prophylactic HPV vaccines. Both of these products are based on viruslike particle technology and in published studies have been effective in phase 1 and 2 clinical trials.[52,63] These have been studies predominately in women without any prior exposure to the types of HPV in the vaccine. The impact of these vaccines for the HIV+ individual is not clear. There will likely not be any immediate impact as most, if not all, HIV+ individuals have already been exposed to numerous types of HPV due to the sexual nature of acquisition of both of these viruses. If these vaccines have wide application in the future, there may be a large impact to decrease the HPV infections and HPV-related disease in the HIV+ population. On the other hand, the HIV+ individuals tend to have multiple HPV infections. Unless vaccines are developed with numerous HPV types in them, the impact of the current formulated products on the HIV+ individual may be small. The impact of HPV vaccines on those women already infected with HPV has not been reported but it is likely to be significantly less than in unexposed women. Future studies to examine vaccine efficacy in HIV+ men and women need to be done.

SUMMARY

HPV infection of both the genital tract and oral cavity of HIV+ men and women is increased. HPV-related pathology is also increased in the HIV+ individuals, usually with further increases seen for those HIV+ individuals with lower CD4 cell counts. Fortunately, the rates of cervical cancer and anal cancer are relatively low and not related to CD4 cell count. Treatment of the HIV+ individual with HPV-related disease is challenging and requires close long-term follow-up to prevent recurrent disease. The mechanism of how HPV and HIV interact is still not known but is more likely to be linked to immune suppression rather than a direct interaction between viruses. The newly developed HPV vaccines will likely have a significant impact on HPV-related disease in immunocompetent individuals. It remains to be seen what impact these vaccine will have on the immune depressed.

REFERENCES

1. Aboulafia, D. M., and Gibbons, S. 2001. Penile cancer and human papillomavirus (HPV) in a Human Immunoeficiency virus (HIV)-infected patient. Cancer Invest **19**:266–72.
2. Arany, I., M. Muldrow, and S. K. Tyring. 2001. Correlation between mRNA levels of IL-6 and TNF alpha and progression rate in anal squamous epithelial lesions from HIV-positive men. Anticancer Res **21**:425–8.
3. Aynaud, O., D. Piron, R. Barrasso, and J. D. Poveda. 1998. Comparison of clinical, histological, and virological symptoms of HPV in HIV-1 infected men and immunocompetent subjects. Sex Transm Infect **74**:32–4.
4. Belafsky, P., R. A. Clark, P. Kissinger, and J. Torres. 1996. Natural history of low-grade squamous intraepithelial lesions in women infected with human immunodeficiency virus. J Acquir Immune Defic Syndr Hum Retrovirol **11**:511–2.
5. Bell, M. C., D. Schmidt-Grimminger, E. Turbat-Herrera, A. Tucker, L. Harkins, N. Prentice, and P. A. Crowley-Nowick. 2000. HIV+ patients have increased lymphocyte infiltrates in CIN lesions. Gynecol Oncol **76**:315–9.
6. Berretta, M., R. Cinelli, F. Martellotta, M. Spina, E. Vaccher, and U. Tirelli. 2003. Therapeutic approaches to AIDS-related malignancies. Oncogene **22**:6646–59.
7. Bollen, L. J., R. Chuachoowong, P. H. Kilmarx, P. A. Mock, M. Culnane, N. Skunodom, T. Chaowanachan, B. Jetswang, K. Neeyapun, S. Asavapiriyanont, A. Roongpisuthipong, T. C. Wright, and J. W. Tappero. 2006. Human papillomavirus (HPV) detection among human immunodeficiency virus-infected pregnant Thai women: implications for future HPV immunization. Sex Transm Dis **33**:259–64.
8. Branca, M., S. Costa, L. Mariani, F. Sesti, A. Agarossi, A. di Carlo, M. Galati, A. Benedetto, M. Ciotti, C. Giorgi, A. Criscuolo, M. Valieri, C. Favalli, P. Paba, D. Santini, E. Piccione, M. Alderisio, M. De Nuzzo, L. di Bonito, and K. Syrjanen. 2004. Assessment of risk factors and human papillomavirus (HPV) related pathogenetic mechanisms of CIN in HIV-positive and HIV-negative women. Study design and baseline data of the HPV-PathogenISS study. Eur J Gynaecol Oncol **25**:689–98.
9. Breese, P. L., F. N. Judson, K. A. Penley, and J. M. Douglas, Jr. 1995. Anal human papillomavirus infection among homosexual and bisexual men: prevalence of type-specific infection and association with human immunodeficiency virus. Sex Transm Dis **22**:7–14.
10. Brinkman, J. A., W. E. Jones, A. M. Gaffga, J. A. Sanders, A. K. Chaturvedi, I. J. Slavinsky, J. L. Clayton, J. Dumestre, and M. E. Hagensee. 2002. Detection of human papillomavirus DNA in urine specimens from human immunodeficiency virus-positive women. J Clin Microbiol **40**:3155–61.
11. Cameron, J. E., D. Mercante, M. O'Brien, A. M. Gaffga, J. E. Leigh, P. L. Fidel, Jr., and M. E. Hagensee. 2005. The impact of highly active antiretroviral therapy and immunodeficiency on human papillomavirus infection of the oral cavity of human immunodeficiency virus-seropositive adults. Sex Transm Dis **32**:703–9.
12. Cannistra, S. A., and J. M. Niloff. 1996. Cancer of the uterine cervix. N Engl J Med **334**:1030–8.
13. Caussy, D., J. J. Goedert, J. Palefsky, J. Gonzales, C. S. Rabkin, R. A. DiGioia, W. C. Sanchez, R. J. Grossman, G. Colclough, S. Z. Wiktor, et al. 1990. Interaction of human immunodeficiency and papilloma viruses: association with anal epithelial abnormality in homosexual men. Int J Cancer **46**:214–9.
14. Chaturvedi, A. K., J. A. Brinkman, A. M. Gaffga, J. Dumestre, R. A. Clark, P. S. Braly, K. Dunlap, P. J. Kissinger, and M. E. Hagensee. 2004. Distribution of human papillomavirus type 16 variants in human immunodeficiency virus type 1-positive and -negative women. J Gen Virol **85**:1237–41.
15. Chaturvedi, A. K., J. Dumestre, A. M. Gaffga, K. M. Mire, R. A. Clark, P. S. Braly, K. Dunlap, T. E. Beckel, A. F. Hammons, P. J. Kissinger, and M. E. Hagensee. 2005. Prevalence of human papillomavirus genotypes in women from three clinical settings. J Med Virol **75**:105–113.

16. Chaturvedi, A. K., L. Myers, A. F. Hammons, R. A. Clark, K. Dunlap, P. J. Kissinger, and M. E. Hagensee. 2005. Prevalence and clustering patterns of human papillomavirus genotypes in multiple infections. Cancer Epidemiol Biomarkers Prev **14**:2439–45.

17. Chen, M. J., M. Y. Wu, J. H. Yang, K. H. Chao, Y. S. Yang, and H. N. Ho. 2005. Increased frequency of genital human papillomavirus infection in human immunodeficiency virus-seropositive Taiwanese women. J Formos Med Assoc **104**:34–8.

18. Chiasson, M. A., T. V. Ellerbrock, T. J. Bush, X. W. Sun, and T. C. Wright, Jr. 1997. Increased prevalence of vulvovaginal condyloma and vulvar intraepithelial neoplasia in women infected with the human immunodeficiency virus. Obstet Gynecol **89**:690–4.

19. Chin-Hong, P. V., and J. M. Palefsky. 2005. Human papillomavirus anogenital disease in HIV-infected individuals. Dermatol Ther **18**:67–76.

20. Chirgwin, K. D., J. Feldman, M. Augenbraun, S. Landesman, and H. Minkoff. 1995. Incidence of venereal warts in human immunodeficiency virus-infected and uninfected women. J Infect Dis **172**:235–8.

21. Clarke, B., and R. Chetty. 2002. Postmodern cancer: the role of human immunodeficiency virus in uterine cervical cancer. Mol Pathol **55**:19–24.

22. Conley, L. J., T. V. Ellerbrock, T. J. Bush, M. A. Chiasson, D. Sawo, and T. C. Wright. 2002. HIV-1 infection and risk of vulvovaginal and perianal condylomata acuminata and intraepithelial neoplasia: a prospective cohort study. Lancet **359**:108–13.

23. Coutlee, F., A. M. Trottier, G. Ghattas, R. Leduc, E. Toma, G. Sanche, I. Rodrigues, B. Turmel, G. Allaire, and P. Ghadirian. 1997. Risk factors for oral human papillomavirus in adults infected and not infected with human immunodeficiency virus. Sex Transm Dis **24**:23–31.

24. Critchlow, C. W., S. E. Hawes, J. M. Kuypers, G. M. Goldbaum, K. K. Holmes, C. M. Surawicz, and N. B. Kiviat. 1998. Effect of HIV infection on the natural history of anal human papillomavirus infection. AIDS **12**:1177–84.

25. Crowley-Nowick, P. A., J. H. Ellenberg, S. H. Vermund, S. D. Douglas, C. A. Holland, and A. B. Moscicki. 2000. Cytokine profile in genital tract secretions from female adolescents: impact of human immunodeficiency virus, human papillomavirus, and other sexually transmitted pathogens. J Infect Dis **181**:939–45.

26. Cu-Uvin, S., J. W. Hogan, D. Warren, R. S. Klein, J. Peipert, P. Schuman, S. Holmberg, J. Anderson, E. Schoenbaum, D. Vlahov, and K. H. Mayer. 1999. Prevalence of lower genital tract infections among human immunodeficiency virus (HIV)-seropositive and high-risk HIV-seronegative women. HIV Epidemiology Research Study Group. Clin Infect Dis **29**:1145–50.

27. Cubie, H. A., A. L. Seagar, G. J. Beattie, S. Monaghan, and A. R. Williams. 2000. A longitudinal study of HPV detection and cervical pathology in HIV infected women. Sex Transm Infect **76**:257–61.

28. Da Costa, M. M., C. J. Hogeboom, E. A. Holly, and J. M. Palefsky. 2002. Increased risk of high-grade anal neoplasia associated with a human papillomavirus type 16 E6 sequence variant. J Infect Dis **185**:1229–37.

29. Daling, J. R., N. S. Weiss, T. G. Hislop, C. Maden, R. J. Coates, K. J. Sherman, R. L. Ashley, M. Beagrie, J. A. Ryan, and L. Corey. 1987. Sexual practices, sexually transmitted diseases, and the incidence of anal cancer. N Engl J Med **317**:973–7.

30. Del Mistro, A., R. Bertorelle, M. Franzetti, A. Cattelan, A. Torrisi, M. T. Giordani, R. Sposetti, E. Bonoldi, L. Sasset, L. Bonaldi, D. Minucci, and L. Chieco-Bianchi. 2004. Antiretroviral therapy and the clinical evolution of human papillomavirus-associated genital lesions in HIV-positive women. Clin Infect Dis **38**:737–42.

31. Delmas, M. C., C. Larsen, B. van Benthem, F. F. Hamers, C. Bergeron, J. D. Poveda, B. Anzen, H. A. van den, F. Meier, J. M. Pena, H. Savonius, D. Sperandeo, B. Suligoi, P. Vernazza, and J. B. Brunet. 2000. Cervical squamous intraepithelial lesions in HIV-infected women: prevalence, incidence and regression. European Study Group on Natural History of HIV Infection in Women. AIDS **14**:1775–84.

32. Demopoulos, B. P., E. Vamvakas, J. E. Ehrlich, and R. Demopoulos. 2003. Non-acquired immunodeficiency syndrome-defining malignancies in patients infected with human immunodeficiency virus. Arch Pathol Lab Med **127**:589–92.

33. DeRossi, S. S., and J. Laudenbach. 2004. The management of oral human papillomavirus with topical cidofovir: a case report. Cutis **73**:191–3.

34. Dolei, A., S. Curreli, P. Marongiu, A. Pierangeli, E. Gomes, M. Bucci, C. Serra, and A. M. Degener. 1999. Human immunodeficiency virus infection in vitro activates naturally integrated human papillomavirus type 18 and induces synthesis of the L1 capsid protein. J Gen Virol **80**(Pt 11):2937–44.

35. Drake, L. A., R. I. Ceilley, R. L. Cornelison, W. L. Dobes, W. Dorner, R. W. Goltz, C. W. Lewis, S. J. Salasche, M. L. Turner, B. J. Lowery, et al. 1995. Guidelines of care for warts: human papillomavirus. Committee on Guidelines of Care. J Am Acad Dermatol **32**:98–103.

36. Duerr, A., P. Paramsothy, D. J. Jamieson, C. M. Heilig, R. S. Klein, S. Cu-Uvin, P. Schuman, and J. R. Anderson. 2006. Effect of HIV infection on atypical squamous cells of undetermined significance. Clin Infect Dis **42**:855–61.

37. Ellerbrock, T. V., M. A. Chiasson, T. J. Bush, X. W. Sun, D. Sawo, K. Brudney, and T. C. Wright, Jr. 2000. Incidence of cervical squamous intraepithelial lesions in HIV-infected women. JAMA **283**:1031–7.

38. Ferenczy, A., F. Coutlee, E. Franco, and C. Hankins. 2003. Human papillomavirus and HIV coinfection and the risk of neoplasias of the lower genital tract: a review of recent developments. CMAJ **169**:431–4.

39. Fontaine, J., C. Hankins, M. H. Mayrand, J. Lefevre, D. Money, S. Gagnon, A. Rachlis, K. Pourreaux, A. Ferenczy, and F. Coutlee. 2005. High levels of HPV-16 DNA are associated with high-grade cervical lesions in women at risk or infected with HIV. AIDS **19**:785–94.

40. Frazer, I. H., G. Medley, R. M. Crapper, T. C. Brown, and I. R. Mackay. 1986. Association between anorectal dysplasia, human papillomavirus, and human immunodeficiency virus infection in homosexual men. Lancet **2**:657–60.

41. Frisch, M., R. J. Biggar, and J. J. Goedert. 2000. Human papillomavirus-associated cancers in patients with human immunodeficiency virus infection and acquired immunodeficiency syndrome. J Natl Cancer Inst **92**:1500–10.

42. Fruchter, R. G., M. Maiman, A. Sedlis, L. Bartley, L. Camilien, and C. D. Arrastia. 1996. Multiple recurrences of cervical intraepithelial neoplasia in women with the human immunodeficiency virus. Obstet Gynecol **87**:338–44.

43. Gagnon, S., C. Hankins, C. Tremblay, K. Pourreaux, P. Forest, F. Rouah, and F. Coutlee. 2005. Polymorphism of human papillomavirus type 31 isolates infecting the genital tract of HIV-seropositive and HIV-seronegative women at risk for HIV infection. J Med Virol **75**:213–21.

44. Gervaz, P., D. Hahnloser, B. G. Wolff, S. A. Anderson, J. Cunningham, R. W. Beart, Jr., A. Klipfel, L. Burgart, and S. N. Thibodeau. 2004. Molecular biology of squamous cell carcinoma of the anus: a comparison of HIV-positive and HIV-negative patients. J Gastrointest Surg **8**:1024–30.

45. Goldie, S. J., K. M. Kuntz, M. C. Weinstein, K. A. Freedberg, M. L. Welton, and J. M. Palefsky. 1999. The clinical effectiveness and cost-effectiveness of screening for anal squamous intraepithelial lesions in homosexual and bisexual HIV-positive men. JAMA **281**:1822–9.

46. Goldie, S. J., M. C. Weinstein, K. M. Kuntz, and K. A. Freedberg. 1999. The costs, clinical benefits, and cost-effectiveness of screening for cervical cancer in HIV-infected women. Ann Intern Med **130**:97–107.

47. Greenspan, D., A. J. Canchola, L. A. MacPhail, B. Cheikh, and J. S. Greenspan. 2001. Effect of highly active antiretroviral therapy on frequency of oral warts. Lancet **357**:1411–2.

48. Greenspan, D., E. M. de Villiers, J. S. Greenspan, Y. G. de Souza, and H. H. zur. 1988. Unusual HPV types in oral warts in association with HIV infection. J Oral Pathol **17**:482–8.

49. Haas, S., T. W. Park, E. Voigt, R. Buttner, and S. Merkelbach-Bruse. 2005. Detection of HPV 52, 58 and 87 in cervicovaginal intraepithelial lesions of HIV infected women. Int J Mol Med **16**:815–9.

50. Hagensee, M. E., L. A. Koutsky, S. K. Lee, T. Grubert, J. Kuypers, N. B. Kiviat, and D. A. Galloway. 2000. Detection of cervical antibodies to human papillomavirus type 16 (HPV-16) capsid antigens in relation to detection of HPV-16 DNA and cervical lesions. J Infect Dis **181**:1234–9.

51. Hankins, C., F. Coutlee, N. Lapointe, P. Simard, T. Tran, J. Samson, and L. Hum. 1999. Prevalence of risk factors associated with human papillomavirus infection in women living with HIV. Canadian Women's HIV Study Group. CMAJ **160**:185–91.

52. Harper, D. M., E. L. Franco, C. M. Wheeler, A. B. Moscicki, B. Romanowski, C. M. Roteli-Martins, D. Jenkins, A. Schuind, S. A. Costa Clemens, and G. Dubin. 2006. Sustained efficacy up to 4.5 years of a bivalent L1 virus-like particle vaccine against human papillomavirus types 16 and 18: follow-up from a randomised control trial. Lancet **367**:1247–55.

53. Harris, T. G., R. D. Burk, J. M. Palefsky, L. S. Massad, J. Y. Bang, K. Anastos, H. Minkoff, C. B. Hall, M. C. Bacon, A. M. Levine, D. H. Watts, M. J. Silverberg, X. Xue, S. L. Melnick, and H. D. Strickler. 2005. Incidence of cervical squamous intraepithelial lesions associated with HIV serostatus, CD4 cell counts, and human papillomavirus test results. JAMA **293**:1471–6.

54. Hawes, S. E., C. W. Critchlow, P. S. Sow, P. Toure, I. N'Doye, A. Diop, J. M. Kuypers, A. A. Kasse, and N. B. Kiviat. 2006. Incident high-grade squamous intraepithelial lesions in Senegalese women with and without human immunodeficiency virus type 1 (HIV-1) and HIV-2. J Natl Cancer Inst **98**:100–9.

55. Heard, I., V. Schmitz, D. Costagliola, G. Orth, and M. D. Kazatchkine. 1998. Early regression of cervical lesions in HIV-seropositive women receiving highly active antiretroviral therapy. AIDS **12**:1459–64.

56. Heard, I., J. M. Tassie, M. D. Kazatchkine, and G. Orth. 2002. Highly active antiretroviral therapy enhances regression of cervical intraepithelial neoplasia in HIV-seropositive women. AIDS **16**:1799–802.

57. Hillemanns, P., T. V. Ellerbrock, S. McPhillips, P. Dole, S. Alperstein, D. Johnson, X. W. Sun, M. A. Chiasson, and T. C. Wright, Jr. 1996. Prevalence of anal human papillomavirus infection and anal cytologic abnormalities in HIV-seropositive women. AIDS **10**:1641–7.

58. Jamieson, D. J., A. Duerr, R. Burk, R. S. Klein, P. Paramsothy, P. Schuman, S. Cu-Uvin, and K. Shah. 2002. Characterization of genital human papillomavirus infection in women who have or who are at risk of having HIV infection. Am J Obstet Gynecol **186**:21–7.

59. King, M. D., D. A. Reznik, C. M. O'Daniels, N. M. Larsen, D. Osterholt, and H. M. Blumberg. 2002. Human papillomavirus-associated oral warts among human immunodeficiency virus-seropositive patients in the era of highly active antiretroviral therapy: an emerging infection. Clin Infect Dis **34**:641–8.

60. Kiviat, N. B., C. W. Critchlow, K. K. Holmes, J. Kuypers, J. Sayer, C. Dunphy, C. Surawicz, P. Kirby, R. Wood, and J. R. Daling. 1993. Association of anal dysplasia and human papillomavirus with immunosuppression and HIV infection among homosexual men. AIDS **7**:43–9.

61. Kobayashi, A., R. M. Greenblatt, K. Anastos, H. Minkoff, L. S. Massad, M. Young, A. M. Levine, T. M. Darragh, V. Weinberg, and K. K. Smith-McCune. 2004. Functional attributes of mucosal immunity in cervical intraepithelial neoplasia and effects of HIV infection. Cancer Res **64**:6766–74.

62. Koutsky, L. 1997. Epidemiology of genital human papillomavirus infection. Am J Med **102**:3–8.

63. Koutsky, L. A., K. A. Ault, C. M. Wheeler, D. R. Brown, E. Barr, F. B. Alvarez, L. M. Chiacchierini, and K. U. Jansen. 2002. A controlled trial of a human papillomavirus type 16 vaccine. N Engl J Med **347**:1645–51.

64. Kreimer, A. R., A. J. Alberg, R. Daniel, P. E. Gravitt, R. Viscidi, E. S. Garrett, K. V. Shah, and M. L. Gillison. 2004. Oral human papillomavirus infection in adults is associated with sexual behavior and HIV serostatus. J Infect Dis **189**:686–98.

65. Kreiss, J. K., N. B. Kiviat, F. A. Plummer, P. L. Roberts, P. Waiyaki, E. Ngugi, and K. K. Holmes. 1992. Human immunodeficiency virus, human papillomavirus, and cervical intraepithelial neoplasia in Nairobi prostitutes. Sex Transm Dis **19**:54–9.

66. Laga, M., J. P. Icenogle, R. Marsella, A. T. Manoka, N. Nzila, R. W. Ryder, S. H. Vermund, W. L. Heyward, A. Nelson, and W. C. Reeves. 1992. Genital papillomavirus infection and cervical dysplasia – opportunistic complications of HIV infection. Int J Cancer **50**:45–8.

67. Lefevre, J., C. Hankins, D. Money, A. Rachlis, K. Pourreaux, and F. Coutlee. 2004. Human papillomavirus type 16 viral load is higher in human immunodeficiency virus-seropositive women with high-grade squamous intraepithelial lesions than in those with normal cytology smears. J Clin Microbiol **42**:2212–5.

68. Leigh, J. 2000. Oral warts rise dramatically with use of new agents in HIV. HIV Clin **12**:7.

69. Levi, J. E., B. Kleter, W. G. Quint, M. C. Fink, C. L. Canto, R. Matsubara, I. Linhares, A. Segurado, B. Vanderborght, J. E. Neto, and L. J. van Doorn. 2002. High prevalence of human papillomavirus (HPV) infections and high frequency of multiple HPV genotypes in human immunodeficiency virus-infected women in Brazil. J Clin Microbiol **40**:3341–5.

70. Lillo, F. B., D. Ferrari, F. Veglia, M. Origoni, M. A. Grasso, S. Lodini, E. Mastrorilli, G. Taccagni, A. Lazzarin, and C. Uberti-Foppa. 2001. Human papillomavirus infection and associated cervical disease in human immunodeficiency virus-infected women: effect of highly active antiretroviral therapy. J Infect Dis **184**:547–51.

71. Lilly, E. A., J. E. Cameron, K. V. Shetty, J. E. Leigh, S. Hager, K. M. McNulty, C. Cheeks, M. E. Hagensee, and P. L. Fidel, Jr. 2005. Lack of evidence for local immune activity in oral hairy leukoplakia and oral wart lesions. Oral Microbiol Immunol **20**:154–62.

72. Lozada-Nur, F., M. Glick, M. Schubert, and I. Silverberg. 2001. Use of intralesional interferon-alpha for the treatment of recalcitrant oral warts in patients with AIDS: a report of 4 cases. Oral Surg Oral Med Oral Pathol Oral Radiol Endod **92**:617–22.

73. Maiman, M., D. H. Watts, J. Andersen, P. Clax, M. Merino, and M. A. Kendall. 1999. Vaginal 5-fluorouracil for high-grade cervical dysplasia in human immunodeficiency virus infection: a randomized trial. Obstet Gynecol **94**:954–61.

74. Mandelblatt, J. S., K. Gold, A. S. O'Malley, K. Taylor, K. Cagney, J. S. Hopkins, and J. Kerner. 1999. Breast and cervix cancer screening among multiethnic women: role of age, health, and source of care. Prev Med **28**:418–25.

75. Marais, D. J., J. M. Best, R. C. Rose, P. Keating, R. Soeters, L. Denny, C. M. Dehaeck, J. Nevin, P. Kay, J. A. Passmore, and A. L. Williamson. 2001. Oral antibodies to human papillomavirus type 16 in women with cervical neoplasia. J Med Virol **65**:149–54.

76. Martin, F., and M. Bower. 2001. Anal intraepithelial neoplasia in HIV positive people. Sex Transm Infect **77**:327–31.

77. Massad, L. S., C. T. Evans, H. D. Strickler, R. D. Burk, D. H. Watts, L. Cashin, T. Darragh, S. Gange, Y. C. Lee, M. Moxley, A. Levine, and D. J. Passaro. 2005. Outcome after negative colposcopy among human

immunodeficiency virus-infected women with borderline cytologic abnormalities. Obstet Gynecol **106**:525–32.

78. Massad, L. S., K. A. Riester, K. M. Anastos, R. G. Fruchter, J. M. Palefsky, R. D. Burk, D. Burns, R. M. Greenblatt, L. I. Muderspach, and P. Miotti. 1999. Prevalence and predictors of squamous cell abnormalities in Papanicolaou smears from women infected with HIV-1. Women's Interagency HIV Study Group. J Acquir Immune Defic Syndr **21**:33–41.

79. Massad, L. S., E. C. Seaberg, D. H. Watts, N. A. Hessol, S. Melnick, P. Bitterman, K. Anastos, S. Silver, A. M. Levine, and H. Minkoff. 2004. Low incidence of invasive cervical cancer among HIV-infected US women in a prevention program. AIDS **18**:109–13.

80. Melbye, M., C. Rabkin, M. Frisch, and R. J. Biggar. 1994. Changing patterns of anal cancer incidence in the United States, 1940–1989. Am J Epidemiol **139**:772–80.

81. Minkoff, H., L. Ahdieh, L. S. Massad, K. Anastos, D. H. Watts, S. Melnick, L. Muderspach, R. Burk, and J. Palefsky. 2001. The effect of highly active antiretroviral therapy on cervical cytologic changes associated with oncogenic HPV among HIV-infected women. AIDS **15**:2157–64.

82. Minkoff, H., J. G. Feldman, H. D. Strickler, D. H. Watts, M. C. Bacon, A. Levine, J. M. Palefsky, R. Burk, M. H. Cohen, and K. Anastos. 2004. Relationship between smoking and human papillomavirus infections in HIV-infected and -uninfected women. J Infect Dis **189**:1821–8.

83. Moscicki, A. B., J. H. Ellenberg, P. Crowley-Nowick, T. M. Darragh, J. Xu, and S. Fahrat. 2004. Risk of high-grade squamous intraepithelial lesion in HIV-infected adolescents. J Infect Dis **190**:1413–21.

84. Munoz, N., and F. X. Bosch. 1989. Epidemiology of cervical cancer. IARC Sci Publ 9–39.

85. Nicol, A. F., A. T. Fernandes, B. Grinsztejn, F. Russomano, J. R. Lapa e Silva, A. Tristao, M. A. Perez, G. J. Nuovo, O. Martinez-Maza, and M. G. Bonecini-Almeida. 2005. Distribution of immune cell subsets and cytokine-producing cells in the uterine cervix of human papillomavirus (HPV)-infected women: influence of HIV-1 coinfection. Diagn Mol Pathol **14**:39–47.

86. Olaitan, A., M. A. Johnson, A. MacLean, and L. W. Poulter. 1996. The distribution of immunocompetent cells in the genital tract of HIV-positive women AIDS **10**:759–64.

87. Palefsky, J. M. 1998. Human papillomavirus infection and anogenital neoplasia in human immunodeficiency virus-positive men and women. J Natl Cancer Inst Monogr15–20.

88. Palefsky, J. M. 1999. Anal squamous intraepithelial lesions: relation to HIV and human papillomavirus infection. J Acquir Immune Defic Syndr **21**(Suppl 1):S42–8.

89. Palefsky, J. M., E. A. Holly, J. T. Efirdc, M. Da Costa, N. Jay, J. M. Berry, and T. M. Darragh. 2005. Anal intraepithelial neoplasia in the highly active antiretroviral therapy era among HIV-positive men who have sex with men. AIDS **19**:1407–14.

90. Palefsky, J. M., E. A. Holly, M. L. Ralston, M. Da Costa, and R. M. Greenblatt. 2001. Prevalence and risk factors for anal human papillomavirus infection in human immunodeficiency virus (HIV)-positive and high-risk HIV-negative women. J Infect Dis **183**:383–91.

91. Palefsky, J. M., H. Minkoff, L. A. Kalish, A. Levine, H. S. Sacks, P. Garcia, M. Young, S. Melnick, P. Miotti, and R. Burk. 1999. Cervicovaginal human papillomavirus infection in human immunodeficiency virus-1 (HIV)-positive and high-risk HIV-negative women. J Natl Cancer Inst **91**:226–36.

92. Pereira, D. B., M. H. Antoni, A. Danielson, T. Simon, J. Efantis-Potter, C. S. Carver, R. E. Duran, G. Ironson, N. Klimas, and M. J. O'Sullivan. 2003. Life stress and cervical squamous intraepithelial lesions in women with human papillomavirus and human immunodeficiency virus. Psychosom Med **65**:427–34.

93. Petignat, P., C. Hankins, S. Walmsley, D. Money, D. Provencher, K. Pourreaux, J. Kornegay, F. Rouah, and F. Coutlee. 2005. Self-sampling is associated with increased detection of human papillomavirus DNA in the genital tract of HIV-seropositive women. Clin Infect Dis **41**:527–34.

94. Piattelli, A., C. Rubini, M. Fioroni, and T. Iezzi. 2001. Warty carcinoma of the oral mucosa in an HIV+ patient. Oral Oncol **37**:665–7.

95. Piketty, C., T. M. Darragh, M. Da Costa, P. Bruneval, I. Heard, M. D. Kazatchkine, and J. M. Palefsky. 2003. High prevalence of anal human papillomavirus infection and anal cancer precursors among HIV-infected persons in the absence of anal intercourse. Ann Intern Med **138**:453–9.

96. Piketty, C., T. M. Darragh, I. Heard, M. Da Costa, P. Bruneval, M. D. Kazatchkine, and J. M. Palefsky. 2004. High prevalence of anal squamous intraepithelial lesions in HIV-positive men despite the use of highly active antiretroviral therapy. Sex Transm Dis **31**:96–9.

97. Piper, M. A., S. T. Severin, S. Z. Wiktor, E. R. Unger, P. D. Ghys, D. L. Miller, I. R. Horowitz, A. E. Greenberg, W. C. Reeves, and S. D. Vernon. 1999. Association of human papillomavirus with HIV and CD4 cell count in women with high or low numbers of sex partners. Sex Transm Infect **75**:253–7.

98. Robinson, W. R., J. Andersen, T. M. Darragh, M. A. Kendall, R. Clark, and M. Maiman. 2002. Isotretinoin for low-grade cervical dysplasia in human immunodeficiency virus-infected women. Obstet Gynecol **99**:777–84.

99. Robinson, W. R., C. A. Hamilton, S. H. Michaels, and P. Kissinger. 2001. Effect of excisional therapy and highly active antiretroviral therapy on cervical intraepithelial neoplasia in women infected with human immunodeficiency virus. Am J Obstet Gynecol **184**:538–43.

100. Schiffman, M., and S. K. Kjaer. 2003. Chapter 2: Natural history of anogenital human papillomavirus infection and neoplasia. J Natl Cancer Inst Monogr14–9.

101. Schlecht, N. F., R. D. Burk, J. M. Palefsky, H. Minkoff, X. Xue, L. S. Massad, M. Bacon, A. M. Levine, K. Anastos, S. J. Gange, D. H. Watts, M. M. Da Costa, Z. Chen, J. Y. Bang, M. Fazzari, C. Hall, and H. D. Strickler. 2005. Variants of human papillomavirus 16 and 18 and their natural history in human immunodeficiency virus-positive women. J Gen Virol **86**:2709–20.

102. Schuman, P., S. E. Ohmit, R. S. Klein, A. Duerr, S. Cu-Uvin, D. J. Jamieson, J. Anderson, and K. V. Shah. 2003. Longitudinal study of cervical squamous intraepithelial lesions in human immunodeficiency virus (HIV)-seropositive and at-risk HIV-seronegative women. J Infect Dis **188**:128–36.

103. Schwartz, S. R., B. Yueh, J. K. McDougall, J. R. Daling, and S. M. Schwartz. 2001. Human papillomavirus infection and survival in oral squamous cell cancer: a population-based study. Otolaryngol Head Neck Surg **125**:1–9.

104. Silverberg, M. J., L. Ahdieh, A. Munoz, K. Anastos, R. D. Burk, S. Cu-Uvin, A. Duerr, R. M. Greenblatt, R. S. Klein, S. Massad, H. Minkoff, L. Muderspach, J. Palefsky, E. Piessens, P. Schuman, H. Watts, and K. V. Shah. 2002. The impact of HIV infection and immunodeficiency on human papillomavirus type 6 or 11 infection and on genital warts. Sex Transm Dis **29**:427–35.

105. Six, C., I. Heard, C. Bergeron, G. Orth, J. D. Poveda, P. Zagury, P. Cesbron, C. Crenn-Hebert, R. Pradinaud, M. Sobesky, C. Marty, M. L. Babut, J. E. Malkin, A. Odier, S. Fridmann, J. P. Aubert, J. B. Brunet, and I. de Vincenzi. 1998. Comparative prevalence, incidence and short-term prognosis of cervical squamous intraepithelial lesions amongst HIV-positive and HIV-negative women. AIDS **12**:1047–56.

106. Snowhite, I. V., W. E. Jones, J. Dumestre, K. Dunlap, P. S. Braly, and M. E. Hagensee. 2002. Comparative analysis of methods for collection and measurement of cytokines and immunoglobulins in cervical and vaginal secretions of HIV and HPV infected women. J Immunol Methods **263**:85–95.

107. Solomon, D., D. Davey, R. Kurman, A. Moriarty, D. O'Connor, M. Prey, S. Raab, M. Sherman, D. Wilbur, T. Wright, Jr., and N. Young. 2002. The 2001 Bethesda System: terminology for reporting results of cervical cytology. JAMA **287**:2114–9.

108. Strickler, H. D., R. D. Burk, M. Fazzari, K. Anastos, H. Minkoff, L. S. Massad, C. Hall, M. Bacon, A. M. Levine, D. H. Watts, M. J. Silverberg, X. Xue, N. F. Schlecht, S. Melnick, and J. M. Palefsky. 2005. Natural history and possible reactivation of human papillomavirus in human immunodeficiency virus-positive women. J Natl Cancer Inst **97**:577–86.

109. Sun, X. W., T. V. Ellerbrock, O. Lungu, M. A. Chiasson, T. J. Bush, and T. C. Wright, Jr. 1995. Human papillomavirus infection in human immunodeficiency virus-seropositive women. Obstet Gynecol **85**:680–6.

110. Swan, D. C., R. A. Tucker, G. Tortolero-Luna, M. F. Mitchell, L. Wideroff, E. R. Unger, R. A. Nisenbaum, W. C. Reeves, and J. P. Icenogle. 1999. Human papillomavirus (HPV) DNA copy number is dependent on grade of cervical disease and HPV type. J Clin Microbiol **37**:1030–4.

111. Theodore, C., N. Androulakis, A. Spatz, C. Goujard, P. Blanchet, and P. Wibault. 2002. An explosive course of squamous cell penile cancer in an AIDS patient. Ann Oncol **13**:475–9.

112. Tornesello, M. L., F. M. Buonaguro, E. Beth-Giraldo, and G. Giraldo. 1993. Human immunodeficiency virus type 1 tat gene enhances human papillomavirus early gene expression. Intervirology **36**:57–64.

113. Uberti-Foppa, C., M. Origoni, M. Maillard, D. Ferrari, D. Ciuffreda, E. Mastrorilli, A. Lazzarin, and F. Lillo. 1998. Evaluation of the detection of human papillomavirus genotypes in cervical specimens by hybrid capture as screening for precancerous lesions in HIV-positive women. J Med Virol **56**:133–7.

114. Unger, E. R., S. D. Vernon, D. R. Lee, D. L. Miller, and W. C. Reeves. 1998. Detection of human papillomavirus in archival tissues. Comparison of in situ hybridization and polymerase chain reaction. J Histochem Cytochem **46**:535–40.

115. Vernon, S. D., W. C. Reeves, K. A. Clancy, M. Laga, M. St Louis, H. E. Gary, Jr., R. W. Ryder, A. T. Manoka, and J. P. Icenogle. 1994. A longitudinal study of human papillomavirus DNA detection in human immunodeficiency virus type 1-seropositive and -seronegative women. J Infect Dis **169**:1108–12.

116. Vernon, S. D., E. R. Unger, and D. Williams. 2000. Comparison of human papillomavirus detection and typing by cycle sequencing, line blotting, and hybrid capture. J Clin Microbiol **38**:651–5.

117. Viscidi, R. P., L. Ahdieh-Grant, B. Clayman, K. Fox, L. S. Massad, S. Cu-Uvin, K. V. Shah, K. M. Anastos, K. E. Squires, A. Duerr, D. J. Jamieson, R. D. Burk, R. S. Klein, H. Minkoff, J. Palefsky, H. Strickler, P. Schuman, E. Piessens, and P. Miotti. 2003. Serum immunoglobulin G response to human papillomavirus type 16 virus-like particles in human immunodeficiency virus (HIV)-positive and risk-matched HIV-negative women. J Infect Dis **187**:194–205.

118. Waddell, K. M., S. Lewallen, S. B. Lucas, C. Atenyi-Agaba, C. S. Herrington, and G. Liomba. 1996. Carcinoma of the conjunctiva and HIV infection in Uganda and Malawi. Br J Ophthalmol **80**:503–8.
119. Walboomers, J. M., M.V. Jacobs, M. M. Manos, F. X. Bosch, J. A. Kummer, K.V. Shah, P. J. Snijders, J. Peto, C. J. Meijer, and N. Munoz. 1999. Human papillomavirus is a necessary cause of invasive cervical cancer worldwide. J Pathol **189**:12–9.
120. Weissenborn, S. J., A. M. Funke, M. Hellmich, P. Mallmann, P. G. Fuchs, H. J. Pfister, and U. Wieland. 2003. Oncogenic human papillomavirus DNA loads in human immunodeficiency virus-positive women with high-grade cervical lesions are strongly elevated. J Clin Microbiol **41**:2763–7.
121. Wilkin, T. J., S. Palmer, K. F. Brudney, M. A. Chiasson, and T. C. Wright. 2004. Anal intraepithelial neoplasia in heterosexual and homosexual HIV-positive men with access to antiretroviral therapy. J Infect Dis **190**:1685–91.
122. Womack, S. D., Z. M. Chirenje, L. Gaffikin, P. D. Blumenthal, J. A. McGrath, T. Chipato, S. Ngwalle, M. Munjoma, and K.V. Shah. 2000. HPV-based cervical cancer screening in a population at high risk for HIV infection. Int J Cancer **85**:206–10.
123. Wright, T. C., Jr., J. T. Cox, L. S. Massad, L. B. Twiggs, and E. J. Wilkinson. 2002. 2001 Consensus Guidelines for the management of women with cervical cytological abnormalities. JAMA **287**:2120–9.
124. zur, H. H., and E. M. de Villiers. 1994. Human papillomaviruses. Annu Rev Microbiol **48**:427–47.

8. POLYOMAVIRUS SV40 AND AIDS-RELATED SYSTEMIC NON-HODGKIN'S LYMPHOMA

REGIS A. VILCHEZ AND JANET S. BUTEL

Department of Molecular Virology and Microbiology and Baylor-UTHouston Center for AIDS Research, Baylor College of Medicine, Houston, TX

Current affiliation: Schering Plough Research Institute, Kenilworth, New Jersey, USA

Department of Molecular Virology and Microbiology and Baylor-UTHouston Center for AIDS Research, Baylor College of Medicine, Houston, TX

INTRODUCTION

Immunosuppression associated with HIV/AIDS substantially increases the risk of developing non-Hodgkin's lymphoma (NHL).[1–7] This AIDS-related malignancy can be divided into two distinct groups, systemic and primary central nervous system NHL.[2,4,7] (Table 1). A meta-analysis of the data from different cohorts in the United States and Europe has shown a significant reduction in primary central nervous system NHL, compared to a small decrease in systemic NHL between the pre- and highly active antiretroviral (HAART) eras.[8]

HAART is recognized to have improved the survival of HIV-infected patients. However, its present efficacy may allow patients to survive long term with mild to moderate immunosuppression, placing them at risk for the development of systemic NHL.[9–13] Indeed, the estimated incidence of AIDS-related systemic NHL is between 6.5 and 17.0 per 1,000 patients in the HAART era,[10,14] and data suggest that some patients have developed lymphomas despite effective HIV suppression and relatively high CD4$^+$ T-cell counts.[10,15–19] Furthermore, outcomes of systemic NHL still remain inferior compared to those achieved in HIV-negative individuals.[20] These observations have led to the hypothesis of different tumor biology of lymphomas among HIV-infected as compared to HIV-uninfected patients. Therefore, it is important to develop a better understanding of the pathogenesis and biology of this opportunistic complication of HIV/AIDS.

Table 1. Features of systemic and primary central nervous system non-Hodgkin's lymphoma in patients with HIV infection during the HAART era[a]

Characteristics	Systemic NHL	Primary CNS NHL	Other
Histology	Diffuse large B cell and Burkitt's	Diffuse large B cell	–
Median CD4 cell count	~200/mm^3	<50/mm^3	–
Epstein-Barr virus	~40%	~100%	–
Simian virus 40[b]	13–46%	ND	–
Human herpesvirus 8	<5%	0	Primary effusion lymphoma (100%) Castleman's disease (100%)

[a]Data summarized from references 10,14–21,23–26,32,37. Abbreviations used: NHL, non-Hodgkin's lymphoma; CNS, central nervous system; HAART, highly active antiretroviral therapy; ND, not determined.
[b]Data from studies reporting the HIV status of patients. Many studies do not provide this information.

Viral etiologies have been proposed for some HIV/AIDS-related NHL.[2] Some systemic NHL have been associated with deficient immune surveillance of Epstein-Barr virus (EBV) and human herpesvirus 8 (HHV8) or related to the chronic antigenic stimulation and defective immune regulation observed in patients with HIV/AIDS.[1,3,4,6] In addition, a number of studies have detected polyomavirus simian virus 40 (SV40) associated with some systemic B-cell lymphomas among HIV-infected and –uninfected patients.[21–33] These and other findings led the Institute of Medicine of the National Academies to recognize that SV40 is an emergent human pathogen and conclude that the present biological evidence indicates that infections with this potent DNA tumor virus could lead to cancer in humans under natural conditions.[34]

Molecular and clinical evidence suggest that patients with iatrogenic and/or acquired immunosuppression are a population at risk for SV40 infections and perhaps SV40-related disease.[21,23,24,26,32,35–37] Importantly, experimental data have shown that alterations in cellular immunity, such as those occurring in patients with HIV/AIDS, may foster conditions favorable for SV40 oncogenesis in humans.[38–40] This chapter examines the features of SV40 as an opportunistic human pathogen and its potential role in AIDS-related systemic NHL. Unresolved questions about SV40 and human cancer are considered.

SV40 AS A POTENTIAL HUMAN PATHOGEN

Nonhuman primates are hosts to many viruses, some of which are known to infect and cause disease in humans.[41] Polyomavirus SV40 as a potential human pathogen has been tied to the development and distribution of early forms of the poliovirus and adenovirus vaccines.[34,42–45] Early lots of both inactivated and live attenuated forms of poliovaccines were contaminated with SV40.[34,44,46] Different adenovirus vaccines distributed to the United States military also contained SV40, but the numbers of subjects exposed to those vaccines were less than those given polio-vaccines.[47] Contamination of the early vaccine formulations occurred because they were prepared in primary cultures of kidney cells from rhesus monkeys, which are frequently infected with SV40.[34,43,44] It is well known that infectious SV40 survived the vaccine inactivation treatments, and some estimates have suggested that up to 30 million people in the United States alone were exposed to live SV40 when administered potentially contaminated killed (Salk) poliovaccines in the late 1950s and early 1960s.[34,44] Millions more were exposed to SV40 worldwide because contaminated poliovaccines of both killed and live, attenuated types were used in many countries.[42,48] Recent data suggest that exposure to SV40-contaminated oral poliovaccines in some countries may have occurred as late as 1978 because the vaccine production method used in the USSR did not completely inactivate the SV40 virus.[49]

SV40 is a potent oncogenic DNA virus.[1,43] In animal models, the profile of cancers induced by SV40 includes brain and bone cancers, malignant mesotheliomas and systemic lymphomas.[43] Numerous studies have shown that the induction of tumors by SV40 is the result of the disruption of different cell cycle control checkpoints.[1,50,51] Over the last decade, studies from different laboratories, using an array of molecular techniques, have demonstrated SV40 markers in primary human brain and bone cancers and malignant mesothelioma.[42,43,45,52,53] More recent studies have indicated that SV40 is associated with systemic NHL among HIV-positive and -negative individuals.[23–33] A meta-analysis of the molecular evidence established a significant excess risk of SV40 with those selected human cancers.[54] It is of interest that the tumors induced by SV40 in laboratory animals are the same as those selected human malignancies now associated with the virus.

SV40 has been detected in different specimens from children and adults not exposed to known contaminated poliovaccines.[24,55–67] This detection of SV40 markers in young persons, as well as the presence of infectious SV40 in clinical samples,[35,68,69] suggests that SV40 causes infections in humans today. However, epidemiological studies have found no evidence of increased cancer risk in populations which had a high likelihood of having received potentially contaminated poliovaccine about 50 years ago.[44,70–75] On this topic, the Institute of Medicine found that epidemiological studies of SV40 were inadequate to evaluate a causal relationship.[34] The discrepancy between the molecular studies linking SV40 with human disease and the epidemiological studies failing to find such an association is the basis of current debate.

BIOLOGY OF SV40

Virus Properties

SV40 is a member of the family *Polyomaviridae*, which includes JC virus (JCV) and BK virus (BKV). These pathogens are small, nonenveloped, icosahedral DNA viruses. JCV and BKV share 72% DNA sequence homology and each shares approximately 70% homology with SV40. Although these viruses have significant sequence homology, clear distinctions can be made at the DNA and protein levels.

The three viruses also can be differentiated serologically by neutralization and hemagglutination inhibition assays.[76–78]

The SV40 genome consists of two coding regions: the so-called early region which codes for the large and small tumor antigens and the late region which encodes the capsid proteins VP1, VP2, and VP3 (Fig. 1). SV40 large tumor antigen (T-ag) is an essential protein required for initiation of viral replication and for viral effects on host cell growth regulation (Fig. 2). T-ag forms complexes with several cellular proteins, including tumor suppressor proteins p53 and pRb.[1,43,50,51] These interactions between virus and cell proteins in infected cells underlie the tumorigenic potential of SV40. However, it is important to recognize that the oncogenic capacity of the virus is the result of the viral replication strategy to stimulate host cells into a state capable of supporting viral replication, rather than oncogenesis being an integral part of the virus life cycle.

Most of the SV40 T-ag sequence is conserved among different viral isolates, but a variable domain at the extreme C-terminus of T-ag (T-ag-C), defined as approximately the last 86 amino acids of the molecule (residues 622–708), has been identified.[79,80] SV40 strains can be distinguished by nucleotide and amino acid differences in this region.[80,81] Whereas the function of the SV40 T-ag variable domain is not known, data from in vitro studies suggest that it is important in viral interactions with the host. Embedded within this T-ag region is a functional domain (amino acids 682–708), defined as the host range/adenovirus (Ad) helper function (hr/hf) (Fig. 2). A fragment of T-ag-C relieves the human Ad replication block in monkey cells[82–85] by an unknown mechanism (hf function). The hr function was identified because T-ag-C deletion mutants exhibited different growth properties in two monkey kidney cell lines.[86–90] Virions produced by the hr/hf mutants failed to assemble properly, due to an inability to add VP1 to the 75S assembly intermediates.[91] Recent studies with new mutant viruses that express truncated SV40 T-ag proteins suggested that the T-ag-C is active in *trans* and is essential for the viral life cycle.[92] However, cellular proteins that interact with the T-ag-C and participate in hr/hf activities remain to be identified.

The functional roles for SV40 small tumor antigen (t-ag) are more elusive. This protein is not essential for virus replication or for transformation or tumor induction. However, studies have shown that SV40 t-ag enhances T-ag-mediated transformation in different cells[93] and plays a role in the induction of telomerase in SV40-infected human mesothelial cells.[94] SV40 t-ag modifies cellular protein phosphatase 2A (PP2A) by complexing with the catalytic subunit and displacing a regulatory subunit of the enzyme.[95]

The origin (*Ori*) of SV40 DNA replication is embedded within the noncoding regulatory region that contains elements controlling transcription and replication. Isolates of SV40 are known to vary in the structure of their regulatory region.[43,80,96,97] Those variants containing two 72-base-pair (bp) enhancer elements, sequence rearrangements, and/or deletions are designated as having complex regulatory regions and those with one enhancer and no rearrangement as having

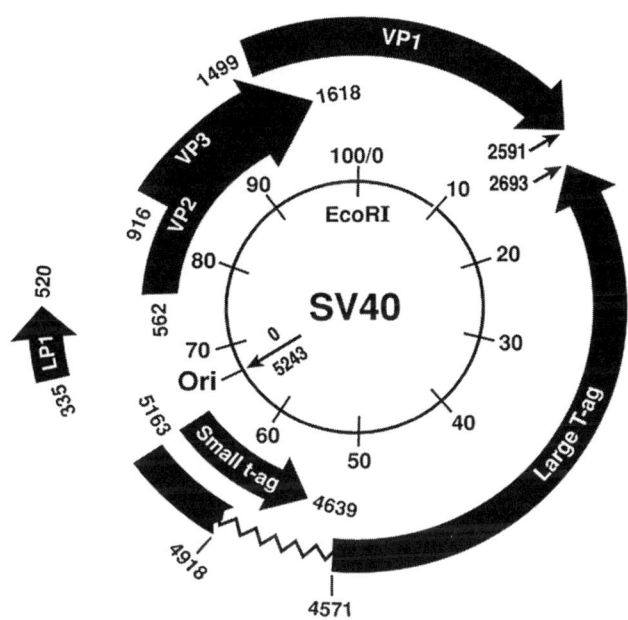

Figure 1. Genetic map of SV40. The circular SV40 DNA genome is represented with the unique EcoRI site shown at map unit 100/0. Nucleotide numbers based on reference strain SV40-776 begin and end at the origin (*Ori*) of viral DNA replication (0/5243). The open reading frames that encode viral proteins are indicated. The beginning and end of each open reading frame are indicated by nucleotide numbers. From Butel and Lednicky.[43]

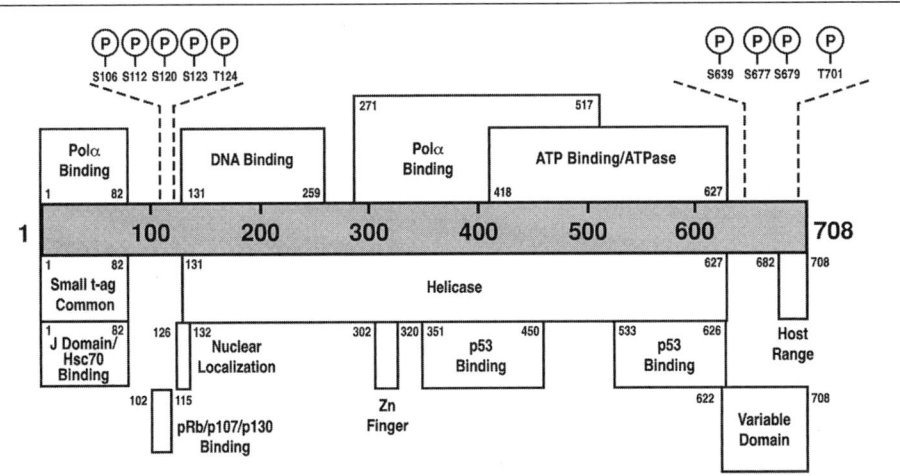

Figure 2. Functional domains of SV40 large T-ag. The numbers given are the amino acid residues. The "variable domain" at the T-ag-C contains amino acid differences among SV40 natural strains and is used for strain identification. From Stewart et al.[79]

simple regulatory region structures.[80,97] Enhancer elements in the regulatory region of SV40 influence the replication of the virus in vitro.[97,98]

Virus Replication

The replication cycle of SV40 provides a basis for considering the potential etiologic role of SV40 in some human malignancies. The major histocompatibility (MHC) class I molecules are the specific cell surface receptors for SV40.[99,100] This initial step in the viral cycle helps explain the broad tropism of SV40 and its ability to infect and induce transformation in many types of cells. The identity of virus receptors provides an important distinction among SV40, JCV, and BKV. JCV uses an N-linked glycoprotein and BKV a glycolipid as their unique host cell receptors.[101] These receptor differences are believed to impact the nature of human infections by these three viruses.

After infection of a cell, SV40 produces both T-ag and t-ag which in turn bind and block the functions of tumor suppressor proteins p53, pRb, p107 and p130/Rb2[43,45,51] (Fig. 2), and PP2A, respectively. These cellular proteins normally function to regulate the cell cycle.[50,51] SV40 T-ag binding sequesters p53, abolishing its function, and allowing cells with genetic damage to survive and enter S phase, leading to an accumulation of genomic mutations that promote genetic instability and tumor growth. pRb binds transcription factor E2F in early G1 of the cell cycle; SV40 T-ag causes dissociation of pRb:E2F complexes, releasing E2F to activate expression of growth-stimulatory genes.[50,51] Studies have shown that different types of human cells can support SV40 replication.[102–104] Importantly, some human cells undergo visible cell lysis in response to SV40 infection, whereas others fail to exhibit cytopathic changes and produce comparatively low levels of virus.[53] Other types of cells may produce infectious virus.

SV40 has been shown to immortalize human primary B lymphocytes in the absence of EBV nuclear antigens[105] and those SV40-immortalized cells expressed the B-cell activation antigen CD23, as well as other B-cell markers. We are evaluating the potential for SV40 to infect human lymphocytes, using a human B-cell line, BJAB.[106] As BJAB cells express the SV40 receptor MHC class I, it was anticipated that the cells would be susceptible to SV40 infection. Following infection with SV40, T-ag expression was measured using intracellular staining and flow cytometry. SV40 T-ag expression peaked in the BJAB cell cultures between days 10 and 14. In addition, recent reports demonstrated that SV40-positive human lymphoblastoid B-cell lines are more tumorigenic in SCID mice than are SV40-negative B-cell lines.[107] Conclusions from these studies are that SV40 can infect and replicate in human B lymphocytes and that various human cell types display differences in susceptibility to infection by SV40.[53,102,103] While the basis for the differences is unknown, SV40 T-ag functions are believed to be important determinants of those distinctions.[108,109] Together, these data highlight that SV40 infections in humans would have the biologic potential to interfere with critical cell regulatory pathways which can lead eventually to the development of cancer.

Oncogenesis in Animal Models

The strong oncogenic capacity of SV40 has been demonstrated in laboratory animal models.[1,34,43,110] The Syrian golden hamster (*Mesocritetus auratus*) is the only small animal model highly susceptible to SV40 infection and carcinogenesis.[43,111–114] The types of malignancies induced by SV40 in the hamster model include brain and bone cancers, malignant mesotheliomas, and lymphomas.[42,43,53] Systemic lymphomas developed among many animals inoculated intravenously,[112,113,115] with the malignancies being of B-cell origin.[116] Animal experiments suggest that host age at the time of infection, virus dose, the route of infection, and the duration of infection are determinants for the development of cancers by SV40. We reported recently that virus strain distinction may represent another factor that influences SV40 infection and oncogenesis.[117]

The classic example of DNA virus strains differing in oncogenic capacity is the human papillomavirus group; of the more than 100 types described, and of which about 30 cause genital infections, only a few types are associated with the development of cervical carcinoma.[118,119] Different natural strains of SV40 have been identified that can be distinguished by nucleotide differences in the regulatory[97] and T-ag-C regions.[80,81,96,120–122] A phylogenetic analysis of SV40 isolates from different sources was conducted to evaluate strain relatedness and the extent of genetic variation.[81] Maximum-parsimony and distance methods revealed distinct SV40 clades (Fig. 3). One clade (clade A) contains strain 776, the historical reference strain of SV40 isolated from the adenovirus vaccine. Clade B contains strains 777 and Baylor, recovered from poliovaccines, and strain SVCPC from human malignancies. The T-ag-C region contains the highest proportion of variable sites in the virus genome and a phylogenetic examination of additional SV40 strains based on just this region indicated that further genetic diversity is likely. An additional clade (clade C) contains strain SVPML-1 (isolated from human brain) and strains associated with some human cancers (Fig. 4). Analysis of additional monkey and human isolates of SV40 is necessary to determine if clade C contains only human-associated SV40 strains.

These findings have provided the rationale for asking whether viruses from different SV40 genogroups vary in biological properties during infection. We have addressed this question in long-term experiments with Syrian golden hamsters in which we compared strains VA45-54, isolated originally from monkey kidney cells, and SVCPC, identified and recovered from human cancers, including AIDS-related systemic NHL.[117] These two strains have nucleotide and amino acid differences at several positions in the T-ag-C, in the small t-ag unique region, and differ in the structure of their regulatory region; they do not belong to the same SV40 genogroup (Figs. 3 and 4).[81] The rate of tumor development was more frequent among animals infected with strain SVCPC than with VA45-54, in experiments of both 8 months (11/22, 50% vs. 4/20, 20%) and 12 months (7/15, 47% vs. 3/13, 23%) duration.

SV40 antibody titers were usually higher in animals with tumors than in those without, but not all animals exposed to large doses of the virus developed detectable levels of long-lasting antibodies.[117] These findings are compatible with

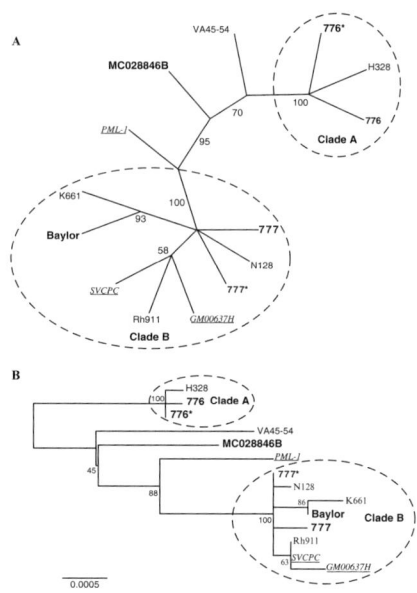

Figure 3. Phylogenetic tree for SV40 based on complete genome sequences of isolates. (A) Unrooted consensus tree of 1,000 bootstrap replicates of the whole-genome data, generated by maximum parsimony. Conventions for labels are as follows: monkey isolates = roman type; vaccine isolates = bold-face type; human isolates = underscored italic type. (B) Consensus tree of 1,000 bootstrap replicates of the whole-genome data, generated by the neighbor-joining method. Distances are proportional to the number of mutational changes, after the Kimura[203] correction has been applied; the scale is in proportion of changes relative to the whole genome. The scale bar (0.0005) represents substitutions per site. Numbers proximal to nodes in tree indicate the proportion of 1,000 bootstrap replications; only bootstrap values above 50% are shown. Two clades (A and B) are identified, with several outlier virus isolates. From Forsman et al.[81]

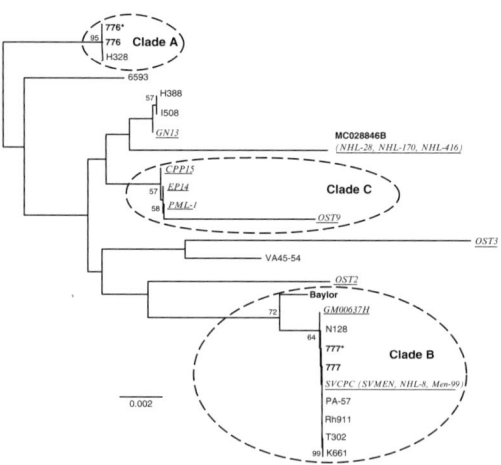

Figure 4. Phylogenetic tree for SV40 based on all available T-ag-C sequences. Consensus tree of 1,000 bootstrap replicates, generated by the neighbor-joining method. Distances are proportional to the number of mutational changes, after the Kimura correction has been applied; the scale is in proportion of changes relative to the T-ag-C gene. Only bootstrap values >50% are shown. Sequences in parentheses are from independent sources, but are the same as the sequence preceding the parentheses. Sequences NHL-8 (SVCPC) and NHL-28 (MC028846B) were from AIDS-related lymphomas. From Forsman et al.[81]

the idea that there may be some differences in virus–host interactions among SV40 strains. One possible factor to be considered is that a faster-replicating virus, such as VA45-54, may be cleared by the host's antiviral immune response more efficiently than a slower-replicating strain, such as SVCPC. Data from the lymphocytic choriomeningitis virus (LCMV) model supports this hypothesis as the speed of replication by LCMV strains, through effects on the host cellular immune response, influences virus persistence within the host.[123] Indeed, a slower-replicating LCMV strain was more apt to evade immune surveillance and persist. Studies are necessary to determine the dynamics of viral infection and disease outcomes displayed by different SV40 genotypes. In addition, studies involving the use of recombinant viruses will be needed to identify the genomic region(s) that contain determinants of tumorigenicity.

Studies in mice on immunity against SV40-transformed cells have documented that both CD4+ and CD8+ cells are important in establishing and maintaining a protective immune response.[38–40] Although CD4+ T cells do not contribute to tumor eradication, they provide help to CD8+ T cells to proliferate and respond against SV40. In addition, recent experimental data suggest that SV40-encoded microRNAs downregulate expression of T-ag, resulting in reduced cytokine production by cytotoxic T cells, and perhaps an increased probability of successful infection due to immune evasion.[124] Thus, the loss of CD4+ T cells in patients with HIV/AIDS may enhance conditions favorable for SV40 infection and disease development.

Virus Transmission

SV40, like polyomaviruses JCV and BKV, establishes persistent infections, often in the kidneys of susceptible hosts.[43] Primary polyomavirus infections occur early in life, and an association with mild respiratory tract disease, mild pyrexia and transient cystitis has been reported for BKV.[125] Symptoms of primary infections by SV40 are not known. Recent data indicate that polyomaviruses BKV and SV40 are frequently present in stool samples from children.[67] SV40 has also been detected on occasion in urine.[35,36,55,68] These findings indicate that fecal-oral and/or urine-oral routes of transmission may occur, although the major route of polyomavirus infection has not been defined.

The mode of transmission of SV40 among humans is unknown and has not yet been directly demonstrated, but we should consider that different routes may be involved. Studies in laboratory animals have indicated that maternal-infant transmission is one possible route of spread.[126] This might represent a pathway for SV40 infections in humans, albeit probably an infrequent one, as there are reports of the detection of DNA sequences and expression of SV40 T-ag in cases of primary brain cancers in infants and young children.[58–62,127] Another possibility is that of zoonotic transmission of SV40 in certain human populations. Laboratory and zoo workers in contact with SV40-infected monkeys and/or tissues from those animals had a prevalence of antibodies to SV40 in the range of 41–55%, suggesting an increased risk for viral infection among this group of workers.[128–130] Another potential condition favorable for transmission of SV40 to humans is the commensal relationship of humans and rhesus monkeys in some communities in Southeast Asia. Indeed, a recent survey of rhesus monkeys indicated that a significant number of animals living in close proximity to people are infected with SV40.[131]

The natural hosts of SV40 are rhesus monkeys (*Macacca mulatta*). SV40 primary infections in these animals may become latent, and the level of virus may be very low in the immunocompetent host. SV40 shedding in the urine and/or feces is a likely means of transmission among monkeys.[46,132,133] Whereas SV40 infections among immunocompetent monkeys appear to be asymptomatic,[134] the virus causes widespread infections among monkeys with simian immunodeficiency virus (SIV);[96,120,121] SV40 has been detected and isolated from multiple organs in animals with SIV/AIDS. These results demonstrate that SV40 can be an opportunistic pathogen in immunosuppressed hosts.

SV40 INFECTIONS IN HUMANS

The prevalence of SV40 infections in humans has not been established, but a battery of evidence indicates that SV40 infections are occurring in different populations today. These include subjects potentially exposed to SV40-contaminated vaccines, as well as individuals not exposed to those vaccines.[24,26,31–36,43–45,52,55–69,80,122,127,135–151] Studies of adult patients with renal disease and of recipients of kidney transplants revealed that CV-1 cells cocultured with urinary cells or peripheral blood mononuclear cells (PBMCs) from those patients developed SV40 cytopathic effects.[35,68] SV40 sequences were detected by polymerase chain reaction (PCR) in kidney biopsies from 56% of patients with focal segmental glomerulosclerosis and SV40 DNA was localized to renal tubular epithelial cell nuclei by in situ hybridization in renal biopsies of those patients. SV40 regulatory region DNA sequences were detected and identified in the allografts of immunocompromised pediatric renal transplant recipients.[56,138] Recently, SV40 markers were detected in the urine and native kidney of a 36-year-old lung transplant patient with polyomavirus nephropathy.[36] In addition, different studies have detected SV40 DNA sequences in PBMCs from various patient populations, including HIV-positive individuals.[32,63,66,68,127,137] These results point out that patients with acquired and/or iatrogenic immunosuppression are at risk for SV40 infections and that the kidney can serve as a reservoir for the virus in humans. However, the natural history and dynamics of SV40 infections in the growing patient population with compromised immunity are not clear.

There are several reasons why determining the prevalence of SV40 in human populations is difficult. The natural history of human infections is not well understood, so the types of specimens to screen by molecular techniques in order to reliably reveal SV40 infection status are not known. Serologic studies in the United States and other countries generally have estimated an SV40 prevalence between 2 and 20%,[43,44,135] but the limitations of the assays used make it difficult to establish the actual prevalence of human infections. A study conducted in the United States showed the presence of SV40 neutralizing antibodies in 16 and 11% of adult HIV-infected and -uninfected individuals, respectively.[139] SV40 neutralizing antibodies were detected in 6% of hospitalized children, with seropositivity increasing with age and significantly associated with kidney transplantation.[138] In Europe, the prevalence of SV40 antibodies was reported as 9 and 4% in subjects tested in Hungary and the Czech Republic, respectively.[151] In the United Kingdom, SV40 seroprevalence was estimated to be 5%.[135] An analysis in Kazakhstan, a Central Asian former member of the USSR, indicated a similar prevalence of SV40 neutralizing antibodies of 5%.[136] Interestingly, 60% of the subjects with SV40 antibodies in the latter study were born between 1969 and the early 1980s, a period in which some oral poliovaccines in the USSR may have been SV40 contaminated.[49] However, SV40 neutralization assays are too laborious to support large population surveys.

Studies using ELISA tests based on virus-like particles (VLPs) (a method more amenable to population screens) have encountered crossreactive polyomavirus antibodies at low serum dilutions in human and macaque sera that have hampered true estimates of SV40 infections.[74,152–157] The SV40-reactive antibodies in human sera usually could be removed by adsorption of sera with VLPs of BKV and/or JCV, leading to the interpretation that there was no clear evidence of recent human infections by SV40.[156] As ELISA tests can detect nonneutralizing antibodies, some crossreactivity is not unexpected because JCV, BKV, and SV40 are partially related. The VLP-based assays are directed against the polyomavirus VP1 proteins, a gene product that shares some sequences (and probably some nonneutralizing epitopes) among the three viruses.[158] Any conclusions about SV40 seroprevalence rates from ELISA-based studies should be viewed with caution as very little is known about the nature and strength of the human immune response to infections with SV40. Antibody responses in humans to SV40 are known to be low titered[159] and, as indicated recently, may wane over time.[160] This difficulty was illustrated in children who received contaminated oral poliovaccine and excreted infectious SV40 in their stools for several weeks after vaccination; no serologic response to the infection was observed.[161] Beyond establishing that past exposure to the virus had occurred, it is unclear how informative any serologic assay would be regarding the pathogenesis of SV40 infections in humans as it is believed that polyomavirus-related diseases result from reactivation of latent infections.[77] Modern molecular biology assays are an excellent and preferred alternative to serology for the analysis of SV40 infections in human populations. In addition, sensitive and specific molecular techniques are recognized to provide insights into the infectious etiology of some human malignancies.[42,162,163] Recently described real-time quantitative PCR assays specific for SV40, BKV, and JCV should facilitate future analyses.[164–166]

ASSOCIATION OF SV40 WITH HUMAN LYMPHOMAS

Molecular Studies in Different Human Populations

Herpesviruses and a human retrovirus have been associated with certain types of human lymphomas. EBV, HHV8, and human T-lymphotropic virus type 1 are associated with primary central nervous system NHL, multicentric Castleman's disease and primary effusion lymphoma, and peripheral T-cell lymphoma, respectively, but are absent from many systemic NHL in both HIV-infected and –uninfected patients.[1,2,9,10,167,168] Studies from our laboratory and others, using different molecular biology techniques, have demonstrated SV40 DNA or T-ag protein in some systemic NHL among HIV-positive and HIV-negative individuals.[21-29,31-33,37,169-171]

Studies in the United States (that were controlled by the inclusion of normal tissues analyzed in parallel) have shown that 19–43% of systemic NHL among adult patients are significantly associated with SV40 DNA sequences[24-26,28,29,32] (Table 2). Positivity rates between 11 and 24% have been reported in studies from other countries,[21,27,31,33,170] whereas other studies found a low frequency or failed to detect SV40 sequences in lymphomas.[22,172-175] These discordances among some studies may occur because of variations in the assays used for analyses[176] or the inclusion of many T-cell neoplasias in the surveys. However, accumulating data suggest that more important than technical issues is that the prevalence of SV40 infections may vary among different geographic regions. Furthermore, the recent recognition of SV40 strains[81] and differences in their biologic properties[117] suggest the theoretical possibility that unrecognized viral strain differences could be responsible for the variable positivity rates reported for SV40 and human lymphomas. Of interest is a recent analysis of stored blood samples from healthy Italian organ donors;[137] 11.2% of donors were positive for SV40 regulatory region sequences. Viral strain differences were noted between individuals born before 1947 and those born after 1958.

Some SV40 sequences associated with systemic NHL have been genotyped[24] (Table 3). The identification of unique SV40 strains established conclusively that the detection of SV40 DNA in human malignancies, including lymphomas, was not

Table 2. Detection of SV40 DNA in lymphomas: geographical differences

Study	Country	SV40 DNA by PCR, no. pos./no. tested (%)		
		Lymphoma	Cancer control	Nonmalignant lymphoid control
Shivapurkar et al., 2002[26]	US	29/68 (43)	5/80 (6)	0/40 (0)
Vilchez et al., 2002[24]	US	64/154 (42)	0/54 (0)	0/186 (0)
Shivapurkar et al., 2004[28]	US	33/90 (36)	–	1/56 (1.2)
Meneses et al., 2005[31]	Costa Rica	30/125 (24)	0/40 (0)	0/51 (0)
Vilchez et al., 2005[25]	US	12/55 (22)	0/5 (0)	0/14 (0)
David et al., 2001[32]	US	15/79 (19)	1/72 (1)	18/115 (16)
Martini et al., 1998[21]	Italy	11/79 (14)	–	3/50 (6)
Chen et al., 2006[33]	Taiwan	13/91 (14)	–	0/106 (0)
Nakatsuka et al., 2003[27]	Japan	14/122 (11)	–	3/64 (4.7)
Capello et al., 2003[172]	Spain, Italy	17/500 (3)	–	0/15 (0)
Daibata et al., 2003[22]	Japan	3/125 (2)	–	0/31 (0)
MacKenzie et al., 2003[173]	UK	0/152 (0)	0/18 (0)	0/27 (0)
Schüler et al., 2006[175]	Germany	0/27 (0)	0/6 (0)	0/55 (0)
Total		241/1667 (14.4)[a]	31/1085 (2.9)[a]	

[a]$p = 0.0001$.

Table 3. Genotyping of SV40 strains found in NHL based on DNA sequences of the C-terminal region of the large T-ag gene[a]

Patient	D.O.B.	HIV+	SV40 strain[b]
1	1936	No	MC028846B
2	1943	No	MC028846B
3	1950	Yes	Unique
4	1965	Yes	MC028846B
5	1965	Yes	SVCPC
6	1971	Yes	Unique (NHL-7)

[a]From Vilchez et al.[24]
[b]Genotyping of SV40 strains described in Forsman et al.[81]

the result of laboratory contamination. Some sequences were like that of an SV40 strain (MC028846B) detected in a contaminated killed poliovaccine from 1955.[177] Another was the same (SVCPC) as sequences detected in other human tumors[57,69,178] and very similar to strain 777, also recovered from a Salk poliovaccine.[81] These findings support the hypothesis that some SV40 infections and morbidity in humans may be linked to past usage of SV40-contaminated vaccines. Significantly, some of the SV40-positive lymphomas in different patient populations developed among individuals born after 1963, the last year that SV40-contaminated poliovaccines are accepted to have been used in different countries.[24,25,31] Indeed, 33% of SV40-positive AIDS-related systemic NHL cases have occurred among individuals born after the period of use of SV40-contaminated vaccines.[25] These observations are similar to published reports involving other selected human cancers,[58–60,62–66] which indicate that it is likely that SV40 is causing infections in humans today.

The frequency of SV40 positivity observed among lymphomas has been higher in NHL from HIV-negative individuals than in those from HIV-infected patients [39 of 75 (52%) vs. 25 of 74 (34%); $p = 0.03$][24] (Table 4). The contribution of EBV to these groups of systemic NHL may affect the reported frequency of SV40, as EBV was associated with 40% of systemic NHL from HIV-infected patients and with only 16% from the HIV-negative group. These findings are similar to EBV rates reported previously.[4] Many studies fail to report the HIV status of the cancer patients.

Among HIV-infected patients with systemic NHL, there was no significant difference in the mean CD4+ T-cell count between patients with EBV-positive and SV40-positive lymphomas ($190/mm^3$ vs. $112/mm^3$, $p = 0.3$) or between virus-positive (EBV and/or SV40) and virus-negative patients ($166/mm^3$ vs. $163/mm^3$, $p = 0.9$). Few systemic NHL appeared to be doubly positive for both SV40 and EBV (7%, 17%),[24,28] showing that SV40 was not randomly associated with lymphoid tumors. NHL comprises a biologically diverse group of hematologic malignancies, but SV40 T-ag sequences have been detected most frequently in diffuse large B-cell lymphomas (DLBCL) among HIV-infected and -uninfected patients.[24] This is the most common histologic type of lymphomas from B cells and accounts for about 45–50% of all NHL cases worldwide.[179]

Table 4. Polyomavirus SV40 and herpesvirus EBV DNA sequences by histologic type of NHL in HIV-infected and -uninfected patients[a]

Histologic type of B-cell neoplasm	HIV-infected patients			HIV-uninfected patients			All cases		
		DNA positive			DNA positive			DNA positive	
	Total tested	SV40 T-ag	EBV LMP-2	Total tested	SV40 T-ag	EBV LMP-2	Total tested	SV40 T-ag	EBV LMP-2
Diffuse large cell	58	19	20	40	25	8	98	44	28
Follicular	1	1	0	25	11	3	26	12	3
Burkitt[b]	13	5	8	7	2	0	20	7	8
Other[c]	2	0	2	3	1	1	5	1	3
Total	74	25	30	75	39	12	149	64	42

From Vilchez et al.[24]
[a]T-cell neoplasms were negative for the presence of EBV and SV40 sequences.
[b]This group included six specimens of variant Burkitt's lymphoma.
[c]This group included the other B-cell neoplasms.

Gene Expression in Virus-Positive Lymphomas

Tumor viruses such as SV40 are known to express the viral transforming protein in some cancer cells.[1] A report from Italy showed the detection of SV40 T-ag staining in 17% of SV40 DNA-positive lymphomas from HIV-infected and -uninfected patients.[21] A study among HIV-uninfected patients from France and Canada reported negative results for SV40 T-ag staining; however, there was no analysis for SV40 DNA sequences on the specimens, complicating the interpretation of those observations.[180] Recently, an age-matched, case-control, blinded study of systemic NHL from patients with HIV/AIDS assessed the expression of SV40 T-ag and phenotypic lymphocyte markers. Immunohistochemistry (IHC) staining showed expression of SV40 T-ag in AIDS-related systemic NHL, whereas none of the control tissue samples were positive for T-ag (12/55, 22% vs. 0/25, 0%; $p = 0.01$). Those samples had been examined previously by PCR for SV40 and EBV DNA sequences, and when the code was broken SV40 T-ag expression was detected only in B-cell lymphoma specimens that contained SV40 DNA sequences.[25] Expression of the viral oncoprotein was present in many malignant lymphoma cells and not in reactive lymphocytes (Fig. 5), but the reaction was not as homogeneous and intense as observed in SV40 hamster tumors and not all cells appeared to be positive. Specimens taken at different time points from a patient with an SV40-positive AIDS-related systemic NHL showed consistent expression of the viral protein.[25] Similar patterns of SV40 T-ag expression have been demonstrated in B-cell lymphomas from HIV-negative individuals[31] (Fig. 6).

These results suggest that some human lymphoma cells express SV40 T-ag, but protein levels seem to be lower than in animal tumors, perhaps reflecting reduced stability of T-ag or increased degradation. It is not known if cells in an

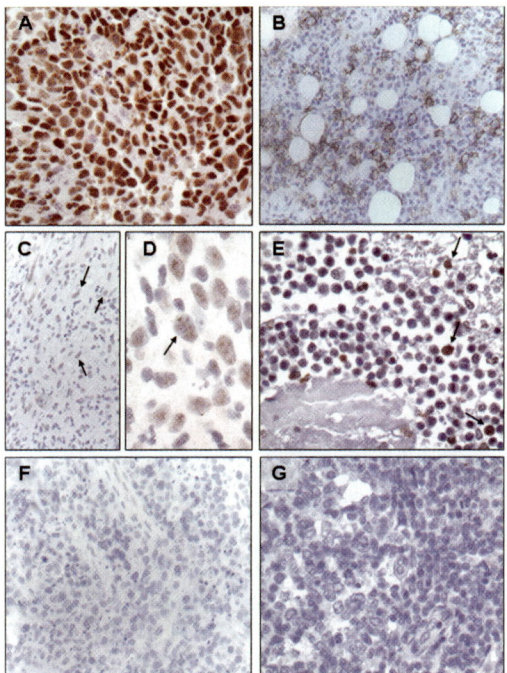

Figure 5. SV40 T-ag expression in AIDS-related non-Hodgkin's lymphoma. (A) SV40-induced hamster tumor stained with PAb101 as a positive control. (B) Expression of CD20 (indicative of B cells) by lymphoma cells. (C, D) Expression of nuclear SV40 T-ag in malignant B cells in an SV40 DNA-positive tumor from a 42-year-old HIV-infected patient, stained with PAb101. (E) Expression of SV40 T-ag in an SV40 DNA-positive tumor from a 29-year-old HIV-infected patient, stained with PAb416. (F) An SV40 DNA-negative diffuse large B-cell lymphoma from a 36-year-old HIV-infected patient reacted with PAb101; no T-ag was detected. (G) Control reaction on an SV40 DNA-negative reactive lymph node specimen from an HIV-infected subject; cells did not react with PAb101. *Arrows* point to representative SV40 T-ag positive cells. A lower proportion of lymphoma cells expressed nuclear SV40 T-ag and the reaction was less intense, as compared to the hamster tumor. Original magnification for all panels (except panel D), 40×. Panel D, 100×. From Vilchez et al.[25]

SV40 DNA-positive lymphoma that appeared negative for T-ag staining expressed low amounts of T-ag that were undetectable by current staining techniques. Alternatively, it may be that the T-ag-negative cells had lost the viral genome because tumor progression and accumulation of mutations in the lymphoma cell had inactivated cellular growth control checkpoints and made functional T-ag dispensable. These questions will need to be addressed by quantitative assays and sensitive in situ hybridization approaches to determine the viral load and the number of malignant cells that contain viral DNA in SV40-positive NHL.

Approximately two thirds of AIDS-related systemic NHL are categorized as DLBCL type, with Burkitt's lymphoma comprising 25% and other histologies a much smaller proportion.[20] Gene expression profiling of DLBCL among HIV-negative patients has shown that this single diagnostic category includes more

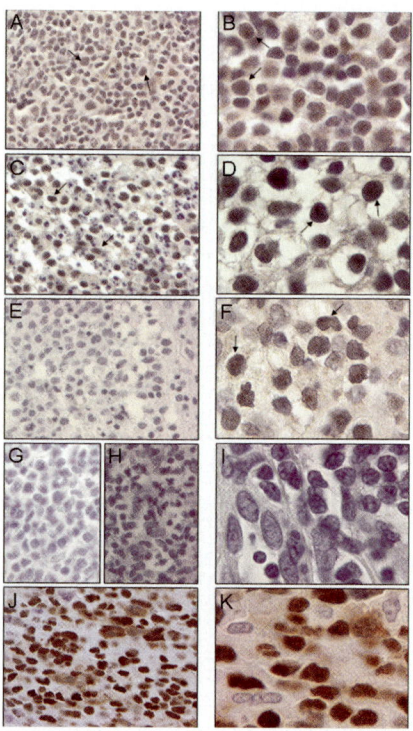

Figure 6. SV40 T-ag expression in NHL from Costa Rica. (A, B) Expression of SV40 T-ag in malignant B cells in an SV40 DNA-positive diffuse large B-cell lymphoma from a 67-year-old female, stained with PAb101. (C, D) Expression of SV40 T-ag in an SV40 DNA-positive diffuse large B-cell lymphoma from a 45-year-old male, stained with PAb101. (E) Same tumor as in panels C and D, stained with negative control mouse serum. (F) Expression of SV40 T-ag in an SV40 DNA-positive diffuse large B-cell lymphoma from a 70-year-old male, stained with PAb101. (G) SV40 DNA-negative lymph node stained with PAb101. (H, I) An SV40 DNA-negative diffuse large B-cell lymphoma from a 29-year-old male reacted with PAb101. (J, K) Expression of SV40 T-ag in an SV40-induced hamster tumor, stained with PAb101. *Arrows* point to T-ag positive cells in panels A–D and F. Original magnification for panels A, C, E, G, H, and J, 40×, panels B, D, F, I, and K, 100×. From Meneses et al.[31]

than one molecularly distinct disease.[181,182] DLBCL consist of at least three distinct gene-expression subgroups, known as germinal center B-cell like (GCB), activated B-cell like (ABC), and type 3 (that behave in a manner similar to the ABC subgroup). Although limited data exist on the phenotypic features of DLBCL in HIV-positive individuals, evidence suggests that tumor pathogenesis is heterogeneous.[15,25,183,184] Examination of B-cell markers by IHC to determine the DLBCL subgroup (GCB-like and non-GCB-like) among HIV-positive individuals showed that a GCB profile was more frequently expressed in SV40-positive than in EBV-related lymphomas (10/12, 83% vs. 6/13, 46%; $p = 0.05$). In contrast, a non-GCB phenotype was more frequent in EBV-positive than in SV40-positive lymphomas (7/13, 54% vs. 2/12, 17%; $p = 0.05$).[25] Significantly, a GCB-cell-like profile also has been associated with SV40-positive DLBCL from HIV-negative individuals.[31]

These results are important as germinal center B cells are poised for apoptosis unless they are rescued by positive stimuli[185] and inhibition of apoptosis is a common feature of tumorigenesis by oncogenic viruses, including SV40.[186–192] Promoter hypermethylation of tumor suppressor genes is a recognized mechanism of gene silencing in some human malignancies, including lymphomas.[150,193–196] Death-associated protein kinase (DAPK) plays a role in apoptosis pathways in B lymphocytes[195,196] and hypermethylation of the *DAPK* gene was observed more frequently in SV40-positive than in SV40-negative NHL.[28] SV40 immortalized human primary B lymphocytes[105] and SV40-positive human lymphoblastoid B-cell lines were more tumorigenic in SCID mice than were SV40-negative B-cell lines.[107] These data suggest that SV40 has the ability to alter human B cells and support the hypothesis that SV40 is functionally important in some AIDS-related systemic NHL.

Epidemiological Studies

In contrast to detection of SV40 markers in specific types of NHL using molecular techniques, epidemiological cohort studies have failed to demonstrate an increased cancer risk in populations that had a high likelihood of having received potentially contaminated poliovaccine,[44,70–73,157,197,198] including patients with HIV/AIDS.[199] However, the epidemiological data have inherent limitations,[34,42,44,200] including a failure to know which individuals received contaminated vaccine, the dosage of infectious SV40 present in particular lots of contaminated vaccine, and who among the exposed vaccinees were successfully infected with SV40. In addition, the comparison (unexposed) groups may have been SV40-infected, either due to person-to-person transmission or, as recent data suggest, because some SV40-contaminated oral poliovaccines were produced and used throughout the world until at least 1978.[49]

Seroepidemiology studies have been carried out using the VLP-based ELISA assays described above. Despite the limitations of the assays noted above, the results have been interpreted to suggest there is no clear evidence of human infections by SV40 or of an association of SV40 with human lymphomas.[74,152,153,155–157,201,202] However, to know how to interpret any serological data it is necessary first to acquire an understanding of the human immune response to SV40 infection.

CONCLUSIONS

A large number of controlled studies provide strong evidence that SV40 is causing infections in humans today. The original source of the major virus exposure in all likelihood was the widespread use of contaminated poliovaccines. Modern sources of exposure are unknown, but probably involve human-to-human transmission via close contact, perhaps by a fecal-oral route or exposure to virus-positive urine. Also unknown are the dynamics of virus replication and tissue distribution patterns within infected hosts. The prevalence of SV40 infections in humans appears to be variable among different populations. This pattern is not unexpected, as usage of contaminated vaccines would have introduced SV40 into different numbers of vaccinees in only certain human populations, and the years hence have not allowed time for the virus to become widely distributed throughout the world. Prospective studies are needed to determine the prevalence of SV40 infections in different human populations and to assess how the virus is transmitted from person to person. Considering that molecular biology approaches provide sensitive and specific approaches to analyze malignancies with a possible infectious etiology, studies using these modern methods should be used to assess the distribution, dynamics, and morbidity of SV40 infections in humans.

The association of SV40 with lymphomas, as with other specific types of human cancer, is strong in some studies, negligible in others, and absent in still others. Such variability is predictable, if virus prevalence differs in the populations being examined. Although decades of in vitro studies have established that SV40 disrupts critical cell cycle control pathways, it remains unknown if these perturbations are sufficient for the virus to induce the development of lymphomas in humans. However, as T-ag is able to exert powerful effects on cell properties, the virus is probably functionally important in at least some SV40-positive human tumors. Models are needed that reproduce key features of SV40 infection and disease in humans. Such models could provide precise evidence of a particular cell pathway in SV40 lymphomagenesis, allow further characterization of the molecular mechanisms of oncogenesis and virus–host interactions, and provide a preclinical system to test therapeutic interventions for this significant and common AIDS-related malignancy.

REFERENCES

1. Butel, J. S. 2000. Viral carcinogenesis: Revelation of molecular mechanisms and etiology of human disease. Carcinogenesis **21**:405–26.
2. Fisher, S. G., and R. I. Fisher. 2004. The epidemiology of non–Hodgkin's lymphoma. Oncogene **23**:6524–34.
3. 1996. IARC Monographs on the Evaluation of Carcinogenic Risks to Humans, Vol. 67. Human Immunodeficiency Virus and Human T Cell Lymphotropic Viruses. Lyon: International Agency for Research on Cancer/World Health Organization.
4. Mueller, N. 1999. Overview of the epidemiology of malignancy in immune deficiency. J Acquir Immune Defic Syndr **21**:S5–10.
5. Engels, E. A., and J. J. Goedert. 2005. Human immunodeficiency virus/acquired immunodeficiency syndrome and cancer: Past, present, and future. J Natl Cancer Inst **97**:407–9.
6. 1992. 1993 revised classification system for HIV infection and expanded surveillance case definition for AIDS among adolescents and adults. MMWR Morb Mortal Wkly Rept **41**(No. RR-17):1–19.
7. Lim, S. T., and A. M. Levine. 2005. Recent advances in acquired immunodeficiency syndrome (AIDS)-related lymphoma. CA Cancer J Clin **55**:229–41.
8. 2000. International Collaboration on HIV and Cancer. Highly active antiretroviral therapy and incidence of cancer in human immunodeficiency virus-infected adults. J Natl Cancer Inst **92**:1823–30.
9. Mitsuyasu, R. 1999. Oncological complications of human immunodeficiency virus disease and hematologic consequences of their treatment. Clin Infect Dis **29**:35–43.
10. Vilchez, R. A., C. A. Kozinetz, J. L. Jorgensen, M. H. Kroll, and J. S. Butel. 2002. AIDS-related systemic non–Hodgkin's lymphoma at a large community program. AIDS Res Hum Retroviruses **18**:237–42.
11. American Cancer Society. 2002. Cancer facts and figures 2002.

12. Richman, D. D., S. C. Morton, T. Wrin, N. Hellmann, S. Berry, M. F. Shapiro, and S. A. Bozzette. 2004. The prevalence of antiretroviral drug resistance in the United States. AIDS **18**:1393–401.

13. Fox, J. M., S. Fidler, and J. Weber. 2006. Resistance to HIV drugs in UK may be lower in some areas. BMJ **332**:179–80.

14. Diamond, C., T. H. Taylor, T. Aboumrad, and H. Anton-Culver. 2006. Changes in acquired immunodeficiency syndrome-related non-Hodgkin lymphoma in the era of highly active antiretroviral therapy: Incidence, presentation, treatment, and survival. Cancer **106**:128–35.

15. Little, R. F., S. Pittaluga, N. Grant, S. M. Steinberg, M. F. Kavlick, H. Mitsuya, G. Franchini, M. Gutierrez, M. Raffeld, E. S. Jaffe, G. Shearer, R. Yarchoan, and W. H. Wilson. 2003. Highly effective treatment of acquired immunodeficiency syndrome-related lymphoma with dose-adjusted EPOCH: Impact of antiretroviral therapy suspension and tumor biology. Blood **101**:4653–9.

16. Vilchez, R. A., C. J. Finch, J. L. Jorgensen, and J. S. Butel. 2003. The clinical epidemiology of Hodgkin lymphoma in HIV-infected patients in the highly active antiretroviral therapy (HAART) era. Medicine **82**:77–81.

17. Tam, H. K., Z. F. Zhang, L. P. Jacobson, J. B. Margolick, J. S. Chmiel, C. Rinaldo, and R. Detels. 2002. Effect of highly active antiretroviral therapy on survival among HIV-infected men with Kaposi sarcoma or non-Hodgkin lymphoma. Int J Cancer **98**:916–22.

18. Gérard, L., L. Galicier, A. Maillard, E. Boulanger, L. Quint, S. Matheron, B. Cardon, V. Meignin, and E. Oksenhendler. 2002. Systemic non-Hodgkin lymphoma in HIV-infected patients with effective suppression of HIV replication: Persistent occurrence but improved survival. J Acquir Immune Defic Syndr **30**:478–84.

19. Navarro, J. T., J. M. Ribera, A. Oriol, C. Tural, F. Millá, and E. Feliu. 2003. Improved outcome of AIDS-related lymphoma in patients with virologic response to highly active antiretroviral therapy. J Acquir Immune Defic Syndr **32**:347–8.

20. Navarro, W. H., and L. D. Kaplan. 2006. AIDS-related lymphoproliferative disease. Blood **107**:13–20.

21. Martini, F., R. Dolcetti, A. Gloghini, L. Iaccheri, A. Carbone, M. Boiocchi, and M. Tognon. 1998. Simian-virus-40 footprints in human lymphoproliferative disorders of HIV$^-$ and HIV$^+$ patients. Int J Cancer **78**:669–74.

22. Daibata, M., Y. Nemoto, M. Kamioka, S. Imai, and H. Taguchi. 2003. Simian virus 40 in Japanese patients with lymphoproliferative disorders. Br J Haematol **121**:190–1.

23. Vilchez, R. A., J. A. Lednicky, S. J. Halvorson, Z. S. White, C. A. Kozinetz, and J. S. Butel. 2002. Detection of polyomavirus simian virus 40 tumor antigen DNA in AIDS-related systemic non-Hodgkin lymphoma. J Acquir Immune Defic Syndr **29**:109–16.

24. Vilchez, R. A., C. R. Madden, C. A. Kozinetz, S. J. Halvorson, Z. S. White, J. L. Jorgensen, C. J. Finch, and J. S. Butel. 2002. Association between simian virus 40 and non-Hodgkin lymphoma. Lancet **359**:817–23.

25. Vilchez, R. A., D. Lopez-Terrada, J. R. Middleton, C. J. Finch, D. E. Killen, P. Zanwar, J. L. Jorgensen, and J. S. Butel. 2005. Simian virus 40 tumor antigen expression and immunophenotypic profile of AIDS-related non-Hodgkin's lymphoma. Virology **342**:38–46.

26. Shivapurkar, N., K. Harada, J. Reddy, R. H. Scheuermann, Y. Xu, R. W. McKenna, S. Milchgrub, S. H. Kroft, Z. Feng, and A. F. Gazdar. 2002. Presence of simian virus 40 DNA sequences in human lymphomas. Lancet **359**:851–2.

27. Nakatsuka, S.-I., A. Liu, Z. Dong, S. Nomura, T. Takakuwa, H. Miyazato, K. Aozasa, and Osaka Lymphoma Study Group. 2003. Simian virus 40 sequences in malignant lymphomas in Japan. Cancer Res **63**:7606–8.

28. Shivapurkar, N., T. Takahashi, J. Reddy, Y. Zheng, V. Stastny, R. Collins, S. Toyooka, M. Suzuki, G. Parikh, S. Asplund, S. H. Kroft, C. Timmons, R. W. McKenna, Z. Feng, and A. F. Gazdar. 2004. Presence of simian virus 40 DNA sequences in human lymphoid and hematopoietic malignancies and their relationship to aberrant promoter methylation of multiple genes. Cancer Res **64**:3757–60.

29. Samaniego, F., S. Wang, D. Young, and S. Wang. 2004. Large T antigen DNA and protein of simian virus 40 in non-Hodgkin's lymphoma. (Abstract #4762). 95th Annual Meeting of the American Association for Cancer Research, March 27–31, 2004, Orlando, FL.

30. Heinsohn, S., S. Golta, H. Kabisch, and U. zur Stadt. 2005. Standardized detection of Simian virus 40 by real-time quantitative polymerase chain reaction in pediatric malignancies. Haematologica **90**:84–9.

31. Meneses, A., D. Lopez-Terrada, P. Zanwar, D. E. Killen, V. Monterroso, J. S. Butel, and R. A. Vilchez. 2005. Lymphoproliferative disorders in Costa Rica and simian virus 40. Haematologica **90**:1635–42.

32. David, H., S. Mendoza, T. Konishi, and C. W. Miller. 2001. Simian virus 40 is present in human lymphomas and normal blood. Cancer Lett **162**:57–64.

33. Chen, P. M., C. C. Yen, M. H. Yang, S. B. Poh, L. T. Hsiao, W. S. Wang, P. C. Lin, M. Y. Lee, H. W. Teng, L. Y. Bai, C. J. Chu, S. C. Chao, A. H. Yang, T. J. Chiou, J. H. Liu, and T. C. Chao. 2006. High prevalence of SV40 infection in patients with nodal non-Hodgkin's lymphoma but not acute leukemia independent of contaminated polio vaccines in Taiwan. Cancer Invest **24**:223–8.

34. Stratton, K., D. A. Alamario, and M. C. McCormick. 2003. Immunization Safety Review: SV40 Contamination of Polio Vaccine and Cancer. Washington, DC: The National Academies Press.
35. Li, R. M., R. B. Mannon, D. Kleiner, M. Tsokos, M. Bynum, A. D. Kirk, and J. B. Kopp. 2002. BK virus and SV40 co-infection in polyomavirus nephropathy. Transplantation **74**:1497–504.
36. Milstone, A., R. A. Vilchez, X. Geiger, A. B. Fogo, J. S. Butel, and S. Dummer. 2004. Polyomavirus simian virus 40 infection associated with nephropathy in a lung-transplant recipient. Transplantation **77**:1019–24.
37. Rizzo, P., M. Carbone, S. G. Fisher, C. Matker, L. J. Swinnen, A. Powers, I. Di Resta, S. Alkan, H. I. Pass, and R. I. Fisher. 1999. Simian virus 40 is present in most United States human mesotheliomas, but it is rarely present in non-Hodgkin's lymphoma. Chest **116**:470S–3S.
38. Utermöhlen, O., C. Schulze-Garg, G. Warnecke, R. Gugel, J. Löhler, and W. Deppert. 2001. Simian virus 40 large-T-antigen-specific rejection of mKSA tumor cells in BALB/c mice is critically dependent on both strictly tumor-associated, tumor-specific CD8$^+$ cytotoxic T lymphocytes and CD4$^+$ T helper cells. J Virol **75**:10593–602.
39. Kennedy, R. C., M. H. Shearer, A. M. Watts, and R. K. Bright. 2003. CD4+ T lymphocytes play a critical role in antibody production and tumor immunity against simian virus 40 large tumor antigen. Cancer Res **63**:1040–5.
40. Tevethia, S. S., and T. D. Schell. 2001. The immune response to SV40, JCV, and BKV. In: Khalili, K., and G. L. Stoner (eds.), Human Polyomaviruses: Molecular and Clinical Perspectives. New York: Wiley-Liss, Inc., pp. 585–610.
41. Brown, D. W. G. 1997. Threat to humans from virus infections of non-human primates. Rev Med Virol **7**:239–46.
42. Vilchez, R. A., C. A. Kozinetz, and J. S. Butel. 2003. Conventional epidemiology and the link between SV40 and human cancers. Lancet Oncol **4**:188–91.
43. Butel, J. S., and J. A. Lednicky. 1999. Cell and molecular biology of simian virus 40: Implications for human infections and disease. J Natl Cancer Inst **91**:119–34.
44. Rollison, D. E. M., and K. V. Shah. 2001. The epidemiology of SV40 infection due to contaminated polio vaccines: Relation of the virus to human cancer. In: Khalili, K., and G. L. Stoner (eds.), Human Polyomaviruses: Molecular and Clinical Perspectives. New York: Wiley-Liss, Inc., pp. 561–84.
45. Arrington, A. S., and J. S. Butel. 2001. SV40 and human tumors. In: Khalili, K., and G. L. Stoner (eds.), Human Polyomaviruses: Molecular and Clinical Perspectives. New York: John Wiley & Sons, pp. 461–89.
46. Shah, K., and N. Nathanson. 1976. Human exposure to SV40: Review and comment. Am J Epidemiol **103**:1–12.
47. Lewis, A. M., Jr. 1973. Experience with SV40 and adenovirus-SV40 hybrids. In: Hellman, A., M. N. Oxman, and R. Pollack (eds.), Biohazards in Biological Research. Cold Spring Harbor: Cold Spring Harbor Laboratory Press, pp. 96–113.
48. 1960. Proceedings of the Second International Conference on Live Poliovirus Vaccines. Papers and discussions held. Scientific Publication No. 50. Washington, DC: Pan American Health Organization.
49. Cutrone, R., J. Lednicky, G. Dunn, P. Rizzo, M. Bocchetta, K. Chumakov, P. Minor, and M. Carbone. 2005. Some oral poliovirus vaccines were contaminated with infectious SV40 after 1961. Cancer Res **65**:10273–9.
50. Sullivan, C. S., and J. M. Pipas. 2002. T antigens of simian virus 40: Molecular chaperones for viral replication and tumorigenesis. Microbiol Mol Biol Rev **66**:179–202.
51. Ahuja, D., M. T. Sáenz-Robles, and J. M. Pipas. 2005. SV40 large T antigen targets multiple cellular pathways to elicit cellular transformation. Oncogene **24**:7729–45.
52. Jasani, B., A. Cristaudo, S. A. Emri, A. F. Gazdar, A. Gibbs, B. Krynska, C. Miller, L. Mutti, C. Radu, M. Tognon, and A. Procopio. 2001. Association of SV40 with human tumours. Semin Cancer Biol **11**:49–61.
53. Gazdar, A. F., J. S. Butel, and M. Carbone. 2002. SV40 and human tumours: Myth, association or causality? Nat Rev Cancer **2**:957–64.
54. Vilchez, R. A., C. A. Kozinetz, A. S. Arrington, C. R. Madden, and J. S. Butel. 2003. Simian virus 40 in human cancers. Am J Med **114**:675–84.
55. Vanchiere, J. A., Z. S. White, and J. S. Butel. 2005. Detection of BK virus and simian virus 40 in the urine of healthy children. J Med Virol **75**:447–54.
56. Butel, J. S., A. S. Arrington, C. Wong, J. A. Lednicky, and M. J. Finegold. 1999. Molecular evidence of simian virus 40 infections in children. J Infect Dis **180**:884–7.
57. Arrington, A. S., M. S. Moore, and J. S. Butel. 2004. SV40-positive brain tumor in scientist with risk of laboratory exposure to the virus. Oncogene **23**:2231–5.

58. Bergsagel, D. J., M. J. Finegold, J. S. Butel, W. J. Kupsky, and R. L. Garcea. 1992. DNA sequences similar to those of simian virus 40 in ependymomas and choroid plexus tumors of childhood. N Engl J Med **326**:988–93.

59. Suzuki, S. O., M. Mizoguchi, and T. Iwaki. 1997. Detection of SV40 T antigen genome in human gliomas. Brain Tumor Pathol **14**:125–9.

60. Zhen, H. N., X. Zhang, X.Y. Bu, Z. W. Zhang, W. J. Huang, P. Zhang, J. W. Liang, and X. L. Wang. 1999. Expression of the simian virus 40 large tumor antigen (Tag) and formation of Tag-p53 and Tag-pRb complexes in human brain tumors. Cancer **86**:2124–32.

61. Weggen, S., T. A. Bayer, A. Von Deimling, G. Reifenberger, D. Von Schweinitz, O. D. Wiestler, and T. Pietsch. 2000. Low frequency of SV40, JC and BK polyomavirus sequences in human medulloblastomas, meningiomas and ependymomas. Brain Pathol **10**:85–92.

62. Malkin, D., S. Chilton-MacNeill, L. A. Meister, E. Sexsmith, L. Diller, and R. L. Garcea. 2001. Tissue-specific expression of SV40 in tumors associated with the Li-Fraumeni syndrome. Oncogene **20**:4441–9.

63. Martini, F., L. Lazzarin, L. Iaccheri, B. Vignocchi, G. Finocchiaro, I. Magnani, M. Serra, K. Scotlandi, G. Barbanti-Brodano, and M. Tognon. 2002. Different simian virus 40 genomic regions and sequences homologous with SV40 large T antigen in DNA of human brain and bone tumors and of leukocytes from blood donors. Cancer **94**:1037–48.

64. Carbone, M., P. Rizzo, A. Procopio, M. Giuliano, H. I. Pass, M. C. Gebhardt, C. Mangham, M. Hansen, D. F. Malkin, G. Bushart, F. Pompetti, P. Picci, A. S. Levine, J. D. Bergsagel, and R. L. Garcea. 1996. SV40-like sequences in human bone tumors. Oncogene **13**:527–35.

65. Mendoza, S. M., T. Konishi, and C. W. Miller. 1998. Integration of SV40 in human osteosarcoma DNA. Oncogene **17**:2457–62.

66. Yamamoto, H., T. Nakayama, H. Murakami, T. Hosaka, T. Nakamata, T. Tsuboyama, M. Oka, T. Nakamura, and J. Toguchida. 2000. High incidence of SV40-like sequences detection in tumour and peripheral blood cells of Japanese osteosarcoma patients. Br J Cancer **82**:1677–81.

67. Vanchiere, J. A., R. K. Nicome, J. M. Greer, G. J. Demmler, and J. S. Butel. 2005. Frequent detection of polyomaviruses in stool samples from hospitalized children. J Infect Dis **192**:658–64.

68. Li, R. M., M. H. Branton, S. Tanawattanacharoen, R. A. Falk, J. C. Jennette, and J. B. Kopp. 2002. Molecular identification of SV40 infection in human subjects and possible association with kidney disease. J Am Soc Nephrol **13**:2320–30.

69. Lednicky, J. A., R. L. Garcea, D. J. Bergsagel, and J. S. Butel. 1995. Natural simian virus 40 strains are present in human choroid plexus and ependymoma tumors. Virology **212**:710–7.

70. Strickler, H. D., P. S. Rosenberg, S. S. Devesa, J. Hertel, J. F. Fraumeni, Jr., and J. J. Goedert. 1998. Contamination of poliovirus vaccines with simian virus 40 (1955–1963) and subsequent cancer rates. J Am Med Assoc **279**:292–5.

71. Olin, P., and J. Giesecke. 1998. Potential exposure to SV40 in polio vaccines used in Sweden during 1957: No impact on cancer incidence rates 1960 to 1993. Dev Biol Stand **94**:227–33.

72. Carroll-Pankhurst, C., E. A. Engels, H. D. Strickler, J. J. Goedert, J. Wagner, and E. A. Mortimer, Jr. 2001. Thirty-five year mortality following receipt of SV40-contaminated polio vaccine during the neonatal period. Br J Cancer **85**:1295–7.

73. Strickler, H. D., J. J. Goedert, S. S. Devesa, J. Lahey, J. F. Fraumeni, Jr., and P. S. Rosenberg. 2003. Trends in U.S. pleural mesothelioma incidence rates following simian virus 40 contamination of early poliovirus vaccines. J Natl Cancer Inst **95**:38–45.

74. Engels, E. A., R. P. Viscidi, D. A. Galloway, J. J. Carter, J. R. Cerhan, S. Davis, W. Cozen, R. K. Severson, S. De Sanjose, J. S. Colt, and P. Hartge. 2004. Case–control study of simian virus 40 and non-Hodgkin lymphoma in the United States. J Natl Cancer Inst **96**:1368–74.

75. Thu, G. O., L. Y. Hem, S. Hansen, B. Moller, J. Norstein, H. Nokleby, and T. Grotmol. 2006. Is there an association between SV40 contaminated polio vaccine and lymphoproliferative disorders? An age–period–cohort analysis on Norwegian data from 1953 to 1997. Int J Cancer **118**:2035–9.

76. Major, E. O. 2001. Human polyomavirus. In: Knipe, D. M., P. M. Howley, D. E. Griffin, R. A. Lamb, M. A. Martin, B. Roizman, and S. E. Straus (eds.), Fields Virology. Philadelphia: Lippincott Williams & Wilkins, pp. 2175–96.

77. Khalili, K., and G. L. Stoner. 2001. Human Polyomaviruses: Molecular and Clinical Perspectives. New York: Wiley-Liss.

78. Shah, K.V. 1996. Polyomaviruses. In: Fields, B. N., D. M. Knipe, P. M. Howley, R. M. Chanock, J. L. Melnick, T. P. Monath, B. Roizman, and S. E. Straus (eds.), Fields Virology. Philadelphia: Lippincott-Raven, pp. 2027–43.

79. Stewart, A. R., J. A. Lednicky, U. S. Benzick, M. J. Tevethia, and J. S. Butel. 1996. Identification of a variable region at the carboxy terminus of SV40 large T-antigen. Virology **221**:355–61.

80. Stewart, A. R., J. A. Lednicky, and J. S. Butel. 1998. Sequence analyses of human tumor-associated SV40 DNAs and SV40 viral isolates from monkeys and humans. J Neurovirol **4**:182–93.

81. Forsman, Z. H., J. A. Lednicky, G. E. Fox, R. C. Willson, Z. S. White, S. J. Halvorson, C. Wong, A. M. Lewis, Jr., and J. S. Butel. 2004. Phylogenetic analysis of polyomavirus simian virus 40 from monkeys and humans reveals genetic variation. J Virol **78**:9306–16.

82. Rabson, A. S., G. T. O'Conner, I. K. Berezesky, and F. J. Paul. 1964. Enchancement of adenovirus growth in African green monkey kidney cell cultures by SV40. Proc Soc Exp Biol Med **116**:187–90.

83. Cole, C. N., L. V. Crawford, and P. Berg. 1979. Simian virus 40 mutants with deletions at the 3′ end of the early region are defective in adenovirus helper function. J Virol **30**:683–91.

84. Grodzicker, T., C. Anderson, P. A. Sharp, and J. Sambrook. 1974. Conditional lethal mutants of adenovirus 2-simian virus 40 hybrids. I. Host range mutants of Ad2+ND1. J Virol **13**:1237–44.

85. Kelly, T. J., Jr., and A. M. Lewis, Jr. 1973. Use of nondefective adenovirus-simian virus 40 hybrids for mapping the simian virus 40 genome. J Virol **12**:643–52.

86. Pipas, J. M. 1985. Mutations near the carboxyl terminus of the simian virus 40 large tumor antigen alter viral host range. J Virol **54**:569–75.

87. Tornow, J., M. Polvino-Bodnar, G. Santangelo, and C. N. Cole. 1985. Two separable functional domains of simian virus 40 large T antigen: Carboxyl-terminal region of simian virus 40 large T antigen is required for efficient capsid protein synthesis. J Virol **53**:415–24.

88. Tornow, J., and C. N. Cole. 1983. Nonviable mutants of simian virus 40 with deletions near the 3′ end of gene *A* define a function for large T antigen required after onset of viral DNA replication. J Virol **47**:487–94.

89. Cole, C. N., and T. P. Stacy. 1987. Biological properties of simian virus 40 host range mutants lacking the COOH-terminus of large T antigen. Virology **161**:170–80.

90. Stacy, T., M. Chamberlain, and C. N. Cole. 1989. Simian virus 40 host range/helper function mutations cause multiple defects in viral late gene expression. J Virol **63**:5208–15.

91. Spence, S. L., and J. M. Pipas. 1994. SV40 large T antigen functions at two distinct steps in virion assembly. Virology **204**:200–9.

92. Poulin, D. L., and J. A. DeCaprio. 2006. The carboxyl-terminal domain of large T antigen rescues SV40 host range activity in *trans* independent of acetylation. Virology **349**:212–21.

93. Hahn, W. C., S. K. Dessain, M. W. Brooks, J. E. King, B. Elenbaas, D. M. Sabatini, J. A. DeCaprio, and R. A. Weinberg. 2002. Enumeration of the simian virus 40 early region elements necessary for human cell transformation. Mol Cell Biol **22**:2111–23.

94. Foddis, R., A. De Rienzo, D. Broccoli, M. Bocchetta, E. Stekala, P. Rizzo, A. Tosolini, J. V. Grobelny, S. C. Jhanwar, H. I. Pass, J. R. Testa, and M. Carbone. 2002. SV40 infection induces telomerase activity in human mesothelial cells. Oncogene **21**:1434–42.

95. Arroyo, J. D., and W. C. Hahn. 2005. Involvement of PP2A in viral and cellular transformation. Oncogene **24**:7746–55.

96. Lednicky, J. A., A. S. Arrington, A. R. Stewart, X. M. Dai, C. Wong, S. Jafar, M. Murphey-Corb, and J. S. Butel. 1998. Natural isolates of simian virus 40 from immunocompromised monkeys display extensive genetic heterogeneity: New implications for polyomavirus disease. J Virol **72**:3980–90.

97. Lednicky, J. A., and J. S. Butel. 2001. Simian virus 40 regulatory region structural diversity and the association of viral archetypal regulatory regions with human brain tumors. Semin Cancer Biol **11**:39–47.

98. Lednicky, J. A., C. Wong, and J. S. Butel. 1995. Artificial modification of the viral regulatory region improves tissue culture growth of SV40 strain 776. Virus Res **35**:143–53.

99. Breau, W. C., W. J. Atwood, and L. C. Norkin. 1992. Class I major histocompatibility proteins are an essential component of the simian virus 40 receptor. J Virol **66**:2037–45.

100. Atwood, W. J., and L. C. Norkin. 1989. Class I major histocompatibility proteins as cell surface receptors for simian virus 40. J Virol **63**:4474–7.

101. Atwood, W. J. 2001. Cellular receptors for the polyomaviruses. In: Khalili, K., and G. L. Stoner (eds.), Human Polyomaviruses: Molecular and Clinical Perspectives. New York: Wiley-Liss, Inc., pp. 179–96.

102. Shein, H. M., and J. F. Enders. 1962. Multiplication and cytopathogenicity of simian vacuolating virus 40 in cultures of human tissues. Proc Soc Exp Biol Med **109**:495–500.

103. O'Neill, F. J., and D. Carroll. 1981. Amplification of papovavirus defectives during serial low multiplicity infections. Virology **112**:800–3.

104. Schneider, C., K. Weißhart, L. A. Guarino, I. Dornreiter, and E. Fanning. 1994. Species-specific functional interactions of DNA polymerase α-primase with simian virus 40 (SV40) T antigen require SV40 origin DNA. Mol Cell Biol **14**:3176–85.

105. Kanki, T. 1994. Immortalization of human primary B lymphocytes by simian virus 40 early region DNA. Hybridoma **13**:327–30.

106. McNees, A. L., and J. S. Butel. 2006. Unpublished data.

107. Dolcetti, R., F. Martini, M. Quaia, A. Gloghini, B. Vignocchi, R. Cariati, M. Martinelli, A. Carbone, M. Boiocchi, and M. Tognon. 2003. Simian virus 40 sequences in human lymphoblastoid B-cell lines. J Virol **77**:1595–7.

108. Deminie, C. A., and L. C. Norkin. 1990. Simian virus 40 DNA replication correlates with expression of a particular subclass of T antigen in a human glial cell line. J Virol **64**:3760–9.

109. Lynch, K. J., S. Haggerty, and R. J. Frisque. 1994. DNA replication of chimeric JC virus–simian virus 40 genomes. Virology **204**:819–22.

110. Carbone, M., R. Stach, I. Di Resta, H. I. Pass, and P. Rizzo. 1998. Simian virus 40 oncogenesis in hamsters. Dev Biol Stand **94**:273–9.

111. Girardi, A. J., B. H. Sweet, V. B. Slotnick, and M. R. Hilleman. 1962. Development of tumors in hamsters inoculated in the neonatal period with vacuolating virus, SV$_{40}$. Proc Soc Exp Biol Med **109**:649–60.

112. Diamandopoulos, G. T. 1972. Leukemia, lymphoma, and osteosarcoma induced in the Syrian golden hamster by simian virus 40. Science **176**:173–5.

113. Diamandopoulos, G. T. 1973. Induction of lymphocytic leukemia, lymphosarcoma, reticulum cell sarcoma, and osteogenic sarcoma in the Syrian golden hamster by oncogenic DNA simian virus 40. J Natl Cancer Inst **50**:1347–65.

114. Cicala, C., F. Pompetti, and M. Carbone. 1993. SV40 induces mesotheliomas in hamsters. Am J Pathol **142**:1524–33.

115. Cicala, C., F. Pompetti, P. Nguyen, K. Dixon, A. S. Levine, and M. Carbone. 1992. SV40 small t deletion mutants preferentially transform mononuclear phagocytes and B lymphocytes *in vivo*. Virology **190**:475–9.

116. Coe, J. E., and I. Green. 1975. B-cell origin of hamster lymphoid tumors induced by simian virus 40. J Natl Cancer Inst **54**:269–70.

117. Vilchez, R. A., C. F. Brayton, C. Wong, P. Zanwar, D. E. Killen, J. L. Jorgensen, and J. S. Butel. 2004. Differential ability of two simian virus 40 strains to induce malignancies in weanling hamsters. Virology **330**:168–77.

118. Lowy, D. R., and P. M. Howley. 2001. Papillomaviruses. In: Knipe, D. M., P. M. Howley, D. E. Griffin, R. A. Lamb, M. A. Martin, B. Roizman, and S. E. Straus (eds.), Fields Virology. Philadelphia: Lippincott Williams & Wilkins, pp. 2231–64.

119. zur Hausen, H. 2002. Papillomaviruses and cancer: From basic studies to clinical application. Nat Rev Cancer **2**:342–50.

120. Ilyinskii, P. O., M. D. Daniel, C. J. Horvath, and R. C. Desrosiers. 1992. Genetic analysis of simian virus 40 from brains and kidneys of macaque monkeys. J Virol **66**:6353–60.

121. Newman, J. S., G. B. Baskin, and R. J. Frisque. 1998. Identification of SV40 in brain, kidney and urine of healthy and SIV-infected rhesus monkeys. J Neurovirol **4**:394–406.

122. Lednicky, J. A., A. R. Stewart, J. J. Jenkins, III, M. J. Finegold, and J. S. Butel. 1997. SV40 DNA in human osteosarcomas shows sequence variation among T-antigen genes. Int J Cancer **72**:791–800.

123. Bocharov, G., B. Ludewig, A. Bertoletti, P. Klenerman, T. Junt, P. Krebs, T. Luzyanina, C. Fraser, and R. M. Anderson. 2004. Underwhelming the immune response: Effect of slow virus growth on CD8$^+$-T-lymphocyte responses. J Virol **78**:2247–54.

124. Sullivan, C. S., A. T. Grundhoff, S. Tevethia, J. M. Pipas, and D. Ganem. 2005. SV40-encoded microRNAs regulate viral gene expression and reduce susceptibility to cytotoxic T cells. Nature **435**:682–6.

125. Dorries, K. 1997. New aspects in the pathogenesis of polyomavirus-induced disease. Adv Virus Res **48**:205–61.

126. Rachlin, J., R. Wollmann, and G. Dohrmann. 1988. Inoculation of simian virus 40 into pregnant hamsters can induce tumors in offspring. Lab Invest **58**:26–30.

127. Martini, F., L. Iaccheri, L. Lazzarin, P. Carinci, A. Corallini, M. Gerosa, P. Iuzzolino, G. Barbanti-Brodano, and M. Tognon. 1996. SV40 early region and large T antigen in human brain tumors, peripheral blood cells, and sperm fluids from healthy individuals. Cancer Res **56**:4820–5.

128. Horvath, L. B. 1972. SV40 neutralizing antibodies in the sera of man and experimental animals. Acta Virol **16**:141–6.

129. Zimmermann, W., S. Scherneck, and E. Geissler. 1983. Quantitative determination of papovavirus IgG antibodies in sera from cancer patients, labworkers and several groups of control persons by enzyme-linked immunosorbent assay (ELISA). Zentralbl Bakteriol Mikrobiol Hyg **254**:187–96.

130. Engels, E. A., W. M. Switzer, W. Heneine, and R. P. Viscidi. 2004. Serologic evidence for exposure to simian virus 40 in North American zoo workers. J Infect Dis **190**:2065–9.

131. Jones-Engel, L., G. A. Engel, J. Heidrich, M. Chalise, N. Poudel, R. Viscidi, P. A. Barry, J. S. Allan, R. Grant, and R. Kyes. 2006. Temple monkeys and health implications of commensalism, Kathmandu, Nepal. Emerg Infect Dis **12**:900–6.

132. Bofill-Mas, S., N. Albiñana-Giménez, P. A. Pipkin, P. D. Minor, and R. Girones. 2004. Isolation of SV40 from the environment of a colony of cynomolgus monkeys naturally infected with the virus. Virology **330**:1–7.

133. Ashkenazi, A., and J. L. Melnick. 1962. Induced latent infection of monkeys with vacuolating SV-40 papova virus: Virus in kidneys and urine. Proc Soc Exp Biol Med **111**:367–72.

134. Sheffield, W. D., J. D. Strandberg, L. Braun, K. Shah, and S. S. Kalter. 1980. Simian virus 40-associated fatal interstitial pneumonia and renal tubular necrosis in a rhesus monkey. J Infect Dis **142**:618–22.

135. Minor, P., P. Pipkin, Z. Jarzebek, and W. Knowles. 2003. Studies of neutralising antibodies to SV40 in human sera. J Med Virol **70**:490–5.

136. Nurgalieva, Z. Z., C. Wong, A. K. Zhangabylov, Z. E. Omarbekova, D. Y. Graham, R. A. Vilchez, and J. S. Butel. 2005. Polyomavirus SV40 infections in Kazakhstan. J Infect **50**:142–8.

137. Paracchini, V., S. Garte, P. Pedotti, F. Poli, S. Frison, and E. Taioli. 2005. Molecular identification of simian virus 40 infection in healthy Italian subjects by birth cohort. Mol Med **11**(1–12):48–51.

138. Butel, J. S., S. Jafar, C. Wong, A. S. Arrington, A. R. Opekun, M. J. Finegold, and E. Adam. 1999. Evidence of SV40 infections in hospitalized children. Hum Pathol **30**:1496–502.

139. Jafar, S., M. Rodriguez-Barradas, D. Y. Graham, and J. S. Butel. 1998. Serological evidence of SV40 infections in HIV-infected and HIV-negative adults. J Med Virol **54**:276–84.

140. Meinke, W., D. A. Goldstein, and R. A. Smith. 1979. Simian virus 40-related DNA sequences in a human brain tumor. Neurology **29**:1590–4.

141. Krieg, P., E. Amtmann, D. Jonas, H. Fischer, K. Zang, and G. Sauer. 1981. Episomal simian virus 40 genomes in human brain tumors. Proc Natl Acad Sci USA **78**:6446–50.

142. Carbone, M., H. I. Pass, P. Rizzo, M. Marinetti, M. Di Muzio, D. J. Mew, A. S. Levine, and A. Procopio. 1994. Simian virus 40-like DNA sequences in human pleural mesothelioma. Oncogene **9**:1781–90.

143. Cristaudo, A., A. Vivaldi, G. Sensales, G. Guglielmi, E. Ciancia, R. Elisei, and F. Ottenga. 1995. Molecular biology studies on mesothelioma tumor samples: Preliminary data on H-ras, p21, and SV40. J Environ Pathol Toxicol Oncol **14**:29–34.

144. Pepper, C., B. Jasani, H. Navabi, D. Wynford-Thomas, and A. R. Gibbs. 1996. Simian virus 40 large T antigen (SV40LTAg) primer specific DNA amplification in human pleural mesothelioma tissue. Thorax **51**:1074–6.

145. Shivapurkar, N., T. Wiethege, I. I. Wistuba, E. Salomon, S. Milchgrub, K. M. Muller, A. Churg, H. Pass, and A. F. Gazdar. 1999. Presence of simian virus 40 sequences in malignant mesotheliomas and mesothelial cell proliferations. J Cell Biochem **76**:181–8.

146. Ramael, M., J. Nagels, H. Heylen, S. De Schepper, J. Paulussen, M. De Maeyer, and C. Van Haesendonck. 1999. Detection of SV40 like viral DNA and viral antigens in malignant pleural mesothelioma. Eur Respir J **14**:1381–6.

147. Dhaene, K., A. Verhulst, and E. Van Marck. 1999. SV40 large T-antigen and human pleural mesothelioma: Screening by polymerase chain reaction and tyramine-amplified immunohistochemistry. Virchows Arch Int J Pathol **435**:1–7.

148. Strizzi, L., G. Vianale, M. Giuliano, R. Sacco, F. Tassi, P. Chiodera, P. Casalini, and A. Procopio. 2000. SV40, JC and BK expression in tissue, urine and blood samples from patients with malignant and non-malignant pleural disease. Anticancer Res **20**:885–9.

149. Procopio, A., L. Strizzi, G. Vianale, P. Betta, R. Puntoni, V. Fontana, G. Tassi, F. Gareri, and L. Mutti. 2000. Simian virus-40 sequences are a negative prognostic cofactor in patients with malignant pleural mesothelioma. Genes Chromosome Cancer **29**:173–9.

150. Toyooka, S., H. I. Pass, N. Shivapurkar, Y. Fukuyama, R. Maruyama, K. O. Toyooka, M. Gilcrease, A. Farinas, J. D. Minna, and A. F. Gazdar. 2001. Aberrant methylation and simian virus 40 Tag sequences in malignant mesothelioma. Cancer Res **61**:5727–30.

151. Butel, J. S., C. Wong, R. A. Vilchez, G. Szücs, I. Dömök, B. Kriz, D. Slonim, and E. Adam. 2003. Detection of antibodies to polyomavirus SV40 in two central European countries. Cent Eur J Public Health **11**:3–8.

152. De Sanjose, S., K. V. Shah, E. Domingo-Domenech, E. A. Engels, A. F. De Sevilla, T. Alvaro, M. Garcia-Villanueva, V. Romagosa, E. Gonzalez-Barca, and R. P. Viscidi. 2003. Lack of serological evidence for an association between simian virus 40 and lymphoma. Int J Cancer **104**:522–4.

153. Rollison, D. E., K. J. Helzlsouer, N. A. Halsey, K. V. Shah, and R. P. Viscidi. 2005. Markers of past infection with simian virus 40 (SV40) and risk of incident non-Hodgkin lymphoma in a Maryland cohort. Cancer Epidemiol Biomarkers Prev **14**:1448–52.

154. Viscidi, R. P., D. E. M. Rollison, E. Viscidi, B. Clayman, E. Rubalcaba, R. Daniel, E. O. Major, and K. V. Shah. 2003. Serological cross-reactivities between antibodies to simian virus 40, BK virus, and JC virus assessed by virus-like-particle-based enzyme immunoassays. Clin Diagn Lab Immunol **10**:278–85.

155. Carter, J. J., M. M. Madeleine, G. C. Wipf, R. L. Garcea, P. A. Pipkin, P. D. Minor, and D. A. Galloway. 2003. Lack of serologic evidence for prevalent simian virus 40 infection in humans. J Natl Cancer Inst **95**:1522–30.

156. Shah, K. V., D. A. Galloway, W. A. Knowles, and R. P. Viscidi. 2004. Simian virus 40 (SV40) and human cancer: A review of the serological data. Rev Med Virol **14**:231–9.

157. Engels, E. A., J. Chen, R. P. Viscidi, K. V. Shah, R. W. Daniel, N. Chatterjee, and M. A. Klebanoff. 2004. Poliovirus vaccination during pregnancy, maternal seroconversion to simian virus 40, and risk of childhood cancer. Am J Epidemiol **160**:306–16.

158. Imperiale, M. J. 2001. The human polyomaviruses: An overview. In: Khalili, K., and G. L. Stoner (eds.), Human Polyomaviruses: Molecular and Clinical Perspectives. New York: Wiley-Liss, Inc., pp. 53–71.

159. Vilchez, R. A., and J. S. Butel. 2004. Emergent human pathogen simian virus 40 and its role in cancer. Clin Microbiol Rev **17**:495–508.

160. Lundstig, A., L. Eliasson, M. Lehtinen, K. Sasnauskas, P. Koskela, and J. Dillner. 2005. Prevalence and stability of human serum antibodies to simian virus 40 VP1 virus-like particles. J Gen Virol **86**:1703–8.

161. Melnick, J. L., and S. Stinebaugh. 1962. Excretion of vacuolating SV-40 virus (papova virus group) after ingestion as a contaminant of oral poliovaccine. Proc Soc Exp Biol Med **109**:965–8.

162. Moore, P. S. and Y. Chang. 1998. Kaposi's sarcoma (KS), KS-associated herpesvirus, and the criteria for causality in the age of molecular biology. Am J Epidemiol **147**:217–21.

163. Fredericks, D. N., and D. A. Relman. 1996. Sequence-based identification of microbial pathogens: A reconsideration of Koch's postulates. Clin Microbiol Rev **9**:18–33.

164. McNees, A. L., Z. S. White, P. Zanwar, R. A. Vilchez, and J. S. Butel. 2005. Specific and quantitative detection of human polyomaviruses BKV, JCV, and SV40 by real time PCR. J Clin Virol **34**:52–62.

165. Pal, A., L. Sirota, T. Maudru, K. Peden, and A. M. Lewis, Jr. 2006. Real-time quantitative PCR assays for the detection of virus-specific DNA in samples with mixed populations of polyomaviruses. J Virol Methods **135**:32–42.

166. Elfaitouri, A., A. L. Hammarin, and J. Blomberg. 2006. Quantitative real-time PCR assay for detection of human polyomavirus infection. J Virol Methods **135**:207–13.

167. Cesarman, E., and D. M. Knowles. 1997. Kaposi's sarcoma-associated herpesvirus: A lymphotropic human herpesvirus associated with Kaposi's sarcoma, primary effusion lymphoma, and multicentric Castleman's disease. Semin Diagn Pathol **14**:54–66.

168. Gerard, L., F. Agbalika, J. Sheldon, A. Maillard, T. F. Schulz, and E. Oksenhendler. 2001. No increased human herpesvirus 8 seroprevalence in patients with HIV-associated non-Hodgkin's lymphoma. J Acquir Immune Defic Syndr **26**:182–4.

169. Vilchez, R. A., and J. S. Butel. 2003. SV40 in human brain cancers and non-Hodgkin's lymphoma. Oncogene **22**:5164–72.

170. Liang, R., and S. Y. Ma. 2004. Higher incidence of simian virus 40 in primary gastric diffuse large B cell lymphoma (DLBCL) than primary nodal disease in Chinese patients, but of no prognostic implication. (Abstract #6671). Annual Meeting of the American Society of Clinical Oncologists.

171. Heinsohn, S. 2004. Role of SV40 in childhood lymphomas: Detection by real time quantitative TaqMan assay. International Symposium on Predictive Oncology and Intervention Strategies, Nice, France, February 7–10, 2004.

172. Capello, D., D. Rossi, G. Gaudino, A. Carbone, and G. Gaidano. 2003. Simian virus 40 infection in lymphoproliferative disorders. Lancet **361**:88–9.

173. MacKenzie, J., K. S. Wilson, J. Perry, A. Gallagher, and R. F. Jarrett. 2003. Association between simian virus 40 DNA and lymphoma in the United Kingdom. J Natl Cancer Inst **95**:1001–3.

174. Hernández-Losa, J., C. G. Fedele, F. Pozo, A. Tenorio, V. Fernández, J. Castellvi, C. Parada, and S. Ramón y Cajal. 2005. Lack of association of polyomavirus and herpesvirus types 6 and 7 in human lymphomas. Cancer **103**:293–8.

175. Schüler, F., S. C. Dölken, C. Hirt, M. T. Dölken, R. Mentel, L. G. Gürtler, and G. Dölken. 2006. No evidence for simian virus 40 DNA sequences in malignant non-Hodgkin lymphomas. Int J Cancer **118**:498–504.

176. Carbone, M., M. A. Rdzanek, J. J. Rudzinski, M. A. De Marco, M. Bocchetta, M. R. Niño, B. Mossman, and H. I. Pass. 2005. SV40 detection in human tumor specimens. Cancer Res **65**:10120–1.

177. Rizzo, P., I. Di Resta, A. Powers, H. Ratner, and M. Carbone. 1999. Unique strains of SV40 in commercial poliovaccines from 1955 not readily identifiable with current testing for SV40 infection. Cancer Res **59**:6103–8.

178. Krieg, P., and G. Scherer. 1984. Cloning of SV40 genomes from human brain tumors. Virology **138**:336–40.

179. Jaffe, E. S., N. L. Harris, H. Stein, and J. W. Vardiman. 2001. World Health Organization classification of tumours. Pathology and genetics of tumours of hematopoietic and lymphoid tissues. Lyon, France: IARC Press.

180. Brousset, P., V. de Araujo, and R. D. Gascoyne. 2004. Immunohistochemical investigation of SV40 large T antigen in Hodgkin and non-Hodgkin's lymphoma. Int J Cancer **112**:533–5.

181. Hans, C. P., D. D. Weisenburger, T. C. Greiner, R. D. Gascoyne, J. Delabie, G. Ott, H. K. Müller-Hermelink, E. Campo, R. M. Braziel, E. S. Jaffe, Z. Pan, P. Farinha, L. M. Smith, B. Falini, A. H. Banham, A. Rosenwald, L. M. Staudt, J. M. Connors, J. O. Armitage, and W. C. Chan. 2004. Confirmation of the

molecular classification of diffuse large B-cell lymphoma by immunohistochemistry using a tissue microarray. Blood **103**:275–82.

182. Rosenwald, A., G. Wright, W. C. Chan, J. M. Connors, E. Campo, R. I. Fisher, R. D. Gascoyne, H. K. Muller-Hermelink, E. B. Smeland, L. M. Staudt, and for the Lymphoma/Leukemia Molecular Profiling Project. 2002. The use of molecular profiling to predict survival after chemotherapy for diffuse large-B-cell lymphoma. N Engl J Med **346**:1937–47.

183. Klein, U., A. Gloghini, G. Gaidano, A. Chadburn, E. Cesarman, R. Dalla-Favera, and A. Carbone. 2003. Gene expression profile analysis of AIDS-related primary effusion lymphoma (PEL) suggests a plasmablastic derivation and identifies PEL-specific transcripts. Blood **101**:4115–21.

184. Hoffmann, C., M. Tiemann, C. Schrader, D. Janssen, E. Wolf, M. Vierbuchen, R. Parwaresch, K. Ernestus, A. Plettenberg, A. Stoehr, G. Fatkenheuer, C. Wyen, M. Oette, and H.-A. Horst. 2005. AIDS-related B-cell lymphoma (ARL): Correlation of prognosis with differentiation profiles assessed by immunophenotyping. Blood **106**:1762–9.

185. Janeway, C. A., P. Travers, M. Walport, and M. Shlomchik. 2001. Immunobiology. New York: Garland Publishing.

186. Cole, S. L., and M. J. Tevethia. 2002. Simian virus 40 large T antigen and two independent T-antigen segments sensitize cells to apoptosis following genotoxic damage. J Virol **76**:8420–32.

187. Cuconati, A., and E. White. 2002. Viral homologs of BCL-2: Role of apoptosis in the regulation of virus infection. Genes Dev **16**:2465–78.

188. Hatzivassiliou, E., and G. Mosialos. 2002. Cellular signaling pathways engaged by the Epstein-Barr virus transforming protein LMP1. Front Biosci **7**:d319–29.

189. Jackson, S., and A. Storey. 2000. E6 proteins from diverse cutaneous HPV types inhibit apoptosis response to UV damage. Oncogene **19**:592–8.

190. Portis, T., J. C. Harding, and L. Ratner. 2001. The contribution of NF-κB activity to spontaneous proliferation and resistance to apoptosis in human T-cell leukemia virus type 1 Tax-induced tumors. Blood **98**:1200–8.

191. Thomson, B. J. 2001. Viruses and apoptosis. Int J Exp Pathol **82**:65–76.

192. Tsai, S. C., K. B. S. Pasumarthi, L. Pajak, M. Franklin, B. Patton, H. Wang, W. J. Henzel, J. T. Stults, and L. J. Field. 2000. Simian virus 40 large T antigen binds a novel Bcl-2 homology domain 3-containing proapoptosis protein in the cytoplasm. J Biol Chem **275**:3239–46.

193. Esteller, M. 2002. CpG island hypermethylation and tumor suppressor genes: A booming present, a brighter future. Oncogene **21**:5427–40.

194. Toyooka, S., M. Carbone, K. O. Toyooka, M. Bocchetta, N. Shivapurkar, J. D. Minna, and A. F. Gazdar. 2002. Progressive aberrant methylation of the *RASSF1A* gene in simian virus 40 infected human mesothelial cells. Oncogene **21**:4340–4.

195. Ng, M. H. L. 2002. Death associated protein kinase: From regulation of apoptosis to tumor suppressive functions and B cell malignancies. Apoptosis **7**:261–70.

196. Reddy, A. N., W. W. Jiang, M. Kim, N. Benoit, R. Taylor, J. Clinger, D. Sidransky, and J. A. Califano. 2003. Death-associated protein kinase promoter hypermethylation in normal human lymphocytes. Cancer Res **63**:7694–8.

197. Engels, E. A., H. A. Katki, N. M. Nielsen, J. F. Winther, H. Hjalgrim, F. Gjerris, P. S. Rosenberg, and M. Frisch. 2003. Cancer incidence in Denmark following exposure to poliovirus vaccine contaminated with simian virus 40. J Natl Cancer Inst **95**:532–9.

198. Rollison, D. E. M., W. F. Page, H. Crawford, G. Gridley, S. Wacholder, J. Martin, R. Miller, and E. A. Engels. 2004. Case–control study of cancer among US Army veterans exposed to simian virus 40-contaminated adenovirus vaccine. Am J Epidemiol **160**:317–24.

199. Engels, E. A., L. H. Rodman, M. Frisch, J. J. Goedert, and R. J. Biggar. 2003. Childhood exposure to simian virus 40-contaminated poliovirus vaccine and risk of AIDS-associated non-Hodgkin's lymphoma. Int J Cancer **106**:283–7.

200. Dang-Tan, T., S. M. Mahmud, R. Puntoni, and E. L. Franco. 2004. Polio vaccines, Simian Virus 40, and human cancer: The epidemiologic evidence for a causal association. Oncogene **23**:6535–40.

201. Engels, E. A. 2005. Does simian virus 40 cause non-Hodgkin lymphoma? A review of the laboratory and epidemiological evidence. Cancer Invest **23**:529–36.

202. Shah, K. V. 2004. Simian virus 40 and human disease. J Infect Dis **190**:2065–9.

203. Kimura, M. 1980. A simple method for estimating evolutionary rates of base substitutions through comparative studies of nucleotide sequences. J Mol Evol **16**:111–20.

9. HIV–HBV AND HIV–HCV COINFECTION AND LIVER CANCER DEVELOPMENT

JIANMING HU AND LAURIE LUDGATE

Department of Microbiology and Immunology, The Pennsylvania State University College of Medicine, Hershey, PA

INTRODUCTION

As antiretroviral therapies continue to improve, patients infected with HIV are living longer, and the health problems that are of primary concern to these patients are changing in recent years. Before the advent of highly active antiretroviral therapy (HAART), HIV-infected patients were most likely to succumb to opportunistic bacterial or fungal infections, secondary to HIV-induced immune suppression. With longer survival times, liver disease and hepatocellular carcinoma (HCC) are becoming increasingly important in these patients.[1,2] In fact, approximately 10–15% of deaths in HIV patients are now due to liver disease. In patients with HIV, most liver disease is due to chronic viral hepatitis. This is not surprising considering that agents causing viral hepatitis, like the hepatitis B virus (HBV) and hepatitis C virus (HCV), are transmitted through similar routes as HIV. An additional complication involving the liver in HIV-infected patients is the fact that many anti-HIV drugs cause hepatotoxicity, which is further exacerbated by viral hepatitis.

CHRONIC HBV AND HCV INFECTIONS

Worldwide, there are over 500 million people who are chronically infected with HBV or HCV, or both. HBV is a small, enveloped DNA virus that belongs to the *Hepadnaviridae* family. HBV is unusual for a DNA virus in that replication of the viral DNA genome is through the reverse transcription of a RNA intermediate.[3] The small, 3.2 kb HBV genome encodes four open reading frames (ORFs) (Fig. 1),

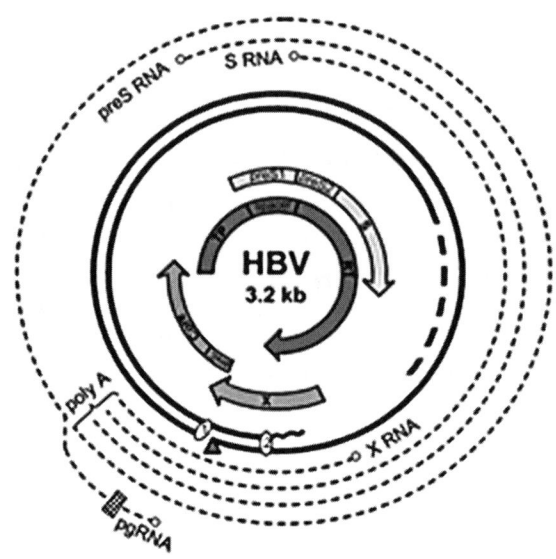

Figure 1. HBV genome organization. *Solid lines*, the partially double stranded, relaxed circular DNA genome; *dotted lines*, viral RNA transcripts; *solid arrows*, encoded proteins. Core, core protein; S, surface protein, RT, reverse transcriptase; X, X protein. Triangle, the RT protein covalently linked to the genome; checked box, the ε RNA packaging signal; ovals, direct repeat 1 and 2, *cis*-acting elements involved in reverse transcription

which are translated to make the viral core (C) protein, the main constituent of the viral nucleocapsid; the reverse transcriptase (RT), the enzyme responsible for DNA replication via reverse transcription; and three envelope glycoproteins. In addition, the HBV X (HBx) protein has a number of pleotropic effects on viral and cellular gene expression, cell signaling, cell cycle, and apoptosis, although the significance of these in viral replication or pathogenesis remains unresolved. HBV is transmitted by contact with blood or other body fluids of an infected person in the same way as HIV. However, HBV is 50–100 times more infectious than HIV.

Of the two billion people who have been infected with HBV, more than 350 million have chronic, lifelong infections.[4,5] In the US alone, there are 1.25 million chronic HBV carriers. In some areas of Asia and Africa, where HBV is endemic, 10–20% of the whole population is chronically infected with HBV (Fig. 2). In these parts of the world, HBV infections are acquired mainly perinatally or early in childhood and have a high (up to 90%) probability of becoming chronic. In contrast, 5–15% of adults who acquire HBV infection will become chronic carriers of the virus. Patients who are chronically infected with HBV are at high risk of premature death from cirrhosis of the liver and HCC, a highly malignant liver cancer. The risk of death from HBV-related liver cancer or cirrhosis is approximately 25% for persons who become chronically infected. Together, these diseases kill about one million persons each year worldwide.

Figure 2. Global prevalence of chronic HBV infection. A map showing the percentage of population chronically infected with HBV in different regions of the world

HCV is an enveloped RNA virus belonging to the *Flaviviridae* family.[6] Like other flaviviruses, the 9-kb long, positive sense, single stranded RNA genome (Fig. 3) of HCV is translated into a polyprotein, which is proteolytically cleaved into the viral structural proteins, including a single capsid (C) protein and two envelope glycoproteins (E1 and E2), and the nonstructural proteins required for viral replication, including two proteases (NS2 and NS3) and the viral RNA-dependent RNA polymerase (NS5b). Like HBV, HCV is transmitted through blood and other body fluids.

HCV frequently causes persistent infection of the liver, although there is no DNA form in its life cycle or latent stage known.[7,8] In fact, the chance of chronic infection with HCV is approximately 55–85% and this varies little with age, in contrast to HBV infections. An estimated 170 million people worldwide are chronically infected with HCV (Fig. 4). Approximately 3.9 million (1.8%) Americans have been infected with HCV, 2.7 million of whom are chronically infected. Approximately 5–20% of chronically infected persons develop liver cirrhosis over a period of 20–30 years, and HCC develops in 1–5% of persons with chronic HCV infection.

COINFECTION OF HBV OR HCV WITH HIV

Worldwide, approximately 40 million people are infected with HIV. Roughly 1.2 million of these people live in the US. In Europe and US, approximately 8–16% of HIV patients are also chronically infected with HBV.[1,9,10] Worldwide, the number of HIV-infected people chronically infected with HBV is estimated at 2–4 million (Fig. 5), with a large proportion of these coming from HBV-endemic regions of Asia and Africa. The risk of developing chronic HBV infection is about threefold

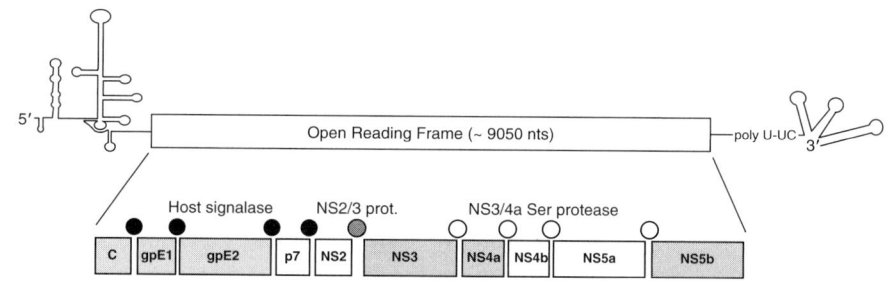

Figure 3. HCV genome organization. The positive sense, single stranded HCV RNA genome encodes a long ORF (ca. 3,000 amino acids). The viral polyprotein is proteolytically cleaved into the structural proteins, C (capsid), gpE1 and gpE2 (envelope glycoprotein 1 and 2), as well as the nonstructural (NS) proteins, NS2 to NS5b, via host (*filled circles*) and viral proteases (*open circles*). The structured 5' and 3' noncoding sequences are important for viral replication and translation

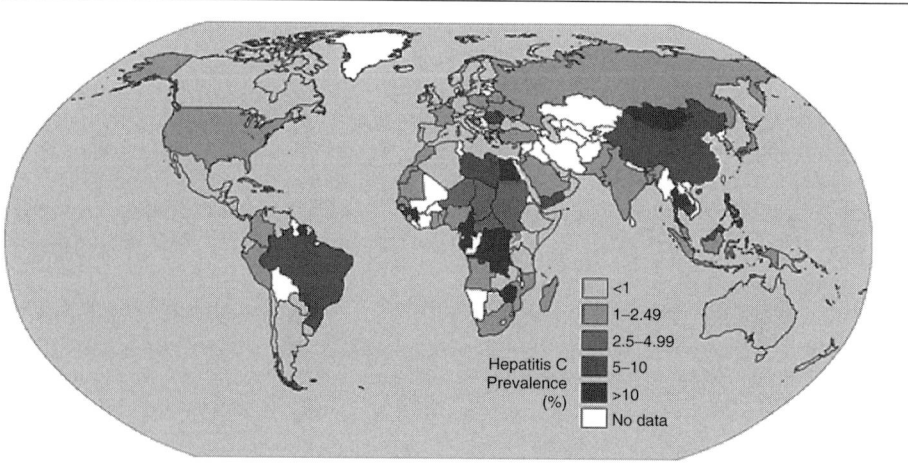

Figure 4. Global prevalence of chronic HCV infection. A map showing the percentage of population chronically infected with HCV in different regions of the world

to sixfold higher in HIV-infected patients than in those who are not infected with HIV, likely due to the fact that HIV-induced immune suppression can reduce the patient's ability to clear HBV.[11] Furthermore, HIV-induced immune suppression may also play a role in the reactivation of latent HBV infections, which are thought to be under immune control following clinical but not virological resolution.[12]

The number of HIV-infected people who are chronically infected with HCV worldwide is approximately 4–5 million (Fig. 5). In developed countries, ~25% of HIV patients also have chronic hepatitis C infection. The number of coinfections varies depending on the route of transmission; for example, the incidence of

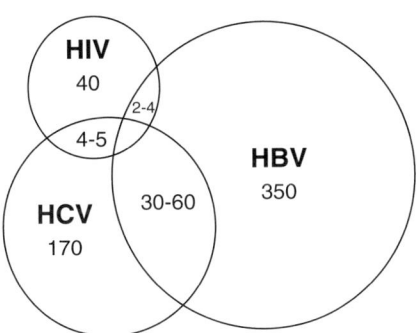

Figure 5. Incidences of chronic HBV, HCV, and HIV coinfections. A Venn diagram depicting coinfection of HIV with HBV or HCV, and of HBV with HCV. Shown are estimated numbers (in millions) of patients singly infected with each virus or doubly infected

HIV/HCV coinfection is higher in populations of injecting drug users (IDU) than in those who were infected by sexual transmission.

HBV–HCV DUAL INFECTIONS AND OCCULT HBV INFECTION

The incidence of HBV–HCV coinfection is not uncommon (Fig. 5), as might be predicted from their shared transmission routes.[13,14] About 10–20% chronic HBV-infected patients are also infected with HCV. In addition, occult HBV infection, defined by the absence of the HBsAg, the main HBV envelope protein, in the serum but the presence of antibodies against the viral core antigen (HBcAg) or HBV DNA, is fairly common.[15] This may be an important factor in the development of HCC in patients with no serologic evidence of HBV or HCV infection, and additionally, may play a role in the development of HCC in chronically infected HCV patients.[16]

HEPATOTROPISM AND LYMPHOTROPISM OF HBV AND HCV

While there is little dispute that HBV and HCV infect the hepatocytes in the liver, it remains controversial as to whether these viruses also infect other cell types, in particular, lymphoid cells. It has been reported that HBV DNA can be detected in peripheral blood mononuclear cells (PBMC) and some HBV isolates may in fact be able to infect PBMC,[17] although the question of true infection or passive endocytosis of virus is still being debated. With respect to HCV, there are many reports of infection of lymphoid cells, particularly B lymphocytes[18] but the level of HCV replication seems to be rather low in general and specific detection problematic. On the other hand, HIV is lymphotropic and infects only a small percentage of CD4+ T cells. Therefore, although the possibility exists that HIV and HCV or HBV may infect the same cell types, true dual infection of the same cell seems unlikely.

HEPATOCELLULAR CARCINOMA

HCC is the most common primary cancer of the liver, accounting for 60–90% of all hepatic malignancies.[19–22] It is the sixth most common cancer among men and the 11th most common cancer in women worldwide. Particularly, in HBV endemic regions of sub-Saharan Africa and Eastern Asia, HCC is the most prevalent cancer and incidences can be as high as 50–150 cases per 100,000 population.[23] In North America and Western Europe where the HBV infection rate is relatively low, HCC incidence is below 10 per 100,000. However, there has been a recent surge of HCC in the developed world, most likely due to the prevalence of HCV infections in these areas.[24,25] Together, chronic HBV and HCV infections are responsible for over 80–90% of all HCC on a global scale, and account for 5% of all human cancer burden.[26] Chronic HBV carriers have been shown to be at a 100–233-fold increased risk for the development of HCC. HBV infection is responsible for the majority of HCC development in the developing world and accounts for 15–20% of HCC in the US. In contrast, HCV accounts for the majority of HCC in the developed world.[27] In addition, as mentioned earlier, coinfection of both HBV and HCV is also common and can further increase the risk of HCC development.

Despite the clear epidemiological evidence that HBV and HCV are responsible for the vast majority of HCC, the mechanism of viral hepatocarcinogenesis remains incompletely understood (Fig. 6).[19–21,28–30] There are several possible mechanisms by which liver cancer may develop in patients with chronic viral hepatitis. The first is by an indirect means: chronic viral infection of the liver produces a state of persistent inflammation, in which cancer is a nonspecific side effect of the immune response against the HBV or HCV infection. Thus, HCC may develop as a result of the continuous damage and regeneration of the liver cells in a mutagenic

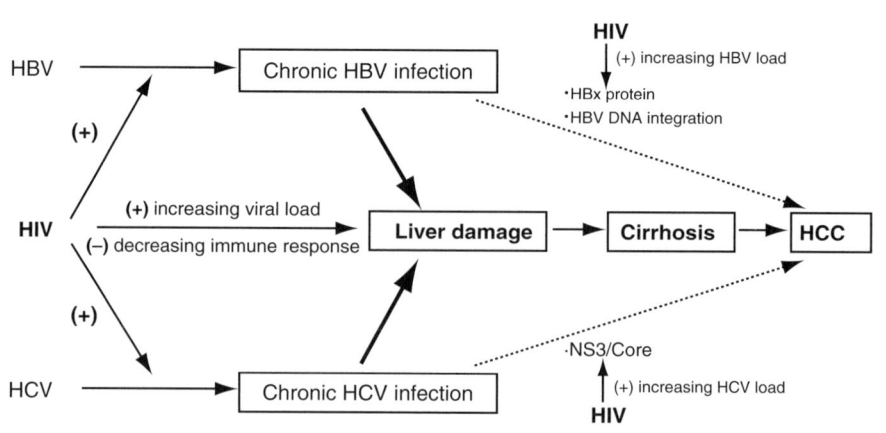

Figure 6. Potential mechanisms HBV- and HCV-induced hepatocarcinogenesis and its enhancement by HIV. *Solid arrows*, likely mechanisms; *dotted arrows*, uncertain mechanisms; (+), enhancing effect; (−), inhibitory effect. See text for details

inflammatory environment, which ultimately leads to the aberrant activation of one or more cellular proto-oncogenes or the inactivation of tumor suppressor genes. Evidence is rather strong in support of this nonspecific carcinogenic mechanism. It is now clear that chronic inflammation and tissue damage over a long period, per se, can be carcinogenic, regardless of the initial trigger events. Thus, not only chronic HBV and HCV infections, but also alcoholic liver damage and metabolic liver damage as a result of genetic mutations including α1-antitrypsin deficiency, Wilson's disease, and hemachromatosis all increase the risk of liver cancer.[22] Indeed, the majority of HCC, regardless of etiology, arises in the background of liver cirrhosis resulting from chronic liver damage.

Another potential mechanism of viral hepatocarcinogenesis involves a more direct role of viral proteins.[20,28,30] Thus, although the different causes of chronic liver damage, as outlined above, can all increase the risk of HCC, chronic HBV or HCV infections present a much greater risk of liver cancer than other hepatic inflammatory disorders. In addition, HCC may develop in chronic HBV infections in the absence of cirrhosis. Although the exact viral proteins and their carcinogenic mechanism remain to be elucidated, the HBx protein and the HCV core, NS3, NS4B and NS5A proteins have been reported to have transforming potential in overexpression systems. As HCC generally only develops after decades of HBV or HCV infections, it seems unlikely that these viruses encode any bona fide oncogenes. Caution is therefore warranted in the interpretation of these overexpression studies out of the context of natural viral infection.

A third molecular mechanism of carcinogenesis, available to HBV but not HCV, is insertional mutagenesis as a result of HBV DNA integration into the host chromosomes.[31] It has been known for decades that HBV DNA does integrate into the host genome during infection although integration is not an obligatory step in viral replication, in contrast to retroviruses. In fact, most HBV-related HCCs harbor HBV DNA integration. HBV DNA has been shown to integrate into the host DNA at multiple sites in a seemingly random fashion, although more recent reports suggest that there may be some preferred sites of integration.[32,33] What is still unclear is the etiological role of DNA integration in HBV carcinogenesis in humans. Elegant studies using the woodchuck hepatitis virus (WHV), a member of the mammalian hepadnaviruses closely related to HBV, have convincingly demonstrated that viral DNA integration, specifically the insertional activation of the cellular *myc* proto-oncogenes, plays a critical role in liver cancer development in chronically WHV-infected woodchucks, virtually all of which develop HCC.[34–36] Reports of potentially similar insertional activation of cellular oncogenes in human liver cancer have appeared[37,38] but its prevalence and true significance in carcinogenesis remain to be clarified.

DOES HIV COINFECTION INCREASE THE RISK OF HCC ASSOCIATED WITH HBV OR HCV INFECTION?

Epidemiological evidence suggests that coinfection with HIV may increase the risk of HCC development in HBV- or HCV-infected patients, although the data so far do not seem definitive.[9,13,39–49] As mentioned above, HIV coinfection can decrease the rate of HBV or HCV clearance and increase the risk of chronic HBV or HCV infection and the risk of reactivation of latent HBV infections (Fig. 6).[9,11] HIV infection increases the viral load of HBV or HCV in chronically infected patients. An inverse correlation between CD4[+] T-cell count and HCV viral load has been noted. For both HBV and HCV-infected patients, HIV coinfection seems to accelerate the progression of liver diseases, leading to increased risk of cirrhosis and possibly liver cancer.[9,50] HIV infection alone has also been reported to increase the risk of HCC, as well as other cancers, presumably as a result of defective immune surveillance against tumorigenesis.[49,51] Although no clear data are available, triple infection with HIV, HBV, and HCV is not uncommon and may present an even greater risk of HCC development than the double infections. The lack of definitive study in this area is not without good reason. It takes years to decades for HCC to develop in HBV- or HCV-infected patients and thus long-term follow-ups are necessary to assess any increased cancer risks in HIV coinfected patients. Before definitive measurements of liver cancer incidences, surrogate markers are sometimes used to predict HCC development. These surrogates include the progression of cirrhosis, which, as mentioned earlier, almost always precedes the development of HCC, and increased viral load, which may predict increased risk of HCC (Fig. 6). In the case of HBV, there is strong evidence now to indicate that HBV viral load is in fact directly correlated with liver disease progression and the risk of HCC development,[52–54] and effective antiviral treatment to decrease HBV replication has been shown to decrease the risk of HCC development.[55–58] However, HCV viral load in the blood does not seem to be correlated directly with liver disease progression.[59,60]

In addition to the difficulties encountered in the epidemiological studies to assess the HCC risk in HIV–HBV or HIV–HCV coinfections, laboratory studies to analyze coinfections are also severely limited by the lack of cell culture systems or convenient animal models that can be infected by both HIV and HBV or HCV. Infection of chimpanzees, which are susceptible to all three viruses, is a possibility, but these studies would be very costly and take years to conduct. As already alluded to earlier, the chance of HIV coinfecting the same cell with either HBV or HCV seems to be remote. Any effect of HIV infection on HCC risk associated with HBV or HCV would be unlikely to be exerted at the level of direct virus–virus interactions. Rather, indirect effects of HIV-mediated immune dysfunction on HBV- or HCV-induced hepatocarcinogenesis are more likely (Fig. 6). In addition to increasing the chance of chronic HBV and HCV infections, which in turn increase the risk of HCC, increased HBV load associated with HIV coinfection could also potentially exacerbate liver damage and thus accelerate disease progression and ultimately cancer development. Recent studies suggest that under conditions of severe immunodeficiency, the normally noncytopathic HBV can damage the infected cells directly with uncontrolled high-level replication.[61] Similarly, although HCV is usually considered to be noncytopathic, HCV replication or HCV proteins may nevertheless directly induce cellular damage such as steatosis,[62] which also occurs more frequently in HCV–HIV coinfections.[63] On the other hand, a decreased immune response against HBV or HCV might actually reduce liver inflammation and damage[49] and, thus, may slow down progression to cirrhosis and cancer (Fig. 6). The increased HBV or HCV load in HIV coinfected patients can, in principle, also influence cancer progression by increasing the expression of HBV or HCV proteins involved in cellular disregulation and transformation and in the case of HBV, the chance of insertional mutagenesis (Fig. 6).

Although HIV is not known to infect hepatocytes, HIV proteins may nevertheless still be able to influence the HBV or HCV-infected cells. For example, the HIV transactivating protein, Tat, which can be present systemically during HIV infection, has been reported to enhance the development of liver cancer[64] and may thus influence the development of HCC in HIV coinfected HBV or HCV patients.

ANTIRETROVIRAL THERAPY-ASSOCIATED HEPATOTOXICITY

Antiretroviral therapy induced liver toxicity is an additional concern in HIV coinfected HCV or HBV patients. One out of eight patients treated with antiretroviral drugs shows hepatotoxicity, a situation that is more likely to occur in HBV- or HCV-infected patients and further exacerbates liver damage accompanying chronic HBV or HCV infections.[1,9,65] Antiretroviral drugs that have shown hepatotoxicity include certain nucleoside analogs (HIV RT inhibitors) and HIV protease inhibitors. A further complication in the treatment of HIV–HBV coinfected patients is the fact that some nucleoside analogs, such as 3TC (lamivudine), are active against both the HIV and HBV RT, and can select for drug-resistant mutants of both viruses. HIV-infected patients are sometimes treated intermittently, in order to prevent the selection of drug-resistant HIV. However, hepatic flare can result when a patient is taken off antiretroviral therapy. This is thought to be due to HBV viral rebound upon drug withdrawal and can lead to an increase in liver damage and subsequent progression to cirrhosis and HCC. It is, therefore, important to ensure that the coinfecting HBV infection continues to be treated while the patient is off the HIV treatment, e.g., using nucleoside analogs specific for the HBV RT but inactive against the HIV RT.

HGV–HIV COINFECTION

An intriguing interaction between HIV and another prevalent human virus, the hepatitis G virus (HGV or GBV-C), may in fact be beneficial to the host.[66,67] Although initially thought to be one of the viruses that can cause hepatitis (hence the name HGV), HGV is not known to cause any human disease but is a relatively common virus that is found worldwide. Like HCV, it is a single-stranded, positive-sense RNA virus that belongs to the family *Flaviviridae*. HGV is transmitted through blood and other body fluids, similar to HIV, HBV, or HCV. Different from HCV, HGV primarily infects lymphocytes. Interestingly, in persons coinfected with HIV and HGV, HGV appears to confer protection against the progression to AIDS.[68,69] This was largely demonstrated before the advent of HAART but has also been shown in the post-HAART era. Furthermore, HIV replication in vitro was shown to be inhibited by coinfection with HGV, suggesting a mechanism involving direct interaction between the viruses.[70] However, not all studies on the effect of HGV coinfection have reported favorable results and the mechanism that provides this putative protection has yet to be elucidated. For the purposes of this review, HGV does not appear to influence liver disease in HBV or HCV coinfections.

SUMMARY

Liver diseases caused by chronic HBV or HCV infection, including cirrhosis and HCC, are emerging as an increasingly important problem faced by millions of HIV-infected patients who are coinfected with HBV or HCV. On one hand, HIV-induced immune suppression enhances the risk of chronic viral hepatitis, increases HBV or HCV load, and may hasten the progression to cirrhosis and liver cancer. On the other hand, significant hepatotoxicity is associated with a number of anti-retroviral drugs, further exacerbating liver damage associated with chronic viral hepatitis. The exact risk of HCC in HIV and HBV or HCV coinfected patients remains to be fully assessed. The elucidation of the multiple virus–virus and virus–host interactions that underlie viral hepatocarcinogenesis and potential HIV enhancement awaits the establishment of appropriate in vitro and in vivo model systems. As millions of HIV-infected patients in the developing countries are gaining access to HAART therapy for their HIV infections, endemic HBV and HCV infections and their associated liver diseases will only become more problematic on a global level. To ameliorate the suffering from HBV- and HCV-induced liver cancer in HIV patients, more effective treatment for chronic HBV and HCV infections are needed. The long time frame of viral hepatocarcinogenesis may afford a window of opportunity to develop and improve such treatment.

REFERENCES

1. Thomas, D. L. 2006. Growing importance of liver disease in HIV-infected persons. Hepatology **43**(2 Suppl 1):S221–9.
2. Cacoub, P., L. Geffray, E. Rosenthal, et al. 2001. Mortality among human immunodeficiency virus-infected patients with cirrhosis or hepatocellular carcinoma due to hepatitis C virus in French Departments of Internal Medicine/Infectious Diseases, in 1995 and 1997. Clin Infect Dis **32**(8):1207–14.
3. Ganem, D., and A. M. Prince. 2004. Hepatitis B virus infection – natural history and clinical consequences. N Engl J Med **350**(11):1118–29.
4. Yim, H. J., and A. S. Lok. 2006. Natural history of chronic hepatitis B virus infection: What we knew in 1981 and what we know in 2005. Hepatology **43**(2 Suppl 1):S173–81.
5. Wright, T. L. 2006. Introduction to chronic hepatitis B infection. Am J Gastroenterol **101**(Suppl 1):S1–6.
6. Lindenbach, B. D., and C. M. Rice, Unravelling hepatitis C virus replication from genome to function. Nature, 2005. 436(7053): p. 933–8.
7. Chen, S. L., and T. R. Morgan. 2006. The natural history of hepatitis C virus (HCV) infection. Int J Med Sci **3**(2):47–52.
8. Seeff, L. B. 1999. Natural history of hepatitis C. Am J Med **107**(6B):10S–15S.
9. Park, J. S., N. Saraf, and D. T. Dieterich. 2006. HBV plus HCV, HCV plus HIV, HBV plus HIV. Curr Gastroenterol Rep **8**(1):67–74.
10. Alter, M. J. 2006. Epidemiology of viral hepatitis and HIV co-infection. J Hepatol **44**(1 Suppl):S6–9.
11. Lascar, R. M., A. R. Lopes, R. J. Gilson, et al. 2005. Effect of HIV infection and antiretroviral therapy on hepatitis B virus (HBV)-specific T cell responses in patients who have resolved HBV infection. J Infect Dis **191**(7):1169–79.
12. Yeo, W., K. C. Lam, B. Zee, et al. 2004. Hepatitis B reactivation in patients with hepatocellular carcinoma undergoing systemic chemotherapy. Ann Oncol **15**(11):1661–6.

13. Sterling, R. K., and M. S. Sulkowski. 2004. Hepatitis C virus in the setting of HIV or hepatitis B virus coinfection. Semin Liver Dis **24**(Suppl 2):61–8.
14. Liu, Z., and J. Hou. 2006. Hepatitis B virus (HBV) and hepatitis C virus (HCV) dual infection. Int J Med Sci **3**(2):57–62.
15. Chemin, I., and C. Trepo. 2005. Clinical impact of occult HBV infections. J Clin Virol **34**(Suppl 1):S15–21.
16. Tamori, A., S. Nishiguchi, S. Shiomi, et al. 2005. Hepatitis B virus DNA integration in hepatocellular carcinoma after interferon-induced disappearance of hepatitis C virus. Am J Gastroenterol **100**(8):1748–53.
17. Michalak, T. I., P. M. Mulrooney, and C. S. Coffin. 2004. Low doses of hepadnavirus induce infection of the lymphatic system that does not engage the liver. J Virol **78**(4):1730–8.
18. Pal, S., D. G. Sullivan, S. Kim, et al. 2006. Productive replication of hepatitis C virus in perihepatic lymph nodes in vivo: Implications of HCV lymphotropism. Gastroenterology **130**(4):1107–16.
19. Brechot, C. 2004. Pathogenesis of hepatitis B virus-related hepatocellular carcinoma: Old and new paradigms. Gastroenterology **127**(5 Suppl 1):S56–61.
20. Branda, M., and J. R. Wands. 2006. Signal transduction cascades and hepatitis B and C related hepatocellular carcinoma. Hepatology **43**(5):891–902.
21. Thorgeirsson, S. S., and J. W. Grisham. 2002. Molecular pathogenesis of human hepatocellular carcinoma. Nat Genet **31**(4):339–46.
22. Fattovich, G., T. Stroffolini, I. Zagni, et al. 2004. Hepatocellular carcinoma in cirrhosis: Incidence and risk factors. Gastroenterology **127**(5 Suppl 1):S35–50.
23. Chen, J. G., J. Zhu, D. M. Parkin, et al. 2006. Trends in the incidence of cancer in Qidong, China, 1978–2002. Int J Cancer.
24. Davila, J. A., R. O. Morgan, Y. Shaib, et al. 2004. Hepatitis C infection and the increasing incidence of hepatocellular carcinoma: A population-based study. Gastroenterology **127**(5):1372–80.
25. El-Serag, H. B. 2004. Hepatocellular carcinoma: Recent trends in the United States. Gastroenterology **127**(5 Suppl 1):S27–34.
26. Parkin, D. M. 2006. The global health burden of infection-associated cancers in the year 2002. Int J Cancer **118**(12):3030–44.
27. Tradati, F., M. Colombo, P. M. Mannucci, et al. 1998. A prospective multicenter study of hepatocellular carcinoma in Italian hemophiliacs with chronic hepatitis C. The Study Group of the Association of Italian Hemophilia Centers. Blood **91**(4):1173–7.
28. Liang, T. J., and T. Heller. 2004. Pathogenesis of hepatitis C-associated hepatocellular carcinoma. Gastroenterology **127**(5 Suppl 1):S62–71.
29. Block, T. M., A. S. Mehta, C. J. Fimmel, et al. 2003. Molecular viral oncology of hepatocellular carcinoma. Oncogene **22**(33):5093–107.
30. Anzola, M. 2004. Hepatocellular carcinoma: Role of hepatitis B and hepatitis C viruses proteins in hepatocarcinogenesis. J Viral Hepat **11**(5):383–93.
31. Cougot, D., C. Neuveut, and M. A. Buendia. 2005. HBV induced carcinogenesis. J Clin Virol **34**(Suppl 1): S75–8.
32. Murakami, Y., K. Saigo, H. Takashima, et al. 2005. Large scaled analysis of hepatitis B virus (HBV) DNA integration in HBV related hepatocellular carcinomas. Gut **54**(8):1162–8.
33. Minami, M., Y. Daimon, K. Mori, et al. 2005. Hepatitis B virus-related insertional mutagenesis in chronic hepatitis B patients as an early drastic genetic change leading to hepatocarcinogenesis. Oncogene **24**(27):4340–8.
34. Hsu, T., T. Moroy, J. Etiemble, et al. 1988. Activation of c-myc by woodchuck hepatitis virus insertion in hepatocellular carcinoma. Cell **55**:627–35.
35. Moroy, T., A. Marchio, J. Etiemble, et al. 1986. Rearrangement and enhanced expression of c-myc in hepatocellular carcinoma of hepatitis virus infected woodchucks. Nature **324**:276–9.
36. Fourel, G., C. Trepo, L. Bougueleret, et al. 1990. Frequent activation of N-myc genes by hepadnavirus insertion in woodchuck liver tumours. Nature **347**(6290):294–8.
37. Dejean, A., L. Bougueleret, K. H. Grzeschik, et al. 1986. Hepatitis B virus DNA integration in a sequence homologous to v-erb-A and steroid receptor genes in a hepatocellular carcinoma. Nature **322**:70–2.
38. Wang, J., X. Chenivesse, B. Henglein, et al. 1990. Hepatitis B virus integration in a cyclin A gene in a hepatocellular carcinoma. Nature **343**(6258):555–7.
39. Benhamou, Y., M. Bochet, V. Di Martino, et al. 1999. Liver fibrosis progression in human immunodeficiency virus and hepatitis C virus coinfected patients. The Multivirc Group. Hepatology **30**(4):1054–8.
40. Colin, J. F., D. Cazals-Hatem, M. A. Loriot, et al. 1999. Influence of human immunodeficiency virus infection on chronic hepatitis B in homosexual men. Hepatology **29**(4):1306–10.
41. Weinig, M., J. G. Hakim, I. Gudza, et al. 1997. Hepatitis C virus and HIV antibodies in patients with hepatocellular carcinoma in Zimbabwe: A pilot study. Trans R Soc Trop Med Hyg **91**(5):570–2.

42. Tswana, S. A., and S. R. Moyo. 1992. The interrelationship between HBV-markers and HIV antibodies in patients with hepatocellular carcinoma. J Med Virol **37**(3):161–4.

43. Sherman, M. 2006. Optimizing management strategies in special patient populations. Am J Gastroenterol **101**(Suppl 1):S26–31.

44. Puoti, M., R. Bruno, V. Soriano, et al. 2004. Hepatocellular carcinoma in HIV-infected patients: Epidemiological features, clinical presentation and outcome. AIDS **18**(17):2285–93.

45. Kramer, J. R., T. P. Giordano, J. Souchek, et al. 2005. The effect of HIV coinfection on the risk of cirrhosis and hepatocellular carcinoma in U.S. veterans with hepatitis C. Am J Gastroenterol **100**(1):56–63.

46. Giordano, T. P., J. R. Kramer, J. Souchek, et al. 2004. Cirrhosis and hepatocellular carcinoma in HIV-infected veterans with and without the hepatitis C virus: A cohort study, 1992–2001. Arch Intern Med **164**(21):2349–54.

47. Garcia-Samaniego, J., M. Rodriguez, J. Berenguer, et al. 2001. Hepatocellular carcinoma in HIV-infected patients with chronic hepatitis C. Am J Gastroenterol **96**(1):179–83.

48. Bruno, R., M. Puoti, P. Sacchi, et al. 2006. Management of hepatocellular carcinoma in human immunodeficiency virus-infected patients. J Hepatol **44**(1 Suppl):S146–50.

49. Smukler, A. J., and L. Ratner. 2002. Hepatitis viruses and hepatocellular carcinoma in HIV-infected patients. Curr Opin Oncol **14**(5):538–42.

50. Hyun, C. B., and W. J. Coyle. 2004. Hepatocellular carcinoma in a patient with human immunodeficiency virus and hepatitis B virus coinfection: An emerging problem? South Med J **97**(4):401–6.

51. Tanaka, T., A. Imamura, G. Masuda, et al. 1996. A case of hepatocellular carcinoma in HIV-infected patient. Hepatogastroenterology **43**(10):1067–72.

52. Chen, C. J., H. I. Yang, J. Su, et al. 2006. Risk of hepatocellular carcinoma across a biological gradient of serum hepatitis B virus DNA level. JAMA **295**(1):65–73.

53. Ohata, K., K. Hamasaki, K. Toriyama, et al. 2004. High viral load is a risk factor for hepatocellular carcinoma in patients with chronic hepatitis B virus infection. J Gastroenterol Hepatol **19**(6):670–5.

54. Hu, J., and D. Nguyen. 2004. Therapy for chronic hepatitis B: The earlier, the better? Trends Microbiol **12**(10):431–3.

55. van Zonneveld, M., P. Honkoop, B. E. Hansen, et al. 2004. Long-term follow-up of alpha-interferon treatment of patients with chronic hepatitis B. Hepatology **39**(3):804–10.

56. Akuta, N., F. Suzuki, Y. Suzuki, et al. 2005. Favorable efficacy of long-term lamivudine therapy in patients with chronic hepatitis B: An 8-year follow-up study. J Med Virol **75**(4):491–8.

57. Liaw, Y. F., J. J. Sung, W. C. Chow, et al. 2004. Lamivudine for patients with chronic hepatitis B and advanced liver disease. N Engl J Med **351**(15):1521–31.

58. Matsumoto, A., E. Tanaka, A. Rokuhara, et al. 2005. Efficacy of lamivudine for preventing hepatocellular carcinoma in chronic hepatitis B: A multicenter retrospective study of 2795 patients. Hepatol Res **33**(5):1299–302.

59. Shindo, M., K. Hamada, Y. Oda, et al. 2001. Long-term follow-up study of sustained biochemical responders with interferon therapy. Hepatology **33**(5):1299–302.

60. Tsuda, N., N. Yuki, K. Mochizuki, et al. 2004. Long-term clinical and virological outcomes of chronic hepatitis C after successful interferon therapy. J Med Virol **74**(3):406–13.

61. Meuleman, P., L. Libbrecht, S. Wieland, et al. 2006. Immune suppression uncovers endogenous cytopathic effects of the hepatitis B virus. J Virol **80**(6):2797–807.

62. Yoon, E. J., and K. Q. Hu. 2006. Hepatitis C virus (HCV) infection and hepatic steatosis. Int J Med Sci **3**(2):53–6.

63. Gaslightwala, I., and E. J. Bini. 2006. Impact of human immunodeficiency virus infection on the prevalence and severity of steatosis in patients with chronic hepatitis C virus infection. J Hepatol.

64. Altavilla, G., A. Caputo, M. Lanfredi, et al. 2000. Enhancement of chemical hepatocarcinogenesis by the HIV-1 tat gene. Am J Pathol **157**(4):1081–9.

65. Pol, S., P. Lebray, and A. Vallet-Pichard. 2004. HIV infection and hepatic enzyme abnormalities: Intricacies of the pathogenic mechanisms. Clin Infect Dis **38**(Suppl 2):S65–72.

66. Stapleton, J. T., C. F. Williams, and J. Xiang. 2004. GB virus type C: A beneficial infection? J Clin Microbiol **42**(9):3915–9.

67. Berzsenyi, M. D., D. S. Bowden, and S. K. Roberts. 2005. GB virus C: Insights into co-infection. J Clin Virol **33**(4):257–66.

68. Xiang, J., S. Wunschmann, D. J. Diekema, et al. 2001. Effect of coinfection with GB virus C on survival among patients with HIV infection. N Engl J Med **345**(10):707–14.

69. Tillmann, H. L., H. Heiken, A. Knapik-Botor, et al. 2001. Infection with GB virus C and reduced mortality among HIV-infected patients. N Engl J Med **345**(10):715–24.

70. Xiang, J., S. L. George, S. Wunschmann, et al. 2004. Inhibition of HIV-1 replication by GB virus C infection through increases in RANTES, MIP-1alpha, MIP-1beta, and SDF-1. Lancet **363**(9426):2040–6.

INDEX

Printed in the United States of America